Commercial Protected Floriculture

Commercial Protected Floriculture

– Author –
Ashish Kaul

Wordpen Academics
Publishing Science That Matter

ISBN: 978-93-90438-04-4 (Hardback)

Published by : **WORDPEN ACADEMICS**
4736/23, Ansari Road, Darya Ganj,
New Delhi – 110 002
Phone: +91-011-35004336
E-mail: info@wordpenacademics.com

Printed at **:** **Replika Press Pvt. Ltd.**

Exclusive Distributor: ***Bio-Green Books, New Delhi***

Preface

The floral industry is one of the major industries in many developing and underdeveloped countries. Floriculture as an industry began in the late 19th century in England, where flowers were grown on a large scale on the vast estates. The present day floral industry is a dynamic, global, fast-growing industry, which has achieved significant growth rates during the past few decades. In the 1950s, the global flower trade was less than US$3 billion. By 1992, it had grown to US$100 billion. In recent years, the floral industry has grown six percent annually, while the global trade volume in 2003 was US$101.84 billion. The floral industry essentially consists of three major components: the growers, the wholesalers and the retailers whose businesses are quite intermingled. The recent trends are more towards eliminating the intermediaries, the wholesalers between the growers and the retailers, so that the flowers are made available at considerably low prices.

Floriculture crops include bedding plants, flowering plants, foliage plants or houseplants, cut cultivated greens, and cut flowers. As distinguished from nursery crops, floriculture crops are generally herbaceous. Bedding and garden plants consist of young flowering plants (annuals and perennials) and vegetable plants. They are grown in cell packs (in flats or trays), in pots, or in hanging baskets, usually inside a controlled environment, and sold largely for gardens and landscaping. Geraniums, Impatiens, and petunias are the best selling bedding plants. Chrysanthemums are the major perennial garden plant in the United States. Flowering plants are largely sold in pots for indoor use. The major flowering plants are poinsettias, orchids, florist chrysanthemums, and finished florist azaleas. Foliage plants are also sold in pots and hanging baskets for indoor and patio use, including larger specimens for office, hotel, and restaurant interiors. Cut flowers are usually sold in bunches or as bouquets with cut foliage. The production of cut flowers is specifically known as the cut flower industry. Farming flowers and foliage employs special aspects of floriculture, such as spacing, training and pruning plants for optimal flower harvest; and post-harvest treatment such as chemical treatments, storage, preservation and packaging. In Australia and the United States some species are harvested from the wild for the cut flower market.

The flowers sold in florist shops typically represent the varieties abundantly available in the season but also include blossoms flown in from around the world. Basic varieties in temperate latitudes include roses, tulips, irises, orchids and lilies.

Fashion sometimes plays a role in floristry; what is considered the flower that everyone needs to have today can change very quickly. Some shops also stock gift baskets, fruits, and chocolates as well as flowers, whereas some shops will purchase these things only when needed for an order. Floral business is seasonal and is heavily influenced by the following holidays and events: Christmas, Valentine's Day, Administrative Professionals' Day, Mothers' Day, All Souls Day, Advent, Easter, weddings and funerals. These occasions make up the largest part of the business, with the sale of house plants and home decor being a smaller, but more constant, part. Flowers for personal enjoyment as well as those selected to celebrate birthdays, anniversaries, thank you's and get well wishes are also a significant portion of a florist's business.

The present book deals with all the important dimensions of Agricultural Management--management process in development organizations, development programmes, human resource planning, management techniques and tools.

— *Author*

Contents

1

Introduction

Floriculture, or flower farming, is a discipline of horticulture concerned with the cultivation of flowering and ornamental plants for gardens and for floristry, comprising the floral industry. The development, via plant breeding, of new varieties is a major occupation of floriculturists.

Floriculture crops include bedding plants, flowering plants, foliage plants or houseplants, cut cultivated greens, and cut flowers. As distinguished from nursery crops, floriculture crops are generally herbaceous. Bedding and garden plants consist of young flowering plants (annuals and perennials) and vegetable plants. They are grown in cell packs (in flats or trays), in pots, or in hanging baskets, usually inside a controlled environment, and sold largely for gardens and landscaping. Geraniums, Impatiens, and petunias are the best selling bedding plants. Chrysanthemums are the major perennial garden plant in the United States.

Flowering plants are largely sold in pots for indoor use. The major flowering plants are poinsettias, orchids, florist chrysanthemums, and finished florist azaleas. Foliage plants are also sold in pots and hanging baskets for indoor and patio use, including larger specimens for office, hotel, and restaurant interiors.

Cut flowers are usually sold in bunches or as bouquets with cut foliage. The production of cut flowers is specifically known as the cut flower industry. Farming flowers and foliage employs special aspects of floriculture, such as spacing, training and pruning plants for optimal flower harvest; and post-harvest treatment such as chemical treatments, storage, preservation and packaging. In Australia and the United States some species are harvested from the wild for the cut flower market.

Floristry

Floristry is the general term used to describe the professional floral trade. It encompasses flower care and handling, floral design or flower arranging, merchandising, and display and flower delivery. Wholesale florists sell bulk flowers and related supplies to professionals in the trade. Retail florists offer fresh flowers and related products and services to consumers.

Floristry is a term can refer to the cultivation of flowers as well as their arrangement, and to the business of selling them. Much of the raw material supplied for the floristry trade comes from the cut flower industry. Florist shops, along with on-line stores are the main flower only outlets, but supermarkets, garden supply stores and filling stations also sell flowers.

Floral Design or floral arts is the art of creating flower arrangements in vases, bowls, baskets or other containers, or making bouquets and compositions from cut flowers, foliages, herbs, ornamental grasses and other plant materials. Often the terms "floral design" and "floristry" are considered synonymous. Florists are people who work with flowers and plants, generally at the retail level. The term is not to be confused with floristics (the study of distribution and relationships of plant species over geographic areas.). Floristry is also not to be confused with horticulture which more broadly relates to the cultivation of flowers and plants.

The craft and Business

The craft of floristry, or being a florist, involves various skills and creativity. A florist should be able to select flowers and other floral supplies and materials that will look good together (based on principles and elements of floral design or market demands), know how to handle and arrange flowers and plants so they will remain fresh as long as possible, and would be desirable for purchase, which also involves knowledge of customers' requirements and expectations. Ability to create a variety of floral designs such as wreaths, bouquets, corsages, boutonnières/'buttonholes', permanent arrangements and other more complicated arrangements is also important.

The flowers sold in florist shops typically represent the varieties abundantly available in the season but also include blossoms flown in from around the world. Basic varieties in temperate latitudes include roses, tulips, irises, orchids and lilies. Fashion sometimes plays a role in floristry; what is considered the flower that everyone needs to have today can change very quickly.

Some shops also stock gift baskets, fruits, and chocolates as well as flowers, whereas some shops will purchase these things only when needed for an order. Floral business is seasonal and is heavily influenced by the following holidays and events: Christmas, Valentine's Day, Administrative Professionals' Day, Mothers' Day, All Souls Day, Advent, Easter, weddings and funerals. These occasions make up the largest part of the business, with the sale of house plants and home decor being a smaller, but more constant, part. Flowers for personal enjoyment as well as those selected to celebrate birthdays, anniversaries, thank you's and get well wishes are also a significant portion of a florist's business.

Education, both formal and informal, is another significant segment of the floristry industry. Established floristry designers and artists impart their craft to students interested in floral design as hobby or career. Courses are generally available in through community colleges, via private post-secondary vocational schools and through professional florist trade associations. The floristry business has a significant market in the corporate and social event world as flowers play a large part in the decor of special events and meetings. Centerpieces, entryways, reception tables, bridal bouquets, wedding chuppahs and stage sets and only a few examples of how flowers are used in the business and social event settings.

Styles of Floristry

Many nations have their own style of floral arranging.

Ikebana

Ikebana is a style of floral design that originated in Japan. Best known for its simplicity of line and form, Ikebana is a design style primarily practiced for personal enjoyment.

English Garden

English Garden style is traditionally an English form of floral design. Stems are placed in a radial fashion and feature abundant use of seasonal flowers and foliages. These designs are often done as low, tufted mounds, or taller vase arrangements that are all sided (360°), and incorporate garden flowers like roses, delphinium and peonies. Many florists that follow this design style do not use Baby's Breath, Carnations and Leatherleaf.

High Style

High Style is a catch-all term to describe contemporary, linear designs that highlights unique forms of both individual floral materials and of the designs themselves. Arrangements generally feature negative space and incorporate asymmetric placement of materials. The style stands in direct contrast to traditional radial arrangements such as English Garden.

High Style designs often incorporate unique, exotic or tropical flowers such as such as Bird of Paradise, Anthurium and Protea but may also employ more common blossoms.

Fresh Flower Sources

There are usually three major sources of fresh flowers for retail florists: local growers, local wholesalers and flower auctions. The majority of these are supplied by specialist horticultural and cut flower suppliers.

Internationally there are hundreds of wholesale flower markets and auctions, the largest of which is located in Aalsmeer, Holland: the Bloemenveiling Aalsmeer. Other major markets include the fledgling Dubai Flower Centre and the Ota Flower Market in Tokyo, Japan.

Within North America, Miami is considered to be the main distribution point for imported fresh flowers. Many local wholesalers purchase fresh flower stock from importers in Miami to resale to local florist in their areas. Wholesale flower districts are present in many North American cities such as New York, Boston and San Francisco. Flower auctions are run using the Dutch clock system and are mainly located in Canada: Toronto, Montreal, and Vancouver.

Shop Layout

Generally, a florist's shop will contain a large array of flowers, sometimes displayed on the street, or will have a large plate glass window to display the flowers. To keep them fresh, the flowers will be inside of a cooler and kept in water, generally in glass or plastic vases or other containers. Most shops have a cooler near the front of the store with large glass doors so that customers can easily view the contents. Some shops also have another cooler out of the customers' view where they keep extra stock and arrangements for customers' orders. Most stores have a back section in which the designers can work on orders with more privacy.

Significance of Particular Flowers

Typically, a florist will organize flowers by season and holiday. Flowers have various different meanings in different cultures. The holidays and events for which flowers are used vary. Poppies are used to remember fallen soldiers only in Great Britain and the Commonwealth countries. The cultural meaning of colours also strongly affects the choice and use of flowers. People often prefer flowers that are associated with their ethnic group or country, and various colours may have special meanings of luck or death or love or other basic human traits. To some, a flower such as a red rose might mean love, but to others it might be considered indecent or simply puzzling. The vastly divergent views on the colour white can also lead to major flower issues. While white represents death in many Asian cultures, it is simultaneously a symbol of purity and innocence in many European and American cultures. Such differences can lead to difficult issues when a bouquet of white lilies, for example, is delivered.

Supermarket Flowers

Cut flowers are widely sold in supermarkets and gas stations. These outlets offer a limited selection, usually in the form of mixed bouquets and roses by the dozen. Flowers purchased at these outlets are generally less expensive than the flowers purchased at a florist shop. However these flowers are not arranged by anyone with professional training or the years of on-the-job training that is standard for the employees of a professional flower shop.

Internet Sales

The World Wide Web has had a significant impact on traditional florists, with the North American market experiencing a more than 20% decline in traditional, independently owned flower shops since 1998. Purchases from shops are slowly being overtaken by on-line flower delivery ordering and floral wire services. However, some independent florists have taken to having their own on-line store, where customers can order on-line as well as on the telephone.

The Flower

A flower, sometimes known as a bloom or blossom, is the reproductive structure found in flowering plants (plants of the division Magnoliophyta, also called angiosperms). The biological function of a flower is to effect reproduction, usually by providing a mechanism for the union of sperm with eggs. Flowers may facilitate outcrossing (fusion of sperm and eggs from different individuals in a population) or allow selfing (fusion of sperm and egg from the same flower). Some flowers produce diaspores without fertilization (parthenocarpy). Flowers contain sporangia and are the site where gametophytes develop. Flowers give rise to fruit and seeds. Many flowers have evolved to be attractive to animals, so as to cause them to be vectors for the transfer of pollen.

In addition to facilitating the reproduction of flowering plants, flowers have long been admired and used by humans to beautify their environment but also as objects of romance, ritual, religion, medicine and as a source of food.

Morphology

Flowering plants are *heterosporous*, producing two types of spores. Microspores are produced by meiosis inside anthers while megaspores are produced inside ovules, inside an ovary. In fact, anthers typically consist of four microsporangia and an ovule is an integumented megasporangium. Both types of spores develop into gametophytes inside sporangia. As with all heterosporous plants, the gametophytes also develop inside the spores (are endosporic).

A flower is a modified stem tip with compressed internodes, bearing structures that are highly modified leaves. In essence, a flower develops on a modified shoot or *axis* from a determinate apical meristem (*determinate* meaning the axis grows to a set size).

Flowers may be directly attached to the plant at their base (sessile—the supporting stalk or stem is highly reduced or absent). The stem or stalk subtending a flower is called a peduncle. If a peduncle supports more than one flower, the stems connecting each flower to the main axis are called pedicels. The apex of a flowering stem forms a terminal swelling which is called the *torus* or receptacle. The parts of a flower are arranged in whorls on the receptacle. The four main whorls (starting from the base of the flower or lowest node and working upwards) are as follows:

- *Calyx*: the outermost whorl consisting of units called*sepals*; these are typically green and enclose the rest of the flower in the bud stage, however, they can be absent or prominent and petal-like in some species.
- *Corolla*: the next whorl toward the apex, composed of units called *petals*, which are typically thin, soft and colored to attract animals that help the process of pollination.
- *Androecium* (from Greek *andros oikia*: man's house): the next whorl (sometimes multiplied into several whorls), consisting of units called stamens. Stamens consist of two parts: a stalk called a filament, topped by an anther where pollen is produced by meiosis and eventually dispersed.
- *Gynoecium* (from Greek *gynaikos oikia*: woman's house): the innermost whorl of a flower, consisting of one or more units called carpels. The carpel or

multiple fused carpels form a hollow structure called an ovary, which produces ovules internally. Ovules are megasporangia and they in turn produce megaspores by meiosis which develop into female gametophytes. These give rise to egg cells. The gynoecium of a flower is also described using an alternative terminology wherein the structure one sees in the innermost whorl (consisting of an ovary, style and stigma) is called a pistil. A pistil may consist of a single carpel or a number of carpels fused together. The sticky tip of the pistil, the stigma, is the receptor of pollen. The supportive stalk, the style, becomes the pathway for pollen tubes to grow from pollen grains adhering to the stigma.

Although the arrangement described above is considered "typical", plant species show a wide variation in floral structure. These modifications have significance in the evolution of flowering plants and are used extensively by botanists to establish relationships among plant species.

In the majority of species, individual flowers have both functional carpels and stamens. These flowers are described by botanists as being *perfect* or *bisexual.* Some flowers lack one or the other reproductive organ and called *imperfect* or *unisexual* If unisex flowers are found on the same individual plant but in different locations, the species is said to be *monoecious.* If each type of unisex flower is found only on separate individuals, the plant is *dioecious.*

Additional discussions on floral modifications from the basic plan are presented in the articles on each of the basic parts of the flower. In those species that have more than one flower on an axis, the collection of flowers is termed an *inflorescence.* Some inflorescences are composed of many small flowers arranged in a formation that resembles a single flower. The common example of this is most members of the very large composite (Asteraceae) group. A single daisy or sunflower, for example, is not a flower but a flower *head*—an inflorescence composed of numerous flowers (or florets).

Many flowers have a symmetry, if the perianth is bisected through the central axis from any point, symmetrical halves are produced—the flower is called regular or actinomorphic, e.g. rose or trillium. When flowers are bisected and produce only one line that produces symmetrical halves the flower is said to be irregular or zygomorphic. e.g. snapdragon or most orchids.

Development

Flowering Transition

The transition to flowering is one of the major phase changes that a plant makes during its life cycle. The transition must take place at a time that is favourable for fertilization and the formation of seeds, hence ensuring maximal reproductive success. To meet these needs a plant is able to interpret important endogenous and environmental cues such as changes in levels of plant hormones and seasonable temperature and photoperiod changes. Many perennial and most biennial plants require vernalization to flower. The molecular interpretation of these signals is through the transmission of a complex signal known as florigen, which involves a variety of genes, including Constans, Flowering Locus C And Flowering Locus T. Florigen is produced in the leaves in reproductively favourable conditions and acts in buds and growing tips to induce a number of different physiological and morphological changes. The first step is the transformation of the vegetative stem primordia into floral primordia. This occurs as biochemical changes take place to change cellular differentiation of leaf, bud and stem tissues into tissue that will grow into the reproductive organs. Growth of the central part of the stem tip stops or flattens out and the sides develop protuberances in a whorled or spiral fashion around the outside of the stem end. These protuberances develop into the sepals, petals, stamens, and carpels. Once this process begins, in most plants, it cannot be reversed and the stems develop flowers, even if the initial start of the flower formation event was dependent of some environmental cue. Once the process begins, even if that cue is removed the stem will continue to develop a flower.

Organ Development

The molecular control of floral organ identity determination is fairly well understood. In a simple model, three gene activities interact in a combinatorial manner to determine the developmental identities of the organ primordia within the floral meristem. These gene functions are called A, B and C-gene functions. In the first floral whorl only A-genes are expressed, leading to the formation of sepals. In the second whorl both A- and B-genes are expressed, leading to the formation of petals. In the third whorl, B and C genes interact to form stamens and in the center of the flower C-genes alone give rise to carpels. The model is based upon studies of homeotic mutants in *Arabidopsis thaliana* and snapdragon, *Antirrhinum majus*. For example, when there is a loss of B-gene function, mutant flowers are produced with sepals in the first whorl as usual, but also in the second whorl instead of the normal petal formation. In the third whorl the lack of B function but presence of C-function mimics the fourth whorl, leading to the formation of carpels also in the third whorl.

Most genes central in this model belong to the MADS-box genes and are transcription factors that regulate the expression of the genes specific for each floral organ.

Flower Specialization and Pollination

Flowering plants usually face selective pressure to optimize the transfer of their pollen, and this is typically reflected in the morphology of the flowers and the behaviour of the plants. Pollen may be transferred between plants via a number of 'vectors'. Some plants make use of abiotic vectors — namely wind (anemophily) or, much less commonly, water (hydrophily). Others use biotic vectors including insects (entomophily), birds (ornithophily), bats (chiropterophily) or other animals. Some plants make use of multiple vectors, but many are highly specialised.

Cleistogamous flowers are self pollinated, after which they may or may not open. Many Viola and some Salvia species are known to have these types of flowers.

The flowers of plants that make use of biotic pollen vectors commonly have glands called nectaries that act as an incentive for animals to visit the flower. Some flowers have patterns, called nectar guides, that show pollinators where to look for nectar. Flowers also attract pollinators by scent and colour. Still other flowers use mimicry to attract pollinators. Some species of orchids, for example, produce flowers resembling female bees in colour, shape, and scent. Flowers are also specialized in shape and have an arrangement of the stamens that ensures that pollen grains are transferred to the bodies of the pollinator when it lands in search of its attractant (such as nectar, pollen, or a mate). In pursuing this attractant from many flowers of the same species, the pollinator transfers pollen to the stigmas—arranged with equally pointed precision—of all of the flowers it visits.

Anemophilous flowers use the wind to move pollen from one flower to the next. Examples include grasses, birch trees, ragweed and maples. They have no need to attract pollinators and therefore tend not to be "showy" flowers. Male and female reproductive organs are generally found in separate flowers, the male flowers having a number of long filaments terminating in exposed stamens, and the female flowers having long, feather-like stigmas. Whereas the pollen of animal-pollinated flowers tends to be large-grained, sticky, and rich in protein (another "reward" for pollinators), anemophilous flower pollen is usually small-grained, very light, and of little nutritional value to animals.

Floral Formula

A *floral formula* is a way to represent the structure of a flower using specific letters, numbers, and symbols. Typically, a general formula will be used to represent the flower structure of a plant family rather than a particular species. The following representations are used:

Ca = calyx (sepal whorl; e. g. Ca^5 = 5 sepals).

Co = corolla (petal whorl; e. g., $Co^{3(x)}$ = petals some multiple of three).

Z = add if *zygomorphic* (e. g., CoZ^6 = zygomorphic with 6 petals).

A = *androecium* (whorl of stamens; e. g., $A^{"}$ = many stamens).

G = *gynoecium* (carpel or carpels; e. g., G^1 = monocarpous).

x: to represent a "variable number",

": to represent "many",

A floral formula would appear something like this:

$Ca^5Co^5A^{10\text{ - }"}G^1$

Several additional symbols are sometimes used.

The four main parts of a flower are generally defined by their positions on the receptacle and not by their function. Many flowers lack some parts or parts may be modified into other functions and/or look like what is typically another part. In some families, like Ranunculaceae, the petals are greatly reduced and in many species the sepals are colourful and petal-like. Other flowers have modified stamens that are petal-like, the double flowers of Peonies and Roses are mostly petaloid stamens. Flowers show great variation and plant scientists describe this variation in a systematic way to identify and distinguish species.

Specific terminology is used to descried flowers and their parts. Many flower parts are fused together; fused parts originating from the same whorl are connate, while fused parts originating from different whorls are adnate, parts that are not fused are free. When petals are fused into a tube or ring that falls away as a single unit, they are sympetalous (also called gamopetalous.) Petals that are connate may have distinctive regions: the cylindrical base is the tube, the expanding region is the throat and the flaring outer region is the limb. A sympetalous flower, with bilateral symmetry with an upper and lower lip, is bilabiate. Flowers with connate petals or sepals may have various shaped corolla or calyx including: campanulate, funnelform, tubular, urceolate, salverform or rotate.

Pollination

The primary purpose of a flower is reproduction. Since the flowers are the reproductive organs of plant, they mediate the joining of the sperm, contained within pollen, to the ovules — contained in the ovary. Pollination is the movement of pollen from the anthers to the stigma. The joining of the sperm to the ovules is called fertilization. Normally pollen is moved from one plant to another, but many plants are able to self pollinate. The fertilized ovules produce seeds that are the next generation. Sexual reproduction produces genetically unique offspring, allowing for adaptation. Flowers have specific designs which encourages the transfer of pollen from one plant to another of the same species. Many plants are dependent upon external factors for pollination, including: wind and animals, and especially insects. Even large animals such as birds, bats, and pygmy possums can be employed. The period of time during which this process can take place (the flower is fully expanded and functional) is called *anthesis*.

Attraction Methods

Plants can not move from one location to another, thus many flowers have evolved to attract animals to transfer pollen between individuals in dispersed populations. Flowers that are insect-pollinated are called *entomophilous*; literally "insect-loving" in Greek. They can be highly modified along with the pollinating insects by co-evolution. Flowers commonly have glands called *nectaries* on various parts that attract animals looking for nutritious nectar. Birds and bees have colour vision, enabling them to seek out "colourful" flowers. Some flowers have patterns, called nectar guides, that show pollinators where to look for nectar; they may be visible only under ultraviolet light, which is visible to bees and some other insects. Flowers also attract pollinators by scent and some of those scents are pleasant to our sense of smell. Not all flower scents are appealing to humans, a number of flowers are pollinated by insects that are attracted to rotten flesh and have flowers that smell like dead animals, often called Carrion flowers including *Rafflesia*, the titan arum, and the

North American pawpaw (*Asimina triloba*). Flowers pollinated by night visitors, including bats and moths, are likely to concentrate on scent to attract pollinators and most such flowers are white. Still other flowers use mimicry to attract pollinators. Some species of orchids, for example, produce flowers resembling female bees in colour, shape, and scent. Male bees move from one such flower to another in search of a mate.

Pollination Mechanism

The pollination mechanism employed by a plant depends on what method of pollination is utilized. Most flowers can be divided between two broad groups of pollination methods:

Entomophilous: flowers attract and use insects, bats, birds or other animals to transfer pollen from one flower to the next. Often they are specialized in shape and have an arrangement of the stamens that ensures that pollen grains are transferred to the bodies of the pollinator when it lands in search of its attractant (such as nectar, pollen, or a mate). In pursuing this attractant from many flowers of the same species, the pollinator transfers pollen to the stigmas—arranged with equally pointed precision—of all of the flowers it visits. Many flowers rely on simple proximity between flower parts to ensure pollination. Others, such as the *Sarracenia* or lady-slipper orchids, have elaborate designs to ensure pollination while preventing self-pollination.

Anemophilous: flowers use the wind to move pollen from one flower to the next, examples include the grasses, Birch trees, Ragweed and Maples. They have no need to attract pollinators and therefore tend not to be "showy" flowers. Whereas the pollen of entomophilous flowers tends to be large-grained, sticky, and rich in protein (another "reward" for pollinators), anemophilous flower pollen is usually small-grained, very light, and of little nutritional value to insects, though it may still be gathered in times of dearth. Honeybees and bumblebees actively gather anemophilous corn (maize) pollen, though it is of little value to them. Some flowers are self pollinated and use flowers that never open or are self pollinated before the flowers open, these flowers are called cleistogamous. Many Viola species and some Salvia have these types of flowers.

Flower-pollinator Relationships

Many flowers have close relationships with one or a few specific pollinating organisms. Many flowers, for example, attract only one specific species of insect, and therefore rely on that insect for successful reproduction. This close relationship is often given as an example of coevolution, as the flower and pollinator are thought to have developed together over a long period of time to match each other's needs.

This close relationship compounds the negative effects of extinction. The extinction of either member in such a relationship would mean almost certain extinction of the other member as well. Some endangered plant species are so because of shrinking pollinator populations.

Fertilization and Dispersal

Some flowers with both stamens and a pistil are capable of self-fertilization, which does increase the chance of producing seeds but limits genetic variation. The extreme case of self-fertilization occurs in flowers that always self-fertilize, such as many dandelions. Conversely, many species of plants have ways of preventing self-fertilization. Unisexual male and female flowers on the same plant may not appear or mature at the same time, or pollen from the same plant may be incapable of fertilizing its ovules. The latter flower types, which have chemical barriers to their own pollen, are referred to as self-sterile or self-incompatible.

Evolution

While land plants have existed for about 425 million years, the first ones reproduced by a simple adaptation of their aquatic counterparts: spores. In the sea, plants—and some animals—can simply scatter out genetic clones of themselves to float away and grow elsewhere. This is how early plants reproduced. But plants soon evolved methods of protecting these copies to deal with drying out and other abuse which is even more likely on land than in the sea. The protection became the seed, though it had not yet evolved the flower. Early seed-bearing plants include the ginkgo and conifers. The earliest fossil of a flowering plant, *Archaefructus liaoningensis*, is dated about 125 million years old. Several groups of extinct gymnosperms, particularly seed ferns, have been proposed as the ancestors of flowering plants but there is no continuous fossil evidence showing exactly how flowers evolved. The apparently sudden appearance of relatively modern flowers in the fossil record posed such a problem for the theory of evolution that it was called an "abominable mystery" by Charles Darwin. Recently discovered angiosperm fossils such as *Archaefructus*, along with further discoveries of fossil gymnosperms, suggest how angiosperm characteristics may have been acquired in a series of steps.

Recent DNA analysis (molecular systematics) shows that *Amborella trichopoda*, found on the Pacific island of New Caledonia, is the sister group to the rest of the flowering plants, and morphological studies suggest that it has features which may have been characteristic of the earliest flowering plants.

The general assumption is that the function of flowers, from the start, was to involve animals in the reproduction process. Pollen can be scattered without bright colours and obvious shapes, which would therefore be a liability, using the plant's resources, unless they provide some other benefit. One proposed reason for the sudden, fully developed appearance of flowers is that they evolved in an isolated setting like an island, or chain of islands, where the plants bearing them were able to develop a highly specialized relationship with some specific animal (a wasp, for example), the way many island species develop today. This symbiotic relationship, with a hypothetical wasp bearing pollen from one plant to another much the way fig wasps do today, could have eventually resulted in both the plant(s) and their partners developing a high degree of specialization. Island genetics is believed to be a common source of speciation, especially when it comes to radical adaptations which seem to have required inferior transitional forms. Note that the wasp example is not incidental; bees, apparently evolved specifically for symbiotic plant relationships, are descended from wasps. Likewise, most fruit used in plant reproduction comes from the enlargement of parts of the flower. This fruit is frequently a tool which depends upon animals wishing to eat it, and thus scattering the seeds it contains. While many such symbiotic relationships remain too fragile to survive competition with mainland organisms, flowers proved to be an unusually effective means of production, spreading (whatever their actual origin) to become the dominant form of land plant life.

While there is only hard proof of such flowers existing about 130 million years ago, there is some circumstantial evidence that they did exist up to 250 million years ago. A chemical used by plants to defend their flowers, oleanane, has been detected in fossil plants that old, including gigantopterids, which evolved at that time and bear many of the traits of modern, flowering plants, though they are not known to be flowering plants themselves, because only their stems and prickles have been found preserved in detail; one of the earliest examples of petrification.

The similarity in leaf and stem structure can be very important, because flowers are genetically just an adaptation of normal leaf and stem components on plants, a combination of genes normally responsible for forming new shoots. The most primitive flowers are thought to have had a variable number of flower parts, often separate from (but in contact with) each other. The flowers would have tended to grow in a spiral pattern, to be bisexual (in plants, this means both male and female parts on the same flower), and to be dominated by the ovary (female part). As flowers grew more advanced, some variations developed parts fused together, with a much more specific number and design, and with either specific sexes per flower or plant, or at least "ovary inferior". Flower evolution continues to the present day; modern flowers have been so profoundly influenced by humans that many of them cannot be pollinated in nature. Many modern, domesticated flowers used to be simple weeds, which only sprouted when the ground was disturbed. Some of them tended to grow with human crops, and the prettiest did not get plucked because of their beauty, developing a dependence upon and special adaptation to human affection.

Symbolism

Many flowers have important symbolic meanings in Western culture. The practice of assigning meanings to flowers is known as floriography. Some of the more common examples include:

- Red roses are given as a symbol of love, beauty, and passion.
- Poppies are a symbol of consolation in time of death. In the United Kingdom, New Zealand, Australia and Canada, red poppies are worn to commemorate soldiers who have died in times of war.
- Irises/Lily are used in burials as a symbol referring to "resurrection/life". It is also associated with stars (sun) and its petals blooming/shining.
- Daisies are a symbol of innocence.

Flowers within art are also representative of the female genitalia, as seen in the works of artists such as Georgia O'Keeffe, Imogen Cunningham, Veronica Ruiz de Velasco, and Judy Chicago, and in fact in Asian and western classical art. Many cultures around the world have a marked tendency to associate flowers with femininity. The great variety of delicate and beautiful flowers has inspired the works of numerous poets, especially from the 18th-19th century Romantic era. Famous examples include William Wordsworth's *I Wandered Lonely as a Cloud* and William Blake's *Ah! Sun-Flower*.

Because of their varied and colourful appearance, flowers have long been a favourite subject of visual artists as well. Some of the most celebrated paintings from well-known painters are of flowers, such as Van Gogh's sunflowers series or Monet's water lilies. Flowers are also dried, freeze dried and pressed in order to create permanent, three-dimensional pieces of flower art. The Roman goddess of flowers, gardens, and the season of Spring is Flora. The Greek goddess of spring, flowers and nature is Chloris.

In Hindu mythology, flowers have a significant status. Vishnu, one of the three major gods in the Hindu system, is often depicted standing straight on a lotus flower. Apart from the association with Vishnu, the Hindu tradition also considers the lotus to have spiritual significance.

Usage

In modern times, people have sought ways to cultivate, buy, wear, or otherwise be around flowers and blooming plants, partly because of their agreeable appearance and smell. Around the world, people use flowers for a wide range of events and functions that, cumulatively, encompass one's lifetime:

- For new births or Christenings
- As a corsage or boutonniere to be worn at social functions or for holidays
- As tokens of love or esteem
- For wedding flowers for the bridal party, and decorations for the hall
- As brightening decorations within the home
- As a gift of remembrance for bon voyage parties, welcome home parties, and "thinking of you" gifts
- For funeral flowers and expressions of sympathy for the grieving
- For worshiping goddesses in Hindu culture it is very common to bring flowers as a gift to temples.

People therefore grow flowers around their homes, dedicate entire parts of their living space to flower gardens, pick wild flowers, or buy flowers from florists who depend on an entire network of commercial growers and shippers to support their trade.

Flowers provide less food than other major plants parts (seeds, fruits, roots, stems and leaves) but they provide several important foods and spices. Flower vegetables include broccoli, cauliflower and artichoke. The most expensive spice, saffron, consists of dried stigmas of a crocus.

Other flower spices are cloves and capers. Hops flowers are used to flavour beer. Marigold flowers are fed to chickens to give their egg yolks a golden yellow colour, which consumers find more desirable. Dandelion flowers are often made into wine. Bee Pollen, pollen collected from bees, is considered a health food by some people. Honey consists of bee-processed flower nectar and is often named for the type of flower, e.g. orange blossom honey, clover honey and tupelo honey.

Hundreds of fresh flowers are edible but few are widely marketed as food. They are often used to add colour and flavour to salads. Squash flowers are dipped in breadcrumbs and fried. Edible flowers include nasturtium, chrysanthemum, carnation, cattail, honeysuckle, chicory, cornflower, Canna, and sunflower. Some edible flowers are sometimes candied such as daisy and rose (you may also come across a candied pansy).

Flowers can also be made into herbal teas. Dried flowers such as chrysanthemum, rose, jasmine, camomile are infused into tea both for their fragrance and medical properties. Sometimes, they are also mixed with tea leaves for the added fragrance.

Plant Growth Substances and Flowering

Plant growth regulators (PGRs) are chemicals that are designed to affect plant growth and/or development.

They are applied for specific purposes to affect specific plant responses. Although there is much scientific information on using PGRs in the greenhouse, it is not an exact science. Achieving the best results with PGRs is a combination of art and science– science tempered with a lot of trial and error and a good understanding of plant growth and development.

Optimizing Results

For best results, PGRs should be handled as production tools, like water and fertilizer. They should not be used as crutches for poor management of other cultural practices. PGRs should be an integrated part of your crop production cycle. They are most effective when applied at the appropriate times to regulate plant growth or development. In other words, growth retardants cannot "shrink" an overgrown plant. They must be applied before the plant is overgrown to avoid plant stretch. When planning PGRs in your production schedule, consider what you want to accomplish with the treatment.

– Do you want to reduce the growth rate of the plant, improve its colour and general condition (toughness)?

If so, you probably want a growth retardant such as B-Nine, Cycocel, A-Rest, Bonzi, or Sumagic.

– Do you want to increase plant branching for enhanced cutting production, or for a more bushy potted plant or hanging basket? If so, you probably want to use a

branching agent or "chemical pincher" such as Atrimmec, Florel® Brand Pistill (Florel), or Off-Shoot-O.

– Do you want to enhance flower initiation or synchronize flowering? If so, you probably want to use Cycocel, Florel, GibGro, or ProGibb.

Answering these questions will indicate which type of PGR you need to use to accomplish your goal. It also will determine the most appropriate timing of the application. Then you will need to select a specific PGR in that class and determine the appropriate dosage and the appropriate application method for the selected application.

Read the Label

Plant growth regulators are classified as pesticides. Therefore, they are subject to all of the same USDA recordkeeping and Worker Protection Standard (WPS) rules as all of your other pesticides. Their use is governed by the manufacturer's label as with other pesticides. The label not only contains information on restrictions, but also much information on how to use the product effectively.

Is the chemical labeled for the crop you wish to treat?

Most of the PGR labels have undergone recent revisions that apply to a broad range of similar crops not specifically listed on the label, with the user taking responsibility for determining appropriate rates. This provides label permission to use the compound on these crops without the manufacturer accepting the responsibility for the rate selection.

Look for information on the effectiveness and on the side effects (phytotoxicity) of the chemical on your specific crop. B-Nine is considered to be a safe, short-term growth retardant with few phytotoxicity problems. However, it has little effect on growth of petunias and may burn treated leaves of kalanchoe.

Begonias are extremely sensitive to Bonzi and Sumagic, and the label warns you to avoid overspray or drift on these crops.

Notice any label warnings regarding the PGR's effect on plant flowering. Many branching enhancers delay flowering. Florel causes flower bud abscission prior to enhancing branching; therefore, it is not recommended within six to eight weeks of marketing. Side effects are frequently affected by the timing of the application; e.g., late applications of Bonzi or Sumagic can delay flowering of pansy, or, late applications of B-Nine or Cycocel can reduce bract size of poinsettias.

Plant growth Regulators for Height Control

Most of the PGRs used in the greenhouse are used to regulate shoot growth of bedding plants, garden mums, poinsettias, and other containerized crops. These PGRs are referred to as "growth retardants." Typical growth retardants are B-Nine, Cycocel, A-Rest, Bonzi, and Sumagic. These PGRs reduce plant height by inhibiting the production of gibberellins, the primary plant hormones responsible for cell elongation. Therefore, their effects are primarily on stem, petiole and flower stalk tissues. Lesser effects are seen in reductions of leaf expansion, resulting in thicker leaves with darker green colour.

Other benefits of using these PGRs in plant production include improved plant appearance by maintaining plant size and shape in proportion with the pot. Plant growth retardants also increase the stress tolerance of plants during shipping and handling and retail marketing, thereby improving shelf-life and extending plant marketability. Remember, growth retardants do not reduce plant size. They reduce the plant's growth rate. You must apply the growth retardant prior to the "stretch." Look for recommendations on the PGR label for time of application.

These recommendations will be given in terms of plant development or plant size as opposed to production time. For example, the Sumagic label specifies that pansies should have attained a minimum height of four inches prior to application. The Bonzi label says that bedding plant plugs should be treated at the one to two true leaf stage and bedding plants (after transplanting) at two inches of new growth or when the plants reach marketable size.

Generally, growth retarding PGRs should be applied just prior to rapid shoot growth. This is usually one to two weeks after transplanting a plug, after the roots are established and as the plant resumes active growth; on pinched plants, it is after the new shoots are visible, just starting to elongate. This is where the art of plant growth regulation is most important. You must learn how your crop grows and when to intervene to obtain the desired results. Remember to note details of crop development in your records of PGR treatments. For example, due to the weather conditions, next year you may need to treat at seven days after transplanting instead of at the ten days after transplanting that you used this year. Gauge when rapid elongation will likely occur and treat to counter it. Many growers use multiple applications of growth retardants to better control plant growth. A single application at a high rate early in the plant production cycle may be excessive if growing conditions are not as good as expected. An early application at a lower rate provides more flexibility, but the tradeoff is in the additional labour involved with a second application if it becomes necessary. Some growers improve crop uniformity by using multiple applications of lower rates to affect small corrections in plant growth. Be aware that excessive rates of Bonzi or Sumagic can cause extremely persistent growth reductions in the flat or even in the landscape. It is always a good idea to evaluate the long-term effects of your treatments by growing some out for yourself and talking with your customers.

Be careful to avoid very late applications, especially of Bonzi or Sumagic as they may delay flower opening on bedding plants. However, drench applications of Bonzi have provided excellent control of poinsettia height very late in the production cycle without causing the reduction in bract size accompanying late spray applications. Learn the art of using PGRs for plant growth regulation.

Plant Growth Regulators for Lateral Branching

Another group of PGRs used in floricultural crops are those that enhance branching, including Florel, Atrimmec and Off-Shoot-O. These PGRs are frequently called "chemical pinchers" because they generally inhibit the growth of the terminal shoots or enhance the growth of lateral buds, thereby increasing the development of lateral branches. They can be used to replace mechanical pinching of many crops. Often this increased branching also will reduce the overall height of the plant. The ethylene released inside the plant by Florel also inhibits internode elongation, keeping treated plants more compact than untreated plants. Florel also affects flowering. If you are looking for enhanced branching, you must have sufficient growth on the plant to allow for sites of lateral development. They cannot enhance lateral branching if there are no laterals on the plant. Again, read the label for details of when to apply for optimum response. You may need to consider combinations of PGRs. For example, if you apply Florel to enhance the branch development of 'Wave' petunias in a hanging basket, you will probably need to follow up with a treatment of a plant growth retardant like Bonzi to control the elongation of those new laterals. Always consider the side effects of treatments.

Plant growth Regulators for Flowering

Plant growth regulators can be used to enhance flowering (GibGro) or to remove flowers (Florel). To improve flowering, GibGro, which contains the growth promoter gibberellic acid, can be used to substitute for all or part of the chilling requirement of some woody ornamentals typically forced in the greenhouse, including azalea. [A broad use label was submitted for EPA approval in 2001 for Pro-Gibb (Valent USA) which would include camellia, hydrangea, and a variety of other floricultural crops.] Special attention must be given to the stage of flower bud development for successful treatment. In addition to overcoming dormancy, these compounds can improve flowering and/or bloom size of camellia, geranium, cyclamen, spathiphyllum, statice, and calla lily. Again, timing is critical since late applications, or excessive rates, may cause excessive plant stretching resulting in weak, spindly stems. Cycocel used to control stem height of hibiscus and geranium also improves early flowering.

Flower removal is especially desirable for stock plants maintained for cuttings of vegetatively propagated ornamentals, like geraniums, fuchsia, begonias, or lantana. Florel (ethephon) is the primary compound used for flower removal. Once ethephon is absorbed by the plant it is converted to gaseous ethylene, a natural plant hormone effective in many plant processes. Ethylene is the primary hormone responsible for flower senescence and fruit ripening. It is the "postharvest" hormone. With proper rates and timing, it will remove unwanted flowers from stock plants or from plugs or young bedding plants.

Flower removal diverts more energy into vegetative growth, increasing the number of laterals available for cuttings on stock plants, and promoting increased branching of plugs and finished plants, which increases fullness in hanging baskets or other containers. Early flower removal also allows synchronization of flowering of a container for a more dramatic appearance or for flowering on a specific marketing date. Since initiation and development of flowers requires time, Florel should not be used on crops within six to eight weeks of marketing.

Application Guidelines

Spray Applications

The pesticide label not only contains information on restrictions but also much information on using the product effectively. The label will identify the target tissue for that PGR – B-Nine is only effective as a foliar spray whereas Bonzi and Sumagic sprays must reach the stems or roots. When making spray applications, look at the physiological development of the plant to see that there is sufficient plant material at the correct stage of growth to make the treatment effective and to accomplish your goal. Generally, there should be sufficient foliage or stems to absorb the PGR. Uptake and effectiveness of a PGR also depend on selecting the application technique that will ensure proper coverage of the target tissue. B-Nine is not soil active and is fairly mobile in the plant. Therefore, a foliar spray application, wetting most of the foliage, will provide a fairly uniform reduction in growth of sensitive crops.

However, the triazoles, Bonzi and Sumagic, are absorbed primarily by stem tissue and then translocated upwards in the plant. Therefore, consistent and complete coverage of the stems is necessary for uniform effects. In other words, if the stem of one lateral receives an inadequate amount of spray, it will grow faster than the others, resulting in a poorly shaped plant, most noticeable in potted crops like poinsettia or chrysanthemum. The triazoles also are very "soil active" which means they may be adsorbed to particles in the media and become available to the plant through root uptake. Therefore, drenching is a very effective application method for these chemicals in crops where it is economically feasible.

The label will provide a recommended application volume for sprays or drenches, especially for chemicals that are soil active. All foliar applications of PGRs should be applied on an area basis, i.e., uniformly spray the area where the plants are located with the recommended volume of solution. Do NOT spray individual plants or spray to reach a subjective target like "spray to glistening." Since every applicator will have a slightly different definition of these goals, there will be no way of recommending appropriate rates or obtaining predictable results. For soil active PGRs, dosage equals the concentration of the solution multiplied by the volume applied in the treated area.

Therefore, to improve predictability, the label-recommended spray application rates are generally set at 2 qt. finished spray per 100 sq.ft., a comfortable walking pace for applicators with hand-held sprayers. Since Bonzi and Sumagic are soil active, precautions should be taken to avoid over-application with sprays.

Spray applications require more attention to detail, because overspray material lands or drips onto the medium. Remember that dosage equals concentration times volume. The effect of Sumagic application volume on growth of vinca (*Catharanthus roseus*) at four weeks after treatment. A 1 ppm spray solution of Sumagic was applied at the label recommended volume of 2 qt. per 100 sq. ft., at 3 qt. per 100 sq. ft., or at twice the label rate, 4 qt. per 100 sq. ft. This high volume application was comparable to the amount of spray you might apply "to runoff." These vinca plants were effectively treated with 0, 1, 1.5, or 2 ppm Sumagic (dose = concentration x volume). Recognizing that stem coverage is necessary for the triazoles, you may need to apply a higher than recommended volume to large or dense plants to obtain adequate coverage. In fact, the Bonzi label recommends 3 qt per 100 sq.ft. for "larger plants with a well developed canopy." Adjust the concentration you apply accordingly. This suggests the importance of record keeping.

Spray Equipment : To assure proper spray volumes, your compressed air sprayer should be equipped with a pressure gauge and regulator and you should consistently use the same nozzle for all PGR applications.

Your sprayer should be calibrated by determining the output of the chemical with the selected nozzle at the selected pressure within a specified time period. Using this information, you can apply a known amount of material to a known area. Spray droplet size also affects response with smaller droplet sizes providing better coverage, but only up to a point. Mist or fog type applicators do NOT provide adequate volume for coverage of plant stems and have not been effective when used with compounds like Bonzi and Sumagic.

Applying Drenches

Drenches have several advantages over sprays. Drenches generally have less effect on flower or bract size and tend to provide longer lasting growth regulation than sprays. Drenches are easier to apply uniformly than sprays because the drench volume is easily measured, and when applied to moist media, it is easy to obtain good distribution of the PGR in the media. Therefore, the resulting growth regulation is frequently more uniform. The label specifies the recommended volumes for drench applications to different size pots or types of media. Read the label. In general, 4 fl. oz. of drench solution is applied to a sixinch "azalea" pot, and that volume is adjusted up or down with pot size to obtain a volume where about 10% of the solution runs out the bottom of the pot when the media is moist.

Remember that the amount of active ingredient applied to plants using soil-active PGRs is a product of the concentration (ppm) of the solution and the volume applied.

Label recommendations for drench applications give solution rates (in ppm) and volume recommendations. In some cases, drench application recommendations are given in terms of milligrams of active ingredient (mg a.i.) per pot. For Bonzi, the label provides mixing directions for mg a.i. solutions for Bonzi, or, you can use the NC State University "PGR Calculator" to obtain solution directions for drench recommendations using this format.

Other methods of applying PGRs directly to the media have been developed and labelled. For example, Bonzi and A-Rest are labelled for chemigation or application through the irrigation system. These are generally limited to flood (sub-irrigation) or drip irrigation, not overhead sprinkler systems. Again, rates vary with the volumes used and method of application. Bonzi applied once by sub-irrigation requires 50% to 75% of the amount of Bonzi that is applied in a typical drench application. Read and exactly follow the label for chemigation applications.

Other Types of PGR Applications: Three other methods of providing a drench type application of soilactive PGRs on a more economical scale are being used by growers. One is media surface application sprays. These are spray applications made to the surface of the media of filled flats or pots. The treatment is applied at normal to high spray volumes, but since it is applied to the media surface it is activated by irrigation and is available to the plant in the root zone. Both Bonzi and Sumagic are labelled for this method of application. Rates are lower than used for sprays, but higher than used for drench applications. A second method is called "sprenches" which is a high volume foliar spray that results in runoff into the media, providing a drench effect. Rates are lower than those recommended for sprays.

A third technique is called "watering in" where the PGR, A-Rest and Bonzi are currently labelled, is injected into the irrigation water and applied in each irrigation at very low rates of active ingredient. All of these application methods use the relationship between rate and volume to provide the desired control and preferred application methods. Again, you must develop techniques that fit your production methods and your growth management preferences.

Beware of Bark: For all media applications, be aware that soil-active PGRs tend to be tied up by bark particles in the media which makes it less available to the plants. Therefore, if your media mix is high in bark you will need to compensate for this unavailability by adjusting your application rates up for any type of drench or media applications.

Growing Conditions: Look also for label recommendations on time of day or condition of the plant for optimum treatment response. Generally, a healthy, unstressed plant growing under low evaporative conditions, e.g., early in the morning or late in the afternoon, is most responsive to treatment. To maximize uptake, the chemical must remain in contact with the leaf long enough to be absorbed. This time varies for the different PGRs. Plants treated with B-Nine or Florel should not be overhead irrigated for at least 12 hours after treatment, but plants treated with Bonzi or Sumagic may be irrigated one hour after treatment. Spraying when the treatment will not dry quickly increases absorption of the active ingredients and increases the effectiveness of the treatment. Read the label for any warnings on how irrigation or environmental conditions will affect plant response to the PGR treatment.

Treat All Recommendations as Starting Rates for Your Own Trials

The multitude of variations possible in application methods, cultivar and species grown, and growing conditions make it impossible to recommend specific rates for all operations. Use the lower label rates for starting your own trials. There are a couple of general rules for using rate recommendations from other sources:

1) Southern growers use higher rates and more frequent applications than Northern growers. Rates for Virginia tend be closer to the Southern rates.
2) Outdoor applications usually require higher rates or more frequent applications than for plants grown inside the greenhouse.

Recordkeeping

Making notes on your application methods and the results of your PGR treatments will allow you to improve the consistency of your own application methods and establish rates and volumes appropriate to your production system. Note the concentration and the volume applied, the stage of development of the crop (number of leaves, approximate height, presence of flowers), and the environmental conditions under which the PGR was applied. It is always helpful to keep a few untreated plants for comparison, especially if you are new to using PGRs.

Costs of PGRs

Also consider the cost of the various plant growth regulators in developing your production program. This area would include about 650 bedding plant flats or 4000 6-inch pots. This table lists only chemical costs. You will need to add your labour and equipment costs to calculate the PGR application costs in your operation. You also will want to consider the costs of multiple applications vs. single applications when determining which PGR to use in a given situation.

Conclusions and Rate Recommendations

Plant growth regulators are valuable production tools that can enhance product quality and marketability while reducing labour for pinching and/or pruning and plant maintenance. They must be used with proper attention to other cultural practices, especially proper fertility and irrigation management. Plant growth regulators cannot correct poor production practices.

Use of Gibberellin in Floriculture

Floriculture, in its broadest sense, covers the cultivation of ornamental flowers and plants with many purposes, which include cultures of cut flowers and seedlings production. This sector is found in the economy of many countries, but mostly in European countries such as Netherland, Italy and Belgian and a few Latin-American countries such as Colômbia and Costa Rica (Castro, 1998). The tendency to cultivate certain species in areas with boundary conditions for adaptation has generated certain difficulties which has led to the search for solutions. Studies then emerged with vegetables regulators about growth plants and the benefits promoted by these substances such as gibberellin. This study aim at improving the quality and quantity of cultivations productivity. Both the quality and other characteristics of flowers can be modified by the application of vegetables regulators, which are organic substances with an important role in growth regulation, acting as a stimulator and an inhibitor, depending on its concentration and others intrinsic plant characteristics. Inside the main group, with the possibility of exogenous utilization, gibberellin was found. Thus, this review aims to detail the use of gibberellin at floriculture.

Gibberellins

In the 1950's, gibberellin was characterized as phytohormones. Phytohormones are organic compounds, nonnutrients, produced by plants and in low concentrations

(10-4 M), they promote, inhibit or modify physiological and morphological processes of vegetables (Davies, 2004). The most important gibberellin is GA1 and most of the others are precursors of GA1, except GA3, GA5 and GA6. It has a complex structure, being chemically isoprenoid.

It is synthesized by a branch in the route of terpenoids, which are synthesized by the route of mevalonic acid and by the route of metileritritolfosfato (MEP). The differences between many gibberellin are the number and the localization of the double connections and the hydroxyls group (Taiz and Zeiger, 2004).

The gibberellin group covers a large number of compounds, where 1/3 are gibberellin with 20 carbons and the others are gibberellin with 19 carbons, being more active with the GA1, GA3, GA4, GA7, GA9 and GA20. Gibberellin determines important physiological changes in plants, such as cell division and expansion, promote the growth of shorts, induce seeds germination that needs cold or light, stimulation of enzyme production such as -amylase in the germination of cereal seeds, induce flowering, partenocarpia, sexual expression, fruit development, senescence and abscission, break of the yolks dormancy, maintenance of apical dominance and promotion of stem elongation. Gibberellins are present throughout the plant and can be detected in leafs, stems, roots, seed, embryos and pollen. In general, the reproductive tissues has high concentration of gibberellin, on the other hand, the roots has lower concentration. High levels of endo genous gibberellin are also found in petals, stamens and in the inflorescence of most species as a function of the growth stage. However, the distribution of hormones fluctuates between the organs according to the reproductive development, showing its important role in stamens development.

The translocations of gibberellins are performed at the same speed as the constituents of the phloem, such as amino-acids and carbohydrates, moving in all directions in the symplast of the plant. The movement of gibberellin, exogenous or endogenous, appears to be non-polarized and can occur at both the phloem and xylem. The biosynthesis of gibberellins can be affected by environmental conditions and in general, the greater production of gibberellins occurs in long days more than in short days (Rodrigues and Leite, 2004). Moreover, the levels of gibberellin can be affected by changes in photoperiod and temperature modifying the activity of specific enzymes in the pathways of their biosynthesis. In general, long day plants and plants which require a cold period are more responsive to exogenous gibberellin, while short and neutral day plants are not responsive at all (Zeevaart, 1971). That way, the induction of flowering can be obtained through artificially reducing the duration of one or more seasons (Hertogh and Le nard, 1993). Commercially, gibberellins is used in Pro-Gibb a product which contains 10% of active principle of gibberellic acid (GA) in its formulation. When the GA became commercially available, it was applied to many plants and extraordinary results were obtained. It was thought that gibberellins could cause very large increase in plant productivity (Rodrigues and Leite, 2004).

Application of Gibberellic Acid (GA)

In the floriculture area, some producers use GA in order to improve the phenotypic characteristics of plants. The GA3 has been used in species of *Hyacinthus, Liatris, Muscari, Iris, Lilium* and *Tulipa* as a substitute for treatments with low temperatures. Although it promotes flowering, its efficiency is considered low compared to treatments with low temperatures, especially for species of the genus *Tulipa*. Treatments with gibberellin application has also been used in *Zantedeschia* to increase the number of flowers. Under non inductive conditions, GA32 and 2,2-dimethil GA4 applications in *Lolium temulentum* promoted flowering, which is already at *Sinapsis alba,* while some gibberellins (GA1, GA3, GA4, GA7, GA9) limited the flowering process, which was also observed in fruit, such as cherry, peach, apricot, almond and citrus. In plants of *Azalea* treated with 1000 mg L-1 GA3, at a minimum temperature of 16°C, significant results were obtained in the uniformity of flowering, without the need for treatment with low temperature for the Hexe and Sweatheart-Supreme cultivars.

When the combination of 100 to 500 mg L-1 (GA3) and 100 mg L-1 of kinetin (adenine-derived molecule that stimulates cell division) was performed and applied in 'red wing cultivar', at intervals of four days, it promoted the anticipation of flowering (Iersel, 1998).

It can also be found, in plants of the same habit of growth such as annual plants at different responses to the application of GA3. It was evidenced by an experiment with *Petunias* and *Impatiens*, where only the latter responded to an application to a GA3 based-product interfering with the stimulation of its flowering (Iersel, 1998). Similar results were observed by Al-khassawneh et al. (2006) on growth and flowering of *Iris nigricans* Dinsm., using plant growth regulators, especially GA3 at tested concentrations (125, 250, 375 e 500 mg L-1).

However, in low concentrations of GA3, changes were not observed in growth and flowering of chrysanthemum 'Faroe' (Vieira, 2008). Several studies have confirmed the earliness of flowering by GA3 application. Chang and Sung (2000) observed in rhododendron (*Rhododendron pulchrum*) that the application of GA3 was effective on the growth of buds and flowers per plant, showing the colours of the buttons 10 days before and anticipating the flowering in nine days against the control. For Aglaonema (*Aglaonema* sp), the appearance of first flower was enhanced by GA3 application in concentrations of 100 and 200 mg L-1, but in a concentration of 400 mg L-1, this event happened five days later. However, the earliness of flowering was with GA3 when compared with controls (Henny, 1983).

Other phenotypic characteristics were observed by other authors. In Better Times rose cultivar, application of GA3 at concentrations of 10 to 100, increased the stem length and fresh weight of cut flowers (Castro, 1998). In bulbous plants, such as cyclamen (Treder et al., 1999), tulip (Rudnicki et al., 1976) and dahlia (Khan and Tewari, 2003), there was an increase in the height of the plants after GA3 application. In anthurium (*Anthurium andreanum*) these applications were not enough to increase the height or to stimulate flowering (Wang, 1999). Taiz and Zeiger (2004) assume that the increase in height of the plant can also be attributed to auxin, since it can cause the synthesis of gibberellins and thereby induce cell elongations. In the literature, it was generally observed that, GA3 caused an increase in the number of flower buds or in the number of flowers or Inflorescences. In *Syngonium podophylum* Schott cv. White Butterfly treated with GA3 (0, 10, 20, 40 e 80 mg L-1) showed that the best average number of flowers produced by the plant (2,4) occurred in treatment with 80 mg L-1, when compared to the control (Henny et al., 1999). Lower value was found in *Hemerocallis hybrida* cv. Graziela Barroso in the third application of GA3, wich showed 2.93 buds per plant (Ottman, 2006). The pre-treatment with GA3 at a concentration of 83 mg L-1 was effective for the production of seedlings serum by rhizomes, due to the formation of trees with greater height (Tavares and Almeida, 2005).

In post-harvest, the application of exogenous plant growth regulators interferes with the senescence of leaves. This can be observed in lily, where application of GA3 in the preservative solutions significantly delayed yellowing and respiration rate, but did not influence the longevity of flowers (Mello, 2001). The addition of 50 ppm of GA to the solution of "pulsing" delayed leaf senescence in lily and aster. The foliar application of 100 mg L-1 GA3, besides increasing the diameter of the stems, inflorescences and floral disc, also increased the vase life of 12 varieties of chrysanthemum grown in greenhouses (pot) (Dehale et al., 1993). This effect was also observed in chrysanthemum 'Gompier-cha', in which application of 100 mg L-1 in plants grown in greenhouses (pot), prolonged the vase life of 16 days compared to untreated controls (Freitas, 2001). The induced absorption of a solution by exposing the stems at an elevated temperature with low concentrations of GA3 (10 e 20 mg L-1), has also shown increased durability in species such as chrysanthemum (Laschi, 1999). However, little is known about the effects of plant growth regulators on the flowers and as such, these effects are contradictory. This result is similar to that of Brackmann et al. (2005) who evaluated the effect of GA3 on three varieties of chrysanthemums and noted the promotion of senescence of both leaves and flowers.

The application of GA3 in the field did not reduce or retard the senescence process in chrysanthemum 'Faroe'. This author also studied the biochemical changes in post-harvest chrysanthemum 'Faroe' submitted to different concentrations of GA3 applied field and observed increase in the level of polyamines.

Some authors assert that the action of regulators may vary according to species and cultivar, which could explain the increased of longevity in post-harvest treated with GA3 (Freitas, 2001). In this review some studies have shown that the efficiency of gibberellin application in flower is related to application technology (product concentration, shape and timing) as well as various factors such as plant age, types of gibberellins, growth and share treated plant, plant species and climatic conditions. In this study we conclude that there are several studies showing the effect on exogenous application of gibberellin in flower, but many other studies show contrary situations. Certainly, relations between some of the components of this system are unclear and difficult to be traced.

2

Commercial Flower Production

The floral industry is one of the major industries in many developing and underdeveloped countries. Floriculture as an industry began in the late 19th century in England, where flowers were grown on a large scale on the vast estates. The present day floral industry is a dynamic, global, fast-growing industry, which has achieved significant growth rates during the past few decades. In the 1950s, the global flower trade was less than US$3 billion. By 1992, it had grown to US$100 billion. In recent years, the floral industry has grown six percent annually, while the global trade volume in 2003 was US$101.84 billion.

The floral industry essentially consists of three major components: the growers, the wholesalers and the retailers whose businesses are quite intermingled. The recent trends are more towards eliminating the intermediaries, the wholesalers between the growers and the retailers, so that the flowers are made available at considerably low prices.

Transportation

Some flowers are sent packed flat in boxes. This enables large amounts of flowers to be packed in small spaces like aircraft holds. Other flowers cannot survive for long periods out of water such as orchids, gerberas (gerber daisies) and water lilies. These are either sent with their own sealed water container (called picks) on each stem end for more expensive or tropical flowers or are transported in buckets of water (This method of transport in water is often referred to as ["Procona]"). The latter method extends the life of flowers and reduces labour time as flowers are ready for sale, but obviously also reduces the amount of flowers that can be transported as they are much heavier than dry-packed flowers and hence air transportation charges are higher.

Flowers take a number of routes to the consumer, depending on where they are grown and how they are to be sold. Some growers cut and pack flowers at their nurseries, sending them directly out to the consumer by mail order. Some flowers are sent to packing companies, who grade the flowers and arrange them in bunches for sale to supermarkets or to deliver by mail order. Some flowers are graded and sleeved by the growers and sold at wholesale flower markets; the wholesalers then sell them on to florists who condition and arrange the flowers for the consumer.

Holland and the History of the Flower Industry

Traditionally, the center of flower production has been near their largest consumers: the developed world, where Japan, Western Europe and North America were both major producers and consumers. The major consumer markets being Germany (22 percent), the United States (15 percent), France (10 percent), the United Kingdom (10 percent), the Netherlands (9 percent), Japan (6 percent), Italy (5 percent), and Switzerland (5 percent).

Holland remains the center of production for the European floral market, as well as a major international supplier to other continents. The flower auction at Aalsmeer is the largest flower market in the world. Since the mid-1970s, the production and distribution of cut flowers in Netherlands has burgeoned. In 1995, Dutch growers produced over 8 billion blooms and the flower auctions collectively traded more than 5.4 billion guilders (about $3.2 billion) in cut flowers and potted plants, contributing over 4 billion guilders annually to the Dutch balance of trade.

New Flower Growing Centres

Experts believe that the production focus has moved from traditional growers to countries where the climates are better and production and labour costs are lower. This has resulted in a paradigm shift in the floral industry. The Netherlands, for instance, has already shifted attention from flower production to flower trading, though it plays an important role still in the development of floricultural genetics. The new centers of production are typically developing countries like Colombia (second largest exporter in the world), Ecuador, Ethiopia, Kenya, and India. Other players in this global industry are Israel, South Africa, Australia, Thailand and Malaysia. New Zealand, due to its position in the Southern Hemisphere, is a common source for seasonal flowers that are typically unavailable in Europe and North America.

In Africa, Kenya is the largest exporter, supplying a large percentage of Europe's flowers, the industry there is represented by the Kenya Flower Council.

In South America, Colombia is the leading flower producer and exporter accounting for 59% of all flowers imported to The United States in 2006. The United States imports 79% of its flowers. Ecuador has become, in recent years, the leading South American rose producer and is well known throughout the world for its high quality, large headed roses due to the high altitude location of its rose farms.INDIA

Cut Flower Crop Production

The term cut flower refers to flowers or flower buds (often with some stem and leaf) that have been cut from the plant bearing it. It is usually removed from the plant for indoor decorative use. Typical uses are in vase displays, wreaths and garlands. Many gardeners harvest their own cut flowers from domestic gardens, but there is a significant commercial market and supply industry for cut flowers in most countries. The plants cropped vary by climate, culture and the level of wealth locally. Often the plants are raised specifically for the purpose, in field or glasshouse growing conditions. Cut flowers can also be harvested from the wild.

The cultivation and practices of raising cut flowers form a part of horticulture. They are often included in that branch of horticulture called floriculture.

Uses

A common use is for floristry, usually for decoration inside a house or building. Typically the cut flowers are placed in a vase. A number of similar types of decorations are used, especially in larger buildings and at events such as weddings. These are often decorated with additional foliage.

In some cultures, a major use of cut flowers is for worship; this can be seen especially in south and southeast Asia.

Sometimes the flowers are picked rather than cut, without any significant leaf or stem. Such flowers may be used for wearing in hair, or in a button-hole. Masses of flowers may be used for sprinkling, in a similar way to confetti.

Garlands (especially in south Asia), and wreaths (in Europe and the Americas) are major derived and value added products.

Longevity of Cut Flowers

Live cut flowers have a limited life. The majority of cut flowers can be expected to last several days with proper care. This generally requires standing them in water in shade. They can be treated in various ways to increase their life.

In most countries, cut flowers are a local crop; because of their limited life after harvest they have to be marketed quickly. In India, much of the product has a shelf life of only a day. Among these are marigold flowers for garlands and temples, which are typically harvested before dawn, and discarded after use the same day.

There is also a market for 'everlasting' or dried flowers, which include species such as Helichrysum bracteatum. These can have a very long shelf life.

Commerce

The largest producers are, in order of cultivated area, China, India, and the United States. The largest importer and exporter by value is the Netherlands, which is both a grower and a redistributor of crops imported from other countries. Most of its exports go to its European neighbours.

In recent decades, with the increasing use of air freight, it has become economic for high value crops to be grown far from their point of sale; the market is usually in industrialised countries. Typical of these is the production of roses in Colombia, mainly for the US market, and production in Kenya and Uganda for the European market. Some countries specialise in especially high value products, such as orchids from Singapore and Thailand.

The total market value in most countries is considerable. It has been estimated at approximately UKL 2 billion in the United Kingdom, of the same order as that of music sales.

Bud

In botany, a bud is an undeveloped or embryonic shoot and normally occurs in the axil of a leaf or at the tip of the stem. Once formed, a bud may remain for some time in a dormant condition, or it may form a shoot immediately.

Overview

The buds of many woody plants, especially in temperate or cold climates, are protected by a covering of modified leaves called *scales* which tightly enclose the more delicate parts of the bud. Many bud scales are covered by a gummy substance which serves as added protection. When the bud develops, the scales may enlarge somewhat but usually just drop off, leaving on the surface of the growing stem a series of horizontally-elongated scars. By means of these scars one can determine the age of any young branch, since each year's growth ends in the formation of a bud, the formation of which produces an additional group of bud scale scars. Continued growth of the branch causes these scars to be obliterated after a few years so that the total age of older branches cannot be determined by this means.

In many plants scales are not formed over the bud, which is then called a naked bud. The minute underdeveloped leaves in such buds are often excessively hairy. Such naked buds are found in shrubs like the Sumac and Viburnums and in herbaceous plants.

In many of the latter, buds are even more reduced, often consisting of undifferentiated masses of cells in the axils of leaves. A terminal bud occurs on the end of a stem and lateral buds are found on the side. A head of cabbage is an exceptionally large terminal bud, while Brussels sprouts are large lateral buds.

Since buds are formed in the axils of leaves, their distribution on the stem is the same as that of leaves. There are alternate, opposite, and whorled buds, as well as the terminal bud at the tip of the stem. In many plants buds appear in unexpected places: these are known as adventitious buds.

Often it is possible to find a bud in a remarkable series of gradations of bud scales. In the buckeye, for example, one may see a complete gradation from the small brown outer scale through larger scales which on unfolding become somewhat green to the inner scales of the bud, which are remarkably leaf-like. Such a series suggests that the scales of the bud are in truth leaves, modified to protect the more delicate parts of the plant during unfavorable periods.

Types of Buds

Buds are often useful in the identification of plants, especially for woody plants in winter when leaves have fallen. Buds may be classified and described according to different criteria: location, status, morphology, and function.

Botanists commonly use the following terms:

- for location:
- o terminal, when located at the tip of a stem (apical is equivalent but rather reserved for the one at the top of the plant);
- o axillary, when located in the axil of a leaf (lateral is the equivalent but some adventitious buds may be lateral too);
- o adventitious, when occurring elsewhere, for example on trunk or on roots (some adventitious buds may be former axillary ones reduced and hidden under the bark, other adventitious buds are completely new formed ones).
- for status:
  - o accessory, for secondary buds formed besides a principal bud (axillary or terminal);
  - o resting, for buds that form at the end of a growth season, which will lie dormant until onset of the next growth season;
  - o dormant or latent, for buds whose growth has been delayed for a rather long time. The term is usable as a synonym of *resting*, but is rather employed for buds waiting undeveloped for years, for example epicormic buds;
  - o pseudoterminal, for an axillary bud taking over the function of a terminal bud (characteristic of species whose growth is sympodial: terminal bud dies and is replaced by the closer axillary bud, for examples beech, persimmon, *Platanus* have sympodial growth).
- for morphology:
  - o scaly or covered, when scales (which are in fact transformed and reduced leaves) cover and protect the embryonic parts;
  - o naked, when not covered by scales;
  - o hairy, when also protected by hairs (it may apply either to scaly or to naked buds).
- for function:
  - o vegetative, if only containing vegetative pieces: embryonic shoot with leaves (a leaf bud is the same);
  - o reproductive, if containing embryonic flower(s) (a flower bud is the same);
  - o mixed, if containing both embryonic leaves and flowers.

Within Zoology

The term bud (as in budding) is used by analogy within zoology as well, where it refers to an outgrowth from the body which develops into a new individual. It is a form of asexual reproduction limited to animals or plants of relatively simple structure. In this process a portion of the wall of the parent cell softens and pushes out. The protuberance thus formed enlarges rapidly while at this time the nucleus of the parent cell divides. One of the resulting nuclei passes into the bud, and then the bud is cut off from its parent cell and the process is repeated. Often the daughter cell will begin to bud before it becomes separated from the parent, so that whole colonies of adhering cells may be formed. Eventually cross walls cut off the bud from the original cell.

Choosing a Site

Most cut flowers prefer a location in full sun throughout the entire day. The field and soil should be well drained. Wind protection is highly desirable for all plants. Windbreaks serve to reduce water stress on plants and help prevent stem breakage and floral damage. It is important to consider any competitive effect which may occur from the roots of any plant used as a windbreak.

The site should have sufficient cold air drainage to avoid recurring early or late season frosts. A source of irrigation water is essential. If animal foraging is a problem, fencing may be required.

Labour efficiency is a critical factor in choosing a site. Flower production requires intensive management. The site should have ready access for production equipment and removal and handling of the harvested product.

Study the proposed site for ease in movement of materials and plants into it to begin production, within the site while producing and harvesting the crop, and of the harvested crop to storage and packing areas. The idea is to maximize efficiency of the labour required for all production and marketing functions.

If a pick-your-own marketing strategy is your choice, a site with easy access is critical. Easy access from public roads to parking areas and from parking to the fields will enhance the consumer's overall shopping experience.

Preparing the Bed

Plants should be grown in beds raised 4–6 inches to maximize drainage. Poorly drained soil should be corrected by placing drain lines 10–12 inches deep under the beds. Check for the existence of a hardpan in the soil.

A deep-rooted cover crop such as alfalfa may help to break up the hardpan. A well-drained soil environment is essential for maximum root development and reduces the potential for root rot problems.

Organic matter should be incorporated into the beds to a depth of 10–12 inches. The addition of organic matter is best if incorporated in the fall, but can be done anytime before planting if the nitrogen status of the soil is monitored. Remember that for beds planted to perennials, it may not be possible to work additional organic matter deep into the soil until the plants are divided or replaced. For these beds, sufficient organic matter should be added initially to provide a soil structure with optimum aeration and drainage.

Always have the soil tested for nutrient content before adding any fertilizer to the planting bed. Production decisions made without adequate information are merely guesses and can be costly. If nitrogen levels are low and large amounts of organic matter have been added, an application of 20–45 pounds of actual nitrogen per acre prior to planting may be appropriate. Marketing strategy, plant growth habit, and labour efficiency determine the ideal size of a planting bed. If a pick-your-own market strategy is the choice, beds should be narrow—about 21D 2–3 feet wide—so consumers easily can pick their flowers without damaging the crop. Bed length should be about 25 feet, with sodded aisles to provide customers with quick and clean access to any product they desire. Traditional production beds are 3–4 feet wide depending on the growth habit of the crop.

Bed width is set to allow maximum light penetration to the center of the bed and to facilitate harvesting. The wider a bed is, the better the ratio of production space to aisle space and the greater the return per acre. Tall, dense flower growth habits reduce the amount of light reaching lower leaves in the center of the bed.

A worker can efficiently reach only 2 feet into a bed to make a proper cut and remove the flower without damage to the crop. Bed length also is set to maximize the area in production versus the area in aisles. The limiting factor to bed length is labour efficiency. The maximum distance a worker should carry harvested flowers is about 50–60 feet. Planting beds should be 100–120 feet long.

What to Grow: The Ideal Crop

An ideal cut flower crop, either fresh or dried, would have the following characteristics:

- low cost of production—materials and labour
- high value and unlimited demand
- high production per square foot of bed space
- extended production and marketing season
- long productive life
- ability to sell fresh and to sell surplus as a dried or preserved floral product
- postharvest vase life of at least 7 days
- resistance to diseases and pests of all types
- resistance to heat and drought stress
- long stems (18 inches)
- easy harvest and handling
- aesthetically pleasing and/or fragrant flowers, foliage, or stems.

No single species or variety of plant material will meet all of these criteria. They are presented here to provide a means of evaluating the relative desirability of producing a crop of a specific plant.

Definitions

Annual: A plant that lives, grows, and dies with a completed life cycle within a single year, usually blooms continually during its life cycle.

Biennial: A plant that normally requires two years to complete its life cycle. It will grow and produce leaves, but produces flowers and seed pods only after subsequently undergoing a cold period, usually the second year.

Perennial: A plant which has a life span of more than two years, but flowers for only a set period during the season.

Fresh Cut Flowers—Annuals

Ageratum houstonianum—Floss flower; in demand for blue flower; a variety to try is 'Blue Horizon.'

Ammi majus—Queen Anne's lace, snowflake; white lacy head can be dyed.

Antirrhinum majus—Snapdragon; tall spike required, varieties to try are Rocket and the Potomacs.

Callistephus chinensis—China aster; many cutting strains available, the Florett Series is an extradouble, large-flowered type.

Caution: aster yellows, a disease transmitted by insects, is devastating and difficult to control.

Celosia cristata—Cockscomb, feather or plume Celosia; for the red crested flower try the Chief

Series. The Sparkler Series is a red feather-type to try. Wheat celosia bear slender white plumes during summer heat and are rose tinted under cool nights of fall.

Centaurea cyanus—Cornflower, bachelors' buttons; frilly buttons of white, pink, and blue.

Centaurea americana—Cornflower; much larger flowers than *C. cyanus*, lilac-pink colour.

Clarkia amoena—Godetia; try the Grace Series; prefers cool temperatures.

Consolida regalis—Larkspur; Giant Imperial Strain is the standard strain; try Qis Series and 'Blue Cloud.'

Dianthus barbatus—Sweet William; old types are biennials, new types act like true annuals; try the new Pride of Park Avenue Series.

Eustoma grandiflorum—Lisianthus, sweet lissies; Yodels are the standard variety but also try the Echo Series and the Heidi Series.

Gypsophila elegans—Annual baby's breath; most commonly grown are the large-flowered white form, but the smaller-flowered pink, rose, and purple also have a market. Recommended for fresh use only.

Helianthus debilis cucumerifolius —Sunflower; relatively small flower of yellow to white and shades of mahogany and rust. Try 'Sunbright' and 'Full Sun,' regularly branched plants which do not produce pollen. In our trials, pollen production has proven to be a quality objectionable to the consumer. Pollen shed from the vase causes a mess.

Iberis amara—Rocket candytuft; fragrant and early bloomer.

Limonium sinuatum—Annual statice; try Excellent Series, Qis Series, and the Turbo Series. Surplus production may be marketed in dried form.

Molucella laevis—Bells of Ireland; apple green "flowers" (calces); surplus production may be used in dried form.

Nigella damascena—Love-in-amist; delicate flowers for fresh use; allow surplus production to set fruit for sale as fresh material, further surplus fruit may be used in dried form.

Scabiosa atropurpurea—Pincushion flower; dense, rounded, rich colored, flower heads.

Zinnia elegans—Zinnia; several flower types and sizes available. Never water overhead; subject to leaf diseases. The dahlia-flowered 'Giant Mammoth,' the cactusflowered 'Zenith' and the State Fair series are reported to be the best of the large flowering zinnias.

Try the Pumila Series, Ruffles and Cut-and-Come-Again for smaller sized flowers.

Fresh Cut Flowers—Perennials

Achillea filipendulina—Yarrow, fernleaf yarrow; try 'Gold Plate' and 'Coronation Gold.' For other colours try 'Jambo' (soft yellow), 'Lilac Improved' (lilac pink), 'Lusaka' (pure white), 'Nakuru' (purple and white), 'Sawa Sawa' (dark purple), and 'Wesersandstein' (light pink).

Artemisia ludoviciana—White sage; grown for the silver-gray foliage which is used fresh or in dried form. 'Silver King' and 'Silver Queen' are standard cultivars.

Asclepias incarnata & *Asclepias tuberosa*—Butterfly flower; rosepurple and neon orange respectively; easy to grow, shippable, and long-lasting cut flowers.

Aster novi-belgii & *Aster ericoides*—Aster; hybrids of both species are good cut flowers; 'Monte Casino' is the standard variety grown, the Master Series is among the best of the new hybrids.

Astilbe—Astilbe, false spirea; requires moist soil in summer; colour range of white, pink, red, and lavender.

Chrysanthemum x superbum—Shasta daisy; most popular from cuttings is 'T.E. Killin' and 'Alaska' from seed; prolific bloomers adaptable to most soils; plants are not long-lived without frequent division.

Echinops bannaticus, Echinops exaltatus, Echinops ritro, Echinops sphaerocephalus—Globe thistle; rich blue to very light blue; excellent producers for either fresh or dried markets; long-lived, but should not be transplanted.

Eryngium amethystinum & *Eryngium planum*—Eryngo, sea holly; easiest of the eryngo to grow; excellent fresh or dried; small silvery-blue to purple-blue flowers; try 'Donau,' 'Blue Star,' 'Blue Diamond,' 'Silver Stone,' and 'Fluella.' Largerflowered types are *E. bourgatti* and *E. x zabelii*. *Eryngium alpinum* is the most difficult to grow, but is the largest-flowered with several great blue cones surrounded by prickly calyx frills.

Gypsophila paniculata—Perennial baby's breath; the standard "filler" of the floral industry; used fresh or dried; suited to dry, light, and slightly alkaline soils. Vegetatively produced plants have large, fully double flowers; 'Perfecta' is the standard. Can be seed grown, but will have smaller flowers and only a portion will have double flowers.

Liatris—Gayfeather, blazing star; one of the longest lasting and finest cut flowers; native to Kansas; well adapted to our climate and soils; long-lived plants. Produce for local markets only; over-supply in national market.

Paeonia—Peony; herbaceous types are among the choicest of fresh and dried cut flowers; extremely long-lived; short harvest season but can be stored cool and dry for up to 12 weeks. Kansas has a competitive advantage in the commercial production of peonies because of climate. Plants require 3–5 years from planting to reach productive potential. Demand currently exceeds supply for both fresh and dried flowers.

Platycodon grandiflorum—Balloon flower; large bellflowerlike blooms follow balloonlike buds; adapts to a wide range of soil types; will tolerate some shade.

Salvia farinacea—Blue salvia, mealycup sage; strong grower in extreme heat; can be used as fresh or dried material; try 'Victoria,' 'Catima,' and 'Blue Bedder.' Salvia is sensitive to methyl-bromide treated soil.

Scabiosa caucasica—Pincushion flower; a traditional florists' cut flower; easy to grow and adaptable to most soils; try 'Fama' and 'Complement,' lilac blue or white seed strains.

Solidago—Goldenrod; hybrids from Holland are better than our native species; easy, trouble-free, and adaptable to most soils. Note: Goldenrod does not cause hayfever symptoms.

Fresh Cut Flowers—Bulbs

Allium—Flowering onion; all species are good fresh cut flowers; easy to grow and adaptable to any well-drained soil. The best species for cutting are: *A. aflatunense, A. caeruleum, A. giganteum, and A. spaerocephalum.*

Gladiolus—Gladioli; staggered planting for continuous harvest; plant new corms each year; dig and sell corms each fall. Standard item for florist and farmers' market sales. Try both standard and "baby" glads; the new hybrid Parigo Series is an intermediate type glad.

Lilium—Lily; the Asiatic and Aurelian hybrids along with the Oriental hybrids *L. auratum*, *L. rubellum*, and *L. speciosum* are best for commercial cut flower production.

Try the white flowered Oriental hybrid 'Casa Blanca.'

Fresh Cut Flowers—Flowering Woody Stems

The stems of *Forsythia*, *Salix*, and *Chaenomeles* can be cut when dormant, held cool and forced into bloom as fresh flowers for late winter and early spring sales.

Dried Materials—Flowers and Miscellaneous Plant Parts

The following list of the top 30 flowers for drying is the result of a membership survey by the Association of Specialty Cut Flower Growers.

Achillea (yarrows)

Ammobium (herb)

Artemisia (silver king, queen, and annual Sweet Annie)

Branches of Plants (myrtle, cedar, willow)

Iberis sempervireas (candytuff)

Carthamus tinctorius (safflower)

Celosia cristata (cockscomb)

Consolida regalis (larkspur)

Daucus carota (Queen Anne's lace)

Echinops (globe thistle)

Eryngium (seaholly)

Eucalyptus

Chrysanthemum parthenium (feverfew)

Gomphrena golbosa (globe amaranth)

Grass and Grains (Wheat, black bearded and Durum; rye, oats, rice, buffalo, quaking, barley, canary, flax, hare's tale, milo, sorghum,)

Gypsophila perfecta & *G. paniculata* (baby's breath)

Helichrysum bracteatum (strawflower)

Helipterum manglesii (Rhodanthe)

Helipterum manglesii (Acroclinum)

Hydrangea Lavandula (lavender)

Lepidium (peppergrass)

Limonium latifolia (latifolia)

Limonium caspicum (caspia)

Limonium sinuatum (annual sinuata)

Limonium suworowii (rattail)

Limonium tataricum (German)

Lunaria (money plant)

Nigella (love-in-a-mist, devil-in-the-bush)

Papaver somniferum (poppy pods)

Rosa (roses)

Tanacetum (tansy)

Xeranthemum (common immortelle)

Zea Mays (corn).

When choosing plants to grow, start with a test plot. Begin small and learn how to grow the plant and determine if it is suitable and economical to produce under your specific conditions. You should be able to produce a high-quality product before expanding production beyond the experimental stage. Keep a journal. You will find little or no information available on many plant materials you may wish to grow. Your experience will be useful for production decisions in subsequent seasons.

Planting

When to Plant

Planting dates depend on your target market and on plant classification—whether it is an annual, biennial, or perennial. In general, the peak demand for the retail florist trade is from fall through Mother's Day. Field production of fresh cut flowers for this market should include planting for maximum harvest in the spring and fall seasons. Farmers markets typically operate from late spring until frost in the fall. Consistent production throughout this period is most desirable.

Annuals

Annuals are planted into the field as soon as the danger of frost is past. Using transplants will bring the crop into flower earlier and may return higher prices early on if you are able to harvest for the Mother's Day market. Sequential plantings may be required to assure a continuous supply of product throughout the market season. Staggered plantings—two weeks apart into July—are common for many annuals. Transplants may be used initially to hit the earliest possible market, with later plantings direct seeded. When choosing cultivars, be careful to select those suitable for cut flower production. Many annuals have been developed for use as bedding plants and are not suited for commercial cut flower production.

Biennials

Biennials should be planted in the fall to ensure an adequate cold treatment before regrowth starts in the spring. The overwinter cold treatment is required for more uniform flowering. If beds or plants are not ready for planting in the fall, some alternatives exist. Placing the plants into cold storage or spraying them with 500–1,000 parts per million of gibberellic acid will substitute for overwintering the plants in the bed. While these alternatives do promote flowering, production quantity and quality are reduced compared with fall-planted stock.

Biennials tend to bloom in a condensed time period. Staggered plantings typically will not result in staggered periods of bloom. Most biennials will bloom at the same time unless subjected to cold storage or gibberellic acid treatments. Some biennials are excellent cut flowers but because of this concentrated, all-at-once bloom habit, few are grown as commercial crops.

Perennials

Perennials are placed into categories based on the presence or absence of storage organs and whether they must be dug in the fall and replanted each spring.

1. cold-hardy plants which have storage organs and can remain in the ground for several years. If allowed to remain in place over winter, they typically will bloom at the same time. Extended bloom periods can be achieved by digging in the fall, placing in 40oF storage, and planting every 2–3 weeks beginning in mid-spring. Examples in this group are liatris and lilies.
2. nonhardy plants which have storage organs and must be dug each year. Continuous bloom periods are relatively easy to achieve by staggering the planting dates from spring through mid-summer. The additional labour required to dig, store, and replant increases their cost of production. The economics of producing each species should be analyzed before extensive production is undertaken. Examples in this group are anemones, dahlias, gladiolus, and ranunculus.
3. cold-hardy plants which have rhizomes or a clump-forming growth habit and may be left undisturbed for a few to several years before being divided. After the first year, plants in this category tend to bloom together, typically within a week or so of the same time each year. Bloom periods can be extended slightly through cultivar selection. Examples of plants in this group are peony and shasta daisy.

How to Plant

In general, transplants should be planted shallow, with the roots placed just below the soil surface. Planting too deep reduces oxygen exchange between the newly developing root system and the atmosphere, slowing growth. Too deep planting encourages the development of root and stem rots.

The soil around the transplants should be firmed to remove large pockets of air but not so much so as to pack the soil. All plants should be watered as soon after planting as possible. Delaying the application of water to the newly planted crop may severely reduce overall production.

Plant Density

A closely spaced crop of annual flowers will usually produce more flowers over the growing season than a crop at wider spacing. The essential criteria in cut flower production is the number of flowers produced per square foot on growth habit of individual species and the growing conditions.

It should be noted that closer spacings produce more flowers per square foot of bed, but also reduce air circulation within the bed.

Poor air circulation could increase the likelihood of foliar diseases. If powdery mildew or leaf spot fungus is a common problem on a species you intend to produce, then a slightly wider spacing may be appropriate.

Crop Support

Some cut flowers may need support to ensure a high percentage of straight stems. A plastic or nylon material in rolls of either a 4- or 6-inch mesh is a popular supporting material. It is laid out horizontally and suspended above the bed by attaching to steel fence posts placed along the edges of the bed.

Other methods exist and any method which produces the desired result with low cost and minimal labour is acceptable. of bed space and not the number of flowers per plant. In general, the closer the spacing of annuals the more flowers produced. A 4- to 6-inch spacing within the row and 6–8 inches between rows is ideal for most annual flowers.

This will vary depending on the growth habit of each species and on the specific environment of the planting site. Perennials left in place each year have the potential to crowd each other to an extent that may reduce overall production in succeeding years. Plants that may be divided every 3 years (shasta daisy) may be planted closer than plants which may be divided every 20 years or more (peony).

In general, perennials are planted at about a 1-foot spacing, within and between rows. Peonies are planted 3 feet apart in a single row or staggered in a double row, with a 3-foot spacing within each row and 2 feet between rows. Plant density will vary depending on growth habit of individual species and the growing conditions.

It should be noted that closer spacings produce more flowers per square foot of bed, but also reduce air circulation within the bed. Poor air circulation could increase the likelihood of foliar diseases.

If powdery mildew or leaf spot fungus is a common problem on a species you intend to produce, then a slightly wider spacing may be appropriate.

Crop Support

Some cut flowers may need support to ensure a high percentage of straight stems. A plastic or nylon material in rolls of either a 4- or 6-inch mesh is a popular supporting material. It is laid out horizontally and suspended above the bed by attaching to steel fence posts placed along the edges of the bed. Other methods exist and any method which produces the desired result with low cost and minimal labour is acceptable.

How to Grow

Specialty cut flowers are a high-dollar crop. Their value is measured in dollars per square foot, not dollars per acre. Withholding or delaying application of any input, which reduces flower production or quality, is false economy.

Watering

Specialty cut flowers grown in the field, for fresh or dried markets, will require more total water than many other field crops.

To maintain floral quality and peak production, the plants must be watered frequently, sometimes daily with some soil types. A water source of sufficient volume and quality should be readily available to the production area. Growers in Kansas should not attempt to produce cut flowers without supplemental watering. Natural rainfall is not sufficient nor reliable. Without supplemental watering, production may not be economically realistic.

Overhead watering is not recommended. It may physically damage the flowers, cause spotting on the petals, splash soil onto the foliage, and promote the spread of disease. Some type of drip irrigation is recommended. It has the advantage of placing the water on the ground where it is needed and not on the flowers or foliage.

The basic principle in designing an irrigation system for cut flowers is to place the water uniformly around each plant. Spacing of irrigation lines depends on soil type and structure.

Coarse, open soils require closer spacing than dense, fine soils. The irrigation system plan should be developed by a competent designer, with the grower having a thorough understanding of the system and its operation.

Irrigation should be scheduled based on the soil moisture status in the root zone. Proper irrigation management provides sufficient but not excessive water to the crop. Water stress will reduce production and quality of a crop.

A consistently saturated soil will reduce growth and promote the development of root rot. The amount and frequency of water required will vary with the weather and stage of maturity of the crop.

Fertilizer

Before initiating any fertilizer program, always test the soil for nutrient content. The increased water requirement of cut flowers creates an increased requirement for fertilization. The application of fertilizer should coincide with crop needs. Higher rates are required in the initial phases of growth while you are building the size of the plants.

Lower rates are required as a crop approaches harvest. For crops with a harvest that extends throughout the growing season, lighter and more frequent application of nitrogen may be necessary to reduce the nonproductive cycle between flushes of bloom. Nitrogen deficiency is the most common nutrient problem. Many growers incorporate 25–45 pounds of actual nitrogen per acre before planting. Minor element deficiencies are common in areas of Kansas with alkaline water or soils.

Weed Control

Weeds must be controlled in the field production of cut flowers. Competition with weeds reduces the quantity and quality of floral production. A bed full of weeds increases the time required to harvest, raising labour costs. Due to limited production of many species of specialty cut flowers, only a few herbicides are labelled for use. Contact your county Extension agent for an update on herbicides labelled for flower production. Weed barriers can be used. They have the added benefit of restricting soil splash onto foliage and flowers from rain and irrigation.

The most common barriers in use are plastics and paper. The major disadvantage with plastic mulches may be in their disposal after removal from the bed. The barriers restrict water evaporation from the soil surface, reducing loss and maintaining a more consistent soil moisture content.

Straw and wood chips have been used as mulches to reduce weed competition, improve water retention, and reduce temperature fluctuations within the soil. Being organic materials, they slowly decompose and increase organic matter in the soil. They are applied in a 2- to 4-inch layer and must be reapplied each year. A hoe and hand weeding are the best methods for control. Availability and cost of labour may be prohibitive in all but the smallest production situations.

Insect Control

Good cultural practices are the best insect control available. A healthy, actively growing plant is more resilient to insect attack. The ideal approach is a preventative program. Control insects early, when they are first detected. Do not wait until a serious infestation occurs. Less chemical can be applied to a smaller area to control these hot spots as they develop than would be required to spray the entire crop. Aphids, leaf hoppers, spider mites, and thrips are the most common insects encountered.

Integrated pest management (IPM) is the process of using all the different methods of controlling insects in an integrated approach. The main goal is to reach an acceptable level of insect control with minimal use of chemical pesticides. Monitoring insect species and keeping records as to insect numbers and location are integral parts of the program.

Biological controls using parasites, predators, and pathogens can help keep insect levels in balance. They are especially useful when the grower specializes in only a few crops. Biologicals tend to be specific to an insect on a crop. It is easier to manage an insect on a crop than it is to manage several insects on a variety of crops. All chemicals should be applied intelligently. Always apply at the lowest possible effective rate. Learn the life cycles of all insects associated with the culture of your crops. Know which stage of each insect's life cycle is susceptible to which chemical spray. Spray only when control is possible.

Disease Control

Foliar fungus diseases are the most serious disease problem on cut flowers. If mildew or other foliar diseases are a recurring problem, it might be wise to increase the spacing between plants slightly to improve air circulation around the foliage. Powdery mildew and black leaf spot are two of the most common foliar diseases. Botrytis attacks flowers during cool damp weather. Improper watering or frequent rains may cause some root rot problems. These can be minimized by planting on raised beds. Occasionally *Fusarium* and *Verticillium* wilt have become problems. Aster Yellows Virus can be devastating on susceptible crops. Manage the crop, Don't neglect it, Practice good sanitation; keep the field free of weeds and plant debris. Adjust planting density for each species to allow sufficient air circulation within the bed. Control leaf hoppers and thrips to reduce the potential for spreading virus. Rogue out diseased plants and remove from the production area.

Harvest

Many authors have stated that flowers should be harvested at the peak of perfection. Although this advice seems straight forward, it is ambiguous considering the differences in flowers and concepts of peak of perfection. For the consumer, the peak of perfection is when the flower is showing its best colour and form, and lasts the longest. For the grower, the wholesale broker and the florist, it is the stage when the flower will hold up in the marketing chain the longest and meet the expectations of the consumer.

Flowers for wholesale markets should be less developed to allow for handling and shipping time. For the longest vase life, flowers should be harvested in the morning before the heat of the day. The flowers are cooler and have more water in them. Less energy will be required to cool the flowers and wilting will be less of a problem.

Flowers should be removed from the field as soon as possible and placed in a cooler at 40oF or a cool place that is less than 60oF until they can be sorted, graded, packaged, and shipped to market. Special attention should be placed on harvest equipment. Knives or shears should be sharp so all stems can be cut cleanly. Dull cutting instruments can crush stems, restricting the flow of water into the stems and shortening the vase life of the flowers.

Knives, shears, and harvest containers should be routinely disinfected before each use. A mild chlorine bleach solution (1:10) is adequate. Harvested flowers should be placed in tepid water (110oF) with a floral preservative added. When using a preservative, put flowers in plastic containers. Preservative chemicals may react with a metal container. Floral preservatives have many components that lengthen the vase life of flowers. First, they acidify the water. Flowers keep longer in acid water (pH 3.5).

Water tests should be conducted to determine the pH of your water. Most water in Kansas is hard or alkaline (pH > 7.0), especially city water. If your water is very hard, it may require additional acid.

Chemically softened water should not be used in any stage of cut flower production or handling. The high level of sodium in chemically softened water can be toxic to flowers. If large-scale production is being considered, installation of a water deionizer is recommended to ensure a reliable source of high-quality water.

Floral preservatives contain chemicals that stop or slow the growth of microorganisms. A low pH also retards microorganism growth. Microorganisms can plug the water-conducting channels—xylem—of the flower and cause decay.

Floral preservatives also provide food in the form of simple sugars, like sucrose. Providing the flower with food lengthens vase life and brightens the flower colour.

Floral preservatives can be made from scratch, but beginners should use commercially prepared products until they become familiar with their crops' handling requirements and water quality.

Postharvest

After flowers are removed from the field and placed in the packing shed, the stems should be cut under water. Stems will accept water as long as the xylem is not blocked. Air bubbles drawn up into the xylem when the stem was cut in the field can block the upward movement of water. The air bubble blockage may be eliminated by removing an inch of stem under water.

Fresh cut flowers are extremely sensitive to temperature and humidity. Because they have such a high surface-area-to-volume ratio, they can lose water and wilt quickly if not kept under low temperature and high humidity conditions. The flower tissue has a high metabolic rate, so cooler temperatures are necessary to slow metabolism. Most flowers should be stored at 32–35 oF and 90–95 percent relative humidity if they are to be kept awhile.

Flowers are sensitive to ethylene gas. Ethylene is produced naturally by flowers, but can come from other sources including decaying plant material, old flowers, ripening fruit, gasoline and propane combustion and welding. Careful selection of a location for the storage area away from motors and welding equipment is important. Sanitation and good inventory management should eliminate decaying plant material and old flowers. Flowers should not be stored in the same facility as ripening fruits and vegetables.

Good air circulation in the storage area will help dilute any ethylene which may be present. Some species are so sensitive to ethylene that you need to do more to protect them. Silver thiosulfate (STS) reduces the harmful effects of ethylene and is available from most suppliers of commercial preservatives. Silver is a heavy metal that can pollute ground-water and soil. There are systems available to recycle the STS and recover the silver.

The stem tips of most spike type flowers will bend upward if laid horizontally. This is known as negative geotropism. Spike type flowers should be stored and shipped upright. Bruising and breaking flowers reduces their aesthetic value and therefore their economic value. Wounds also allow entry of many disease organisms and increase ethylene production by the flowers. Careful handling to prevent damage is imperative to ensure a long-lived, high-quality flower.

Pulsing

Pulsing is a chemical treatment of flowers to prolong the vase life. The treatment is not continuous and typically lasts for 2–12 hours. Silver thiosulfate (STS), and 5–10 percent sucrose solutions are chemicals often pulsed into flowers. These solutions are put in the holding water and the flowers are held at various temperatures for a certain time period.

Different species and different cultivars require different pulse treatments. Although pulse treatments have not been determined for many specialty cut flowers, there are too many different treatments to cover in this publication.

Marketing and Selling

Flowers can be marketed to a flower wholesaler, a retail florist, or directly to the public at a farmers market or farm market. Each market has advantages and disadvantages. Direct to the public is the easiest market to enter. Sorting, handling, and packaging usually are minimal as is the capital investment. A wide variety of species and cultivars may be sold with success. The disadvantage is the volume of sales may limit the income.

Wholesalers will accept large volumes of flowers. You will have to package, sort, and grade flowers to their specifications, and they may not accept some specialty flowers. They set the price for your flowers. It takes awhile for the grower and wholesaler to develop a good working relationship. Retail florists buy from wholesalers.

If you market directly to the florist, you cut out the middleman and can get a higher wholesale price for your flowers. Florists may not be as concerned with packaging as wholesalers. Both florists and wholesalers will insist on high quality. Florists may be more interested in specialty flowers than wholesalers. You will have to deal with several florists to sell the same volume you would to a wholesaler. The market or markets you choose to enter must provide sufficient profit margins for you to stay in business.

Ornamental Plant

Ornamental plants are plants that are grown for decorative purposes in gardens and landscape design projects, as house plants, for cut flowers and specimen display. The cultivation of these forms a major branch of horticulture.

Garden Plants

Most commonly ornamental garden plants are grown for the display of aesthetic features including: flowers, leaves, scent, overall folige texture, fruit, stem and bark, and aesthetic form. In some cases, unusual features may be considered to be of interest, such as the prominent and rather vicious thorns of *Rosa sericea* and *cacti*. In all cases, their purpose is for the enjoyment of gardeners, visitors, and/or the public.

Trees

Similarly certain trees may be called ornamental trees. This term is used when they are used as part of a garden or landscape setting, for instance for their flowers, their texture, form and shape, and other aesthetic characteristics. In some countries trees in 'utilitarian' landscape use such as screening and roadside plantings are called *amenity trees*.

Cultivation

For plants to be considered as ornamental, they may require specific work and activity by a gardener. For instance, many plants cultivated for topiary and bonsai would only be considered as ornamental by virtue of the regular pruning carried out on them by the gardener, and they may rapidly cease to be ornamental if the work was abandoned.

Ornamental plants and trees are distinguished from utilitarian and crop plants, such as those used for agriculture and vegetable crops, and for forestry or as fruit trees. This does not preclude any particular type of plant being grown both for ornamental qualities in the garden, and for utilitarian purposes in other settings. Thus lavender is typically grown as an ornamental plant in gardens, but may also be grown as a crop plant for the production of lavender oil.

The Term

The term ornamental plant is used here in the same sense that it is generally used in the horticultural trades. The term largely corresponds to 'garden plant', though the latter is much less precise, as any plant may be grown in a garden.

3

Bedding Plants Production for Floriculture

Bedding plants are plants (usually annuals) massed with others to produce the maximum in visual appeal. With an eye to the five basic elements of landscape design (colour, scale, line, form and texture), a landscape designer skillfully arranges each bedding plant in relation to the accompanying annuals, perennials, shrubs and trees.

Bedding plants are showy plants which are grown in containers and greenhouses and designed to be transplanted when they are mature. Gardeners use bedding plants to create a display of flowers or foliage quickly. Some gardeners like to grow their own bedding plants, while others purchase them from nurseries when they have need for colourful, interesting plants. Typically, displays of bedding plants start appearing in the spring, and nurseries will sell them into the summer months.

Annuals are commonly used as bedding plants, but it's also perfectly acceptable to use perennials or biennials for a longer lasting display. The plants are generally chosen because they have showy flowers or foliage. The term "bedding" comes from the fact that these plants are often arranged in flower beds, but bedding plants can also be grown in hanging containers, ornamental pots, and window boxes just as easily.

Marigolds, sweet William, petunias, primroses, snapdragons, salvias, pansies, violets, cyclamen, chrysanthemums, sage, lantana, alyssum, strawberries, and many others can be used as bedding plants. Gardeners can array these plants in an assortment of ways, and by pinching flowers back as they grow old, a gardener can often keep them in bloom for weeks or months, while maintaining a neat appearance in the process.

Growing bedding plants from seed is often very easy, with the assistance of a greenhouse to start seedlings. Seeds are readily available from nurseries and garden supply stores, along with containers for growing starts. Once the last chance of frost is over, the plants can be established in the locations where the gardener wishes to grow them. Gardeners who want some colour in the winter can establish winter bedding plants such as flowering kale, primulas, and ornamental cabbage in the fall, giving the plants a chance to adjust to cooler temperatures so that they will grow in the winter. One of the nice things about bedding plants is that they are not designed to be permanent additions to the garden, which means that the look and feel of a garden can be changed radically every year or every other season. Mature ornamental plants can also greatly enliven a home or garden, and only a few hours of work is required to prepare the soil and set the plants out. Gardeners can also experiment with things like planting colourful plants in fanciful patterns, or playing with colour to make the garden more visually interesting and engaging.

Growing and Marketing Bedding Plants

Growing bedding plants can be a relatively easy way to start a horticultural business or add to an agricultural business. Bedding plants can be a profitable cash crop from spring through fall, but they do require planning and attention to achieve a consistent, high-quality crop. Historically, bedding plants were defined as herbaceous plants sold to home gardeners for "bedding out" in the outdoor landscape in the spring to provide seasonal colour. However, from a greenhouse production point of view, bedding plants are a het-erogeneous group of plants started under controlled conditions that share a common production methodology. Bedding plants include a wide range of plant species and cultivars that may have multiple applications. These include herbaceous annuals and peren-nials, biennials, herbs, ground covers, vegetables, small fruits (strawberry), and a few woody species that die to the ground in the fall.

When consumers buy bedding plants, they prob-ably don't see the same thing you see. To many grow-ers, bedding plants are small plants to be planted in the landscape. To many consumers, these plants are instant color to perk up the home and garden. Planting bedding plants gives consumers an oppor-tunity to dig in the soil around the home and spend some time outside in the spring. Most consumers want to put little or no effort into maintaining the plants once they are planted. Householders want plants that produce color all summer long even if there is too much or too little rain or too much heat. Many consumers look for something different from their neighbors' plants. Others look for those plants that performed well last year. Consumers often remember which plants did poorly, and they make an effort not to buy them again the next year. So, think consumer and think instant color when planning bedding plant production.

Market Period

By far, the largest market period for bedding plants occurs in four to six weeks during the spring. In Alabama, this occurs from late March to early May. American consumers demand that bedding plants be in bloom at the time of purchase. Most bedding plants are sold in small containers or market packs for transplanting into the garden. This poses several challenges for the grower. The large assortment of species and cultivars tests the knowledge and skill of the grower to meet environmental and cultural requirements. Likewise, scheduling production for a wide variety of plants to meet market dates with top quality requires an intimate knowledge of the needs and timing for each plant type. Lastly, peak sales are often driven by the arrival of warm weather that encourages large-scale customer purchases. Therefore, spring sales may be "late" or "early" depending on the weather. The grower may be faced with the problem of holding a crop past its peak or pushing a crop for early sales.

Traditionally, few bedding plants have been grown during the summer. Two market changes, however, have created a demand for large-container bedding plants for late spring and summer. These are the use of bedding plants by landscape firms in large commercial installations such as corporations and large, private institutions and a late start on gardens by consumers. In each of these cases, the customer demands fully-grown bedding plants for instant colour, usually in 4- to 6-inch pots.

Because of the economic dependence on spring sales, much marketing and promotion have gone into fall bedding plant sales. Development of improved cultivars of cold-hardy annuals such as pansy and ornamental kale has resulted in the fall's becoming a more important season for bedding plant sales, especially in the Southeast. Although not as large as spring sales, this market has grown tremendously over the past 10 to 15 years.

Industry Size and Growth

Bedding-plant production is an important component of the floriculture industry in Alabama and all the United States. In 1998, the USDA reported nationwide bedding-plant production at $1.8 billion, up more than 4 percent from 1997. Strong growth was evident in the value of bedding-plants produced in the past 10 years with an average increase of more than 8 percent per year. In 1997, Alabama producers grew annual bedding-plant flats valued at $38.5 million. Bedding- and garden-plant production comprised 72 percent of all reported floral crops produced in Alabama in 1997 (valued at $53.6 million). Bedding-plant production comprised 54 percent of the total United States floral crop production (valued at $3.9 billion) in 1998. Bedding-plant production continues to grow at a strong pace both in Alabama and across the nation.

Establishing a Market

Each year, many small businesses spring up selling bedding plants by the roadside. The consumer's demand for plants in the spring seems enormous, and the industry continues to grow at a strong pace. Many different kinds of businesses want to get some of the market and sell bedding plants. Bedding-plant sales often seem to be a sure thing, but finding a market before planting the first seedling or plug is the smartest business strategy for any bedding-plant grower.

Start by planning to grow a small number of the most popular bedding plants. In successive years, plan to increase the number gradually if sales were what was projected. If this is the first season growing bedding plants, check out the competition and the products they sold in the past year. Check the prices of the competition and make some estimates of how much it will cost to grow bedding plants. Many established growers in the area are often willing to share marketing information. Work out a plan on paper to produce plants and sell them at a profit.

Plan Each Step of Production

Follow five steps each year to plan a bedding plant production schedule.

1. Decide who makes up your market.
2. Decide when to market your plants.
3. Select which kinds of plants and which cultivars of those plants to grow.
4. Determine how you will produce them.
5. Calculate at what price to sell them for a profit.

First, decide who makes up your market. Many kinds of consumers use bedding plants throughout the year. Homeowners are perhaps the largest market because they want a variety of bedding plants. Landscape contractors, landscape maintenance firms, garden centres, home centres, supermarkets, and discount stores also buy bedding plants. Talk with these potential customers several months to a year before they will need the plants. Many of these businesses will make a contract with a grower to produce a certain number, size, and type of bedding plants well ahead of their need. They will also need some bedding plants only a few weeks before sales.

Second, decide when customers want to buy plants. Consumer demand for bedding plants is greatest in spring, yet there is also some demand for bedding plants in the summer months and in the fall. Determine when plants should be ready for sale, then plan from that market date back to a planting date. Bedding plants sell best when several flowers are in colour. Be sure to plan for enough time to pro-duce flowers on the plants, especially if that's what customers specify.

Third, determine which plants and which cultivars of those plants to produce. Several factors should be considered when deciding which bedding plants to grow. Two prime considerations are popularity among consumers and performance in the garden, especially heat tolerance. You may not be able to grow every-thing that the customer wants. Some crops, such as begonias, are expensive to grow because they take a very long period of time to produce. Others are not heat tolerant, a big factor to consider in Alabama. When shipping to out-of-state markets, find out which plants work well in particular climates. What have homeowners been buying? Impatiens have been the best selling variety of bedding plant for 10 years in a row. Consider what people bought last year when planning for next year. Consumers' past purchases are a very good indication of what they will want to buy next year. Plan, too, for production of a few of the good-performing but unusual plants. The newer items often sell well because some people want plants just a little different from their neighbors' plants.

Selecting Heat Tolerant Varieties

Anticipate the stress that summer heat and drought will put on plants outdoors and consider growing those cultivars that perform well in the South. Customers may buy plants that are not heat tolerant this year because they are attractive, but they won't be back for more next year because the plants performed poorly. Several universities and botanical gardens throughout the South conduct outdoor trials of new bedding plant cultivars and types for garden performance. Results from these trials can be obtained by contacting individual institutions. Many of these are available on the World Wide Web.

Trials conducted at Auburn University over the past five years provide recommendations for annual plants on a seasonal basis. Viola x writ-trockiana (pansy), Viola tricolour (Johnny jump up), Brassica oleracea (ornamental cabbage and kale), and Papaver nudicaule (Iceland poppy) are some of the best annuals to plant in the fall for colour in the landscape from fall through spring. Bellis perennis (English daisy), Calendula officinalis (Pot marigold), and Primula x polyantha (primrose) are some other annuals that could be planted in the fall for colour in the early spring. Annuals that could be planted for colour in the spring to last through early summer include Antirrhinum majus (snapdragon), Dianthus chinensis (China pinks), Lobelia erinus (edging lobelia), Lobularia maritima (sweet alyssum), and Petunia x hybrida (hybrid petunia). Annuals that show better than average heat or drought tolerance in Alabama landscapes are Begonia x semperflorens-cultorum (wax begonia), Catharanthus roseus (annual vinca), Coleus x hybridus (coleus), Gomphrena globosa (globe amaranth), Melampodium palu-dosum (medallion flower), Portulaca grandiflora (moss roses), Salvia farinacea (mealycup sage), and Zinnia linearis (narrow-leaf zinnia). Consider plant-ing some of the annuals that perform well under cooler conditions for the transition from summer to fall. These include Tagetes erects (African marigold) and Petunia x hybrida (petunia).

There are three All America Selections Gardens in or very near Alabama. Two have Display Garden status, meaning that plants are not judged but rather are used to display new varieties in comparison to proven favorites. Display Gardens are located at the Birmingham Botanical Gardens and at Bellingrath Gardens near Mobile. The closest official Trial Garden is at Callaway Gardens in Pine Mountain, Georgia. Gardeners and growers are welcome to view the All America Display and Trial Gardens to gather information for their own gardens and businesses.

All America Selections can also be used as promotional tools when advertising products to consumers. Some consumers know that the symbol "AAS" means they are buying a variety of plant that per-formed well in nationwide tests. Others may learn to see it as the mark of a good product to look for in the future. You can help educate customers and increase the satisfaction they get from plants by helping them to select AAS plants for their gardens.

Cultivar Series

Cultivars of many bedding plant species are marketed by seed companies as groups called series. These are often breeding lines with similar genetic backgrounds. Cultivars within a series often have similar characteristics such as plant size and shape, environmental tolerance, and cultural requirements. However, the individuals within a series have different flower colours and different colour patterns. Selecting a variety of colours from the same series allows the grower to provide the market with a variety of colours to choose from while keeping cultural requirements simple.

Getting Started

A new business will need some type of protective structure in which to grow bedding plants. Any number of structures can be constructed or pur-chased to suit this need. The level of sophistication required in these structures will depend, in part, on whether or not you plan to grow your own seedlings. Containers, media, fertilizer, growth regulators, and pesticides are also costs to plan for when getting started. If this is your first horticultural venture, ask your county Extension agent for publication ANR-691, "Starting a Greenhouse Business," for additional details on starting up.

Production Structures

Even mild Alabama winters are not warm enough to produce most spring bedding plants out-side. A greenhouse structure with a heat source is usually required; however, crops with short production times may be produced outside, in cold frames, or in greenhouses without a heat source.

A greenhouse constructed of double-layer plastic, with the plastic layers 2 to 4 inches apart, can make a lightweight structure ideal for bedding-plant pro-duction. The plastic is stretched over an aluminum frame to form a Quonset structure. The ends of the house should have plastic covers that can be easily removed when the temperature climbs in the spring. Ventilation to bring cool air inside on warm days is just as important as heating on cool days.

Propagation Schemes

The majority of bedding plants are propagated from seed, usually sown from late December into March. Ten to 15 years ago most bedding plant seeds were sown and germinated in open containers, usually wooden flats, or sown directly in the final container. Once the seedlings were large enough to handle (usually 2 mature leaves), they were pricked out by hand and transplanted to market flats for finishing.

Therefore, most of the production time was spent in the final container. However, this method had several problems: 1) a lot of time and labour was required for transplanting, 2) diseases could spread easily from seedling to seedling in an open tray, and 3) damage to sensitive roots caused transplant shock. Seedlings were essentially uprooted in the transplanting process and had to start over developing a root system.

The development of plug technology in the 1970s and 1980s resulted from efforts to overcome the problems associated with the traditional method. As the name suggests, these little plants are ready to be "plugged" into a pot or cell pack to produce a mature plant in an average of 3 to 5 weeks less time than it takes to produce the crop from seed. A plug is the individual cell in a plug flat. A plug flat is a plastic waffle-like tray (11" x 22" in size) composed of 50 to 800 individual cells, each 2- to 5-cm deep. Plug flats are named and sold based on the number of cells available. Bedding-plant seeds are sown directly in plug flats filled with media by a mechanical seeder, germinated in the flats, and grown to a trans-plantable stage. Plug technology has the following advantages:

1) Reduces the time and labour required for trans-planting because it is easier and faster to remove one seedling as a unit
2) Maximizes greenhouse space utilization be-cause seedlings are perfectly spaced and remain in the plug tray longer
3) Increases seedling and final crop uniformity because faster growing seedlings in a plug tray crowd early and growth slows while slower seedlings have a chance to catch up
4) Reduces transplant shock because roots are not disturbed and the seedling continues growth as soon as it is transplanted
5) Reduces the spread of disease organisms be-cause each seedling is isolated
6) Reduces crop duration because no transplant shock occurs
7) Mechanizes almost every phase of plug pro-duction
8) Provides an easier and less expensive way to ship seedlings than traditional wooden trays provide.

The main disadvantages of the plug system are that the initial cost for equipment such as automatic seeders, flat fillers, and environmental control facilities for germination is high, and extreme skill and careful attention to detail are required for successful plug production.

Beyond the large differences between the traditional and plug methods during germination and initial seedling stages, finishing techniques in the final containers are essentially the same.

Start from Seeds or Buy Plugs?

The majority of bedding-plant growers utilize plugs for finish production. However, a major pro-duction decision is whether to buy plugs from specialist plug growers or to invest in the equipment and skill needed to grow plugs in-house. Much of this decision involves comparing the cost of buying plugs versus producing plugs. The advantage of buying plugs, especially for the new grower, is the elimination of the germination and early seedling growth stages, which can be costly and have greater risk of crop loss.

Many small- to medium-size growers essentially choose to be bedding-plant finishers, buying plugs from specialist plug growers and dealing only with transplanting and finishing. This avoids the cost of equipment, cost of skilled plug growers, germination problems, and other problems associated with the seedling stage. Specializing only in finishing crops results in the ability to tightly schedule the production area for rapid crop turnover.

Many larger growers may grow plugs not only for their own production but also for sale to other growers. In-house plug production is done to reduce the cost per seedling unit, control plug quality, and grow cultivars that can be difficult to find from specialist plug growers. Often, sales of plug flats to other growers is necessary to justify the cost of equipment and skill required for plugs.

Plug Production

Seed

Producing plug-grown bedding plants has resulted in a demand for high quality seeds that germinate uniformly. Clearly, plug growers desire a high germination rate and seedling survival not only for in-house production but for customers as well. The combination of mechanical seeders and market demand has placed pressure on seed companies to consider seed quality as part of the overall effort to breed new cultivars. Several seed treatments are used in an effort to improve seedling stands in plug flats. These are the following:

Refined Seed: These are cleaned seeds that have been physically separated by size, shape, weight, or density. Typically, individual seed with poor germination and survival may be smaller than normal, have a different shape, or weigh less than normal. Mechanical techniques have been developed to re-move unusual seeds from a lot of seed.

Primed Seed: These seeds have been soaked in a solution with a high osmotic potential so that seed water uptake proceeds to a certain point, then stops. Thus, pregermination metabolic activities proceed but emergence of the radicle does not occur. The seeds are then removed for the solution, dried, packaged, and labelled with a sow by date. The main benefit of primed seed is that germination is more uniform over a wider range of environmental conditions.

Seed Pellets: These small seeds can be covered with a thick sugar coating to facilitate sowing by a mechanical seeder. The coating may be colored for visibility on the plug medium.

Plug Medium

A medium for plug flats, like other container media, must have good aeration and be well drained. However, because of the small soil volume of each cell, the medium components must have a small particle size. Peat, vermiculite, and perlite are the most commonly used components, but finer grades are employed. In addition, a plug medium must have a high water holding capacity and a high ability to hold fertilizer because of the small soil volume and the fact that seedlings may be grown in the medium for 4 to 8 weeks. Like many container mixes, medium pH for plug mixes is adjusted using dolomitic limestone, and superphosphate and micronutrients are added, though at much lower rates. Little else is added to a plug medium because soluble salts levels considered low for many pot plants are too high and may inhibit germination of seed. Many commercial companies produce special bagged mixes designed for plugs.

Seed Germination

Generally, temperature, oxygen, light, and moisture must be controlled for optimum seed germination. Temperature is probably the most important factor regulating the rate of seed germination (time from sowing to radical emergence). Germination rate is invariably low when the temperature is lower than optimum for a given species. As temperature rises, germination rate increases and remains steady within a range optimum for the species. Above optimum, the rate declines again. Specific germination temperature information can be found in citations under the additional reading section of this publication. Oxygen must be available to the seed to carry out the high respiration rates needed for germination. Excessive moisture or poorly drained media can reduce the oxygen available to seed and reduce germination percentage. Most bedding-plant seed have a relatively high humidity requirement during the time of water uptake, usually in the 90 to 100 percent range. However, constantly covering the seed in a thick film of water can limit the oxygen supply to seed for many species and inhibit germination. The seed of some bedding-plant species require light for germination so these should be sown on the medium surface without seed covering. Others benefit from some covering, usually with sifted vermiculite, either because germination is inhibited by light or light plays no role and germination is enhanced by the extra moisture.

Growth Retardants

Growth retardants are often applied to plug-grown seedlings to decrease internode elongation and to strengthen the stems. Though many label directions recommend application 2 to 3 weeks after sowing, it's best to apply growth retardants at the correct stage of development. For most species, this is at the second true leaf stage. Growth retardants are generally applied at a reduced rate compared to larger plants and, if required, reapplied as needed. Seedlings are more susceptible to phytotoxicity and stunting from incorrect application of growth retardants. The most effective application temperatures are 55 to 70 degrees F. Therefore, reduce the rate and apply only under cool conditions.

Germination Facilities

Facilities for seed germination and early growth of seedlings in plugs must provide for more exacting environmental control than is normally found in most greenhouses. The germination area must provide a high degree of control over temperature, light, and humidity; usually two or three different environments must be established to meet the needs of a wide variety of bedding-plant species. In a greenhouse, this can be accomplished by using several small green-house sections with above average heating, cooling, and sophisticated environmental control systems.

The best way to provide a high degree of humidity control and avoid excessive moisture around the seed is to reduce the moisture droplet size far below that which is usually employed by conventional mist used for propagation of vegetative cuttings. Fog systems have been developed that work far better than conventional mist and convert 90 percent of the water droplet size to between 5 and 20 microns. When a system like this is employed, ventilation must be restricted and carefully controlled to keep the environment even.

Beyond the greenhouse, many plug growers have developed and use specialized germination rooms. These rooms are constructed with treated lumber and heavy insulation and are moisture-tight on the inside to reduce temperature fluctuations. Ultra-fine fog systems are used to achieve a high degree of control over germination humidity. Light is provided from fluorescent lamps that are turned on for 16 hours per day. Air conditioners may be used in those rooms designed for cool-season crops such as pansy. Plug flats are placed on shelves spaced 2 feet apart. Seeds generally only remain in the germination room to the end of stage 1; any longer and the seedlings may stretch from insufficient light.

Finishing

Growing-on or finishing refers to the stage (stage 4) of production from transplanting into the final container until the crop reaches a marketable stage.

Bedding Plant Containers

Most bedding plants are produced and sold in market flats. A market flat is composed of a sheet of market packs (called an insert) inserted in a plastic flat. An insert is a thermo formed, polystyrene plastic sheet of pull-apart packs. The packs are composed of 1 to 9 cells, each containing a single plant, de-signed for customers to pull-apart from the insert to purchase as a unit. The insert rests in a plastic flat to provide support and is the single unit for shipping. The choice of which container type and size to use for bedding plants is largely determined by the ultimate finish size of the plant species and cultivar grown, market demands, and constraints imposed by mechanization.

Each species has a characteristic height and spread that influences the practical container size to use. Usually, a given species may be grown in a range of sizes and, within this range, finishing times will vary. But if the container size is too small, plants will crowd and lose lower foliage before flowering. Conversely, if the container size is too large, the plant will flower before filling the container to an aesthetically acceptable extent. Probably, a far more important factor is to meet the demands of the market to which the product is sold and make a profit. The entry of mass marketing has had a profound effect on the choice of container sizes and, therefore, on the growers themselves. Rather than leaving the choice of container size to the grower, mass markets have sought to impose container size standards that fit within their marketing strategies. Because of customer specification and tight competition, the choice of containers is often very limited. This usually means smaller container sizes, a shorter finishing time, and a tighter profit margin.

Growers who market to traditional retail outlets, however, strive to choose container sizes that differentiate the product from the competition in terms of quality and appearance. The key is for customers to be able to identify your produce by appearance and good garden performance. Larger soil volume combined with quality growing techniques should produce a larger, more vigorous plant that is ready to provide quick gratification in the home garden or commercial landscape.

Given a choice, select the largest feasible market pack to accommodate the root system of the bedding plants to be grown. Larger market packs mean fewer plants per flat, and since most plants are priced by the flat, the number of cells per flat will affect costs and profits. Inserts come in 24, 36, 48, or 56 cells to the flat. Plants grown in a larger-sized cell will generally last longer in the sales area than plants grown in small-sized cells.

Standard Market Flats

The 11- by 21-inch plastic flat is still the standard for the industry. Inserts come in a wide variety of combinations of packs per flat and cells per pack. In catalogs, inserts are often designated by a number. The first part of the number indicates the number of packs and the last part, the number of cells per pack, (e.g., a 1204 insert has 12 packs each with 4 cells for a total of 48 plants in a flat). The most widely used flat is the 12 packs per flat, each pack with 2, 3, 4, or 6 cells per pack. Mass markets have popularized the 18 packs per flat with 1 to 4 cells per pack, while growers for garden centres often use the 8 packs per flat with 4, 6, or 9 cells per pack.

Jumbo Flats

One of the main growing areas of the flat market is the jumbo pack that has 6 packs per flat with 4 or 6 cells per flat. The flat size remains the same as standard flats, but the cells are 31D 8 inches deep instead of the 23D 8 inches for standard packs providing a 25 to 30 percent increase in soil volume. Jumbo packs are used by growers seeking longer shelf-life, reduced losses in the retail area, and better quality for the customer.

Slimline Flats

For the large grower who markets to chain out-lets, slimline flats offer a way to reduce production costs when the market resists a price increase. These flats are 81D 2 inches wide (rather than 11 inches) by 20 or 21 inches long. Therefore, they take up 15 per-cent less space than standard flats and more flats can be grown per unit greenhouse area.

Bedding plants marketed in containers larger than market flats are becoming more popular each year for both homeowners and the landscape con-tractor markets. Some bedding plants grown in 4- or 6-inch pots will produce a colourful show in several weeks and bring a premium price in the market. Even larger-sized colour pots, containers with multiple plants of different species, or cultivars mixed in a colourful display, are becoming more popular. Showing flower colour is essential to marketing these larger products. Start growing larger-sized products by transplanting market-pack-grown plants to large containers early in the spring.

Growing Media

Soilless mixes are the most often used growing media for bedding plants. These mixes provide anchorage that enables the plant to support itself and regulate the supply of water, oxygen, and nutrients to the roots. No one specific medium can be singled out as the best, and quality bedding plants can be grown in a wide variety of media combinations. The type of medium a grower chooses depends on personal preference, type of greenhouse, type of irrigation, crop type, and cost. Some soilless mixes work better for some types of plants than do others, but many growers use one mix for all bedding plants. The first decision involving growing media is often whether to purchase medium or mix your own at the greenhouse operation. On face value, commercially bagged mixes seem expensive but once all the costs are considered, mixing your own formulation may be out of the question for most small growers. Like many economic decisions in the greenhouse, carefully consider all the costs of both approaches. For a commercial bag mix consider the cost per bag and shipping costs. For mixing your own formulation, consider each component's cost, management time, labour, office expenses, equipment costs and depreciation, and costs of special structures for the mixing facility and component storage. Many large growers mix their own formulation because the cost of commercially bagged mixes would be prohibitive. One other option is to purchase bulk mixes from local formulators. These companies will mix a medium to your specification and deliver by truck at a lower cost than you would pay for commercially bagged mixes.

Regardless, make sure that the medium comes from a reliable source so that there are few weed seeds and potential for insect or disease problems. Fertilizer and other components should be well blended to ensure a medium that is uniform through-out the batch and from bag to bag. The media should be about half peat moss and about half of some aggregate, such as perlite or vermiculite. This will help the media to hold water yet allow the excess water to drain adequately.

The pH of the media (and it is a good idea to have your media tested each year!) should be between 5.5 and 6.5 to ensure that the nutrients you apply in fertilizing can be absorbed by the plants. A pH outside this range will tie-up the nutrition applied in fertilizer, the plants will become deficient in those nutrients, and you will have plants with yellow or brown leaves. Some elements become toxic to the plant at a pH below 5.0.

Transplanting, Production Temperature, and Spacing

Plug technology has enabled growers to auto-mate even the transplant stage of bedding-plant pro-duction. However, most growers still transplant by hand today. Transplant seedlings into slightly moist media. The media should fill the cells of the market packs to but not over the rim. Seedlings should be inserted into the new medium at about the same level they were growing in the plug tray.

Immediately after transplanting, the plants should be watered thoroughly and moved to the green-house. As a preventive measure, some plants may be watered with a drench of fungicide to prevent damping-off diseases. Root zone heating will help the seedling to establish a good root system in the new container. Monitor the temperature of the environment, which should be maintained at 75 to 80 degrees F day and 65 to 68 degrees F night temperatures. Water and fertilize when necessary, allowing the top of the soil media to dry between watering.

Spacing of plants in market flats is determined by how many cells fit into a flat. Spacing can be regulated on larger-sized pots, such as 4- or 6-inch pots or hanging baskets. Plants can be grown pot to pot and spaced slightly apart in the last 2 to 3 weeks of production.

Fertilizers

Moderation is the key to successful bedding-plant fertilization. Bedding plants should be provided all 16 essential elements either from the liquid fertilization program and/or incorporation at the time the medium is mixed. Most growers use a combination of both. Water-soluble fertilizers come in a wide range of formulation. Many growers successfully use a 20-10-20 formulation; however, 20-10-20 is an acid-reaction fertilizer that can reduce the medium pH over time. If the pH of the media is too low, apply a basic-reaction fertilizer such as 15-0-15. Many growers will alternate between these two fertilizer formulations to maintain pH. Fertilizers especially for soilless mixes should have a good supply of micro-nutrients and be low in ammoniacal nitrogen, a cause of ammonium toxicity during cool weather. As an average, 150 to 200 parts per million (ppm) nitrogen in the irrigation water works well. Reduce the concentration and frequency of fertilization at the beginning and end of the production cycle.

Growth Retardants

Several bedding plants may begin to stretch, developing long, thin, spindly stems a few weeks after transplanting, especially under low-light conditions in early spring. To keep them shorter, sturdier, and greener some growers apply a growth retardant. Several growth retardants are approved for use on bedding plants. Growth retardants decrease the length of the stem (shorten the internode between leaves), thus developing a shorter, sturdier plant. Some growth retardants have an added benefit of greening up the foliage. Ancymidol (A-Rest) can be used as a foliar spray applied 2 to 4 weeks after transplanting at a rate of 30 to 130 ppm. Diaminozide (B-Nine SP) can be applied as a foliar spray 2 to 4 weeks after transplanting at a rate of 2,500 to 5,000 ppm. Bonzi, Cycocel, and Sumagic are also labelled for application to bedding plants. Plants differ in their sensitivity to these chemicals, so be sure to check the label before applying them. Because no chemical growth retardants are approved for application to vegetable bedding plants (or fruit such as strawberry), the only chemical alternative is water, which is a powerful growth retardant. When plants are produced with sufficient but limited water, the result is a shorter, sturdier plant. Withholding excessive amounts of water and fertilizer, combined with sufficient sunlight and cooler temperatures, helps to reduce plant size and increase the quality of vegetable bedding plants.

Scheduling

Scheduling bedding plant crops is essential to greenhouse profitability. While quality and reason-able price are important, crops also must be ready when the market demands. Demand in the market, climate, geographic location, season of the year, and bedding plant type/cultivar all influence the dates for sowing, transplanting, and flowering. Despite the availability of a lot of information on bedding plant scheduling, careful crop records detailing finishing times over a period of years for a particular location are the best guide. Variations in microclimate within a city, even a particular greenhouse can make a difference of a week or so. Scheduling bedding plant crops is often done by assigning the weeks of the year week numbers. The first full week in January is numbered week #1, then each week is numbered consecutively through December (weeks 1 to 52). Schedule a crop as follows:

1. Determine the week number of the sales date.
2. Count back the number of weeks required in the finish container. This is the transplant week number.
3. Count back the number of weeks required in the plug flat. This is the sowing week number. Example: Celosia requires 5 weeks in a 406-plug flat and 5 weeks in a 1204-market flat to finish. The tar-get sales date is week #15. The transplant week number will be week #10 and the sowing week number will be #5.

Problems?

Nutritional Problems

One nutritional problem associated with low medium pH is micronutrient (usually iron and manganese) toxicity. This problem frequently appears as stippling (spotting) or chlorosis (yellowing) on the older leaves of some plants, especially geraniums. Symptoms are similar on marigolds where toxicity looks like tiny bronze spots on the older leaves. Keep the media pH at 5.8 to 6.2 to prevent the problem. If these symptoms appear on growing plants, don't guess about the soil pH. A foliar analysis (test of the leaves to determine their nutrient contents) is the only way to determine for sure that it is micronutrient toxicity and not some other cause.

To treat plants with micronutrient toxicity, a heavy application of a basic-reaction fertilizer such as 200 ppm nitrogen from calcium nitrate will help to remove excess soluble micronutrients and increase the pH. If the pH of the medium is below 5.0, an application of lime should help. Finely ground dolomitic lime at the rate of 1 pound per 100 gallons can be applied to plants in containers. Hydrated lime mixed at the rate of 1 pound per 100 gallons is an alternative if used carefully. The lime should be mixed with water and allowed to settle overnight. Only the clear solution should be applied, followed by clear water to rinse the foliage.

When nutrition problems occur in your bedding plants, be sure to have the plants and media tested. Media analyses can sometimes be conducted by the county Extension agent or can be sent to the Soil Testing Laboratory at Auburn University. In the Soil Testing Laboratory, the mineral nutrition in the medium can be analyzed for approximately $10. The amount of nutrition in the plants (a plant analysis) can be determined for about the same price. It's a good idea to have the pH of the medium and the water supply tested yearly. Water with a low pH applied to a medium with good pH can lead to nutritional problems. A small investment in yearly testing can help eliminate many nutritional problems.

Insect Pests

It is important to regularly check crops for insect and disease problems. Problems identified early are easier and more economical to control. Several pre-cautions can be taken to reduce the likelihood of insect problems. Before using any insecticide or miticide, read the label directions carefully to be sure it is safe to use on specific crops and to double-check the recommended rate. Check with the county Extension office if you have any questions. Eliminate weeds under the benches and around greenhouses to eliminate a major source of insect pests. Weeds are a great hiding and feeding place for many insects, especially whiteflies and aphids, before infesting crop plants. These insects will also feed on weeds while crop plants are being sprayed with in-secticide only to return to crop plants later to feast. Several herbicides are registered to control weeds around greenhouses. Diquat liquid herbicide is registered for use under greenhouse benches for many types of weeds. Roundup (glyphosate) and Surflan (oryzalin) liquid herbicides are registered for use as a spray to control several species of weeds growing under greenhouse benches.

Disease Control

One of the greatest disease problems, root rot, is caused by a group of fungi. Symptoms of this kind of disease are a general lack of vigor or growth, slow growth, and especially black roots or a blackening of the crown. These symptoms can be caused by any one or more of these fungi: Pythium, Phytophthora, Rhizoctonia, or Thielaviopsis. It is important for growers to watch crops carefully, especially at the earliest stages for these symptoms. Different fungi causing rot problems require different treatments. It is essential that the disease be diagnosed by a plant disease clinic in order to determine the proper treatment. Experience from treating past outbreaks will help you more quickly recognize and treat new occurrences. Refer to the product label for the correct amount to apply and read the label directions care-fully to be sure that it is safe to apply the chemical to a specific crop. Banrot 40W can be used as a soil or media drench at seeding or transplanting time as a preventive application for many bedding plants. It can help to prevent Pythium, Phytophthora, Rhizoctonia, Fusarium, and Thielaviopsis root rots. Truban 3G or Terrazole 5G (etridiazole) or Banrot 8G (etridiazole) can be used as a dry soil mix as a preventive mea-sure against Pythium and Phytophthora for many ornamental bedding plants. Subdue 2E can be applied as a soil drench to many ornamental bedding plants as a treatment for Pythium and Phytophthora caused diseases. Rhizoctonia on ornamental bedding plants can be controlled with soil drenches of Chipco 26019 50W (iprodione), or Terraclor 75W (pcnb).

Finishing the Crop

Depending upon the weather and production method, plants should be marketable according to the schedule. Help keep plants in top quality by cooling them in the evening to a minimum of 65 degrees F. Cooler evenings with warmer days help the plants maintain a good size and foliage colour. Do not fertilize heavily toward the end of the production time to help prepare the plants for sale. Reduce either the frequency of fertilizer or the concentration in the last 2 or 3 weeks of the production cycle. Reducing the night temperature and reducing the amount of fertilizer will produce a bedding plant that lasts longer when it is taken from the greenhouse. Lengthening the post-harvest life of the bedding plant makes it look better in the sales area and keeps it looking good until the consumer can plant it.

Costs and Pricing

A 1999 USDA report showed that the average wholesale price of Alabama-grown bedding plant flats was $7.74 per flat in 1998. Geraniums were $6.18, Impatiens were $6.77, New Guinea Impatiens were $10.38, petunias were $6.61, and vegetables were $8.98 per flat. A grower's price will vary from these prices because the overhead or expenses of running the business will vary from business to business. Some growers make the mistake of looking at only the cost of materials to produce the bedding plants or the direct costs of production. For example, there was an average difference of only $0.43 per flat to grow bedding plants from seed versus growing them from plugs in 1998. The plugs are more expensive to produce but can save money in the long run because they save time and space for the grower.

Post-Production Care in the Sales Area

It is difficult to imagine a grower putting considerable time, money, and effort into producing a premium quality crop only to neglect it during market time. Bedding plants need some care during the short period of time they are in the sales area. Several measures can be taken to help keep crops looking their best. High-quality bedding plants sold within 3 days after removal from the greenhouse have a good post-harvest life. Plants sold after 5 days tend to show signs of wilting or decline. If the sales area is cool, plants can remain at a consistent level of quality for up to 10 days. Warmer temperatures cause the plants to dry out quickly and decline in quality faster. Plants given reduced amounts of fertilizer and water in the last several weeks of the production cycle have a longer shelf life than those kept on a consistent level of water and fertilizer. In the sales area, shade the bedding plants somewhat from direct sun. This helps to prevent them from dry out and to keep good flower colour. Approximately 60 to 80 percent shade is recommended.

Wetting agents incorporated into the growing media keep moisture in the soil and are intended to extend the shelf life of the plant. Some do a fair job, others may burn the roots of plants in production. Experiment with some of these products to see what works best for your soil mix. Raise plants off the ground by placing them on benches. Not only will it be easier for consumers to reach plants, but this will keep plants out of puddles and away from soil-borne diseases. Air can also better circulate around the plants raised off the ground.

Growing bedding plants to produce beautiful colours for consumers' homes is as much an art as a science. Practice growing crops should improve the quality and reduce the number of problems encountered each year.

4

The World Floriculture Industry: Prospects for Growth

Floriculture has emerged as a profitable agribusiness option the world over in recent years, particularly in the developing nations. World floriculture trade totals more than US$ 6,800 million (value of imports during 1995). The Netherlands alone account for nearly half of this trade and countries like Colombia, Italy, Denmark, USA, Belgium, Israel, Costa Rica, Germany, Canada, France, Spain, Kenya and Ecuador all have a share of more than US$ 100 million. Asian countries are also entering the international trade and making their presence felt. Thailand, Malaysia, China, Singapore and India are leading the Asian contribution. However, while their present share of the trade appears insignificant, the growth of these countries in recent years has been very impressive. For cut flowers with a trade of US$ 3,700 million Thailand has a share of 2%. India which now has a share of only 0.3%, had a growth rate of 563% in the 1991-95 period. The other country with impressive growth rate has been Malaysia (259%) over the same period.

Among the growing markets (cut flowers, cut foliage, live plants and cuttings) Germany with a share of US$ 2,000 million was the world's biggest market, followed by USA with a share of US$ 831 million. Japan, though seventh in ranking with a trade worth US$ 314 million, emerged as the fastest growing export market during the 1991-95 period indexing a growth of 162%. Consumption in these three countries are characterized by the domination of middle class population.

The world production of floriculture products at a very conservative estimate is valued at $ 20 billion. Japan with a production value worth US$ 3,736 million leads, and with the Netherlands (US$ 3,558 million) and the USA (US$ 3,270 million) account for nearly half of the world total.

In terms of total area in production, Asia and Pacific with an area of 134,000 ha cover nearly 60% of the total world area. A significant feature of the production area statistics is that in most Asian countries, only a small share of the area contributes to production for world trade (exports). This is because either only a small percentage of the area is under protected cultivation (as in China and India with large total areas) or the production is mainly for the domestic market (e.g. Japan). Global production of floriculture products will continue to expand, especially in new producing countries in Asia, Africa and Latin America, though the traditional large producing countries like the Netherlands will also work towards higher productivity per worker and per unit area. In general, however, the trend in production of cut flowers which is going to be very labour intensive, will tend to be located in areas with lower labour and other production costs. The production strategies will require reduced costs so that the profitability improves. For the purchaser it is not important where the flowers come from. They are mainly interested in price, quality, freshness and variety.

For the growing flower exporting countries of Asia, one of the major barrier in increasing profitability is import duty. Around the world, duties are used to protect the local flower industry from unfair competition. The only exception being Japan, among the leading floriculture markets, which charges no duty as yet. But the total imports of cut flowers into Japan are less than 7% of their trade. It remains to be seen, if Japan will continue to have no non-tariff barriers or restrictions on trade, when China, assisted by Dutch technology and quality propagating material from around the world, increases its exports to Japan, particularly considering that China itself has a duty of nearly 53% (including 13% VAT).

For Asian countries, another hampering factor is the soaring cost of air freight, which may take away the competitive advantage of low cost land, labour and year round growing climate. Various international carriers no longer prefer flowers because they have alternative and more lucrative 'exotic' cargo such as computer chips.

The cut flower trade worth $ 3.7 billion is likely to increase by around 10% annually. The share of the developing countries of the total trade has consistently been around 20%. During the last five years there has been a consistent growth in the number of producing countries particularly among the developing nations in Asia, Africa and Latin America. In general, the world supply is increasing more than demand, resulting in reduced prices per stem. With a highly perishable product like flowers, even a small over supply can have a dramatic effect on prices. This will require a greater attention to be paid to identifying crops ideally suited for a particular location, as well as product diversification. One must critically analyse the strengths, weaknesses, opportunities and threats before embarking on expansion of the production/ trade programmes. Greater attention to high value crops, though traded in lower volumes, will increasingly become more important.

Orchids, while not a major crop in most regions, have a special place in floriculture. Another crop demanding better attention in world trade has been Anthurium. Its world trade is second only to orchids among tropical flowers. Its popularity is increasing, particularly in Asia.

Most Asian countries have potential for growing high value crops and need to devote larger areas mainly to meet the huge growing demand in Asian markets, where their freight rates will also be lower. Countries in Asia also require identification and commercialisation of certain native tropical plants to cater to niche markets in Europe and the USA.

As importers are trying to import from the cheapest country, the Asian growers could easily lose their large European market, in view of the strong competition from growers from Africa and Latin America, who may have lower freight costs and no or little import duty to pay. In addition the growers in Holland, with the advantage of producing right in the heart of the most developed market in the world, with low or no transportation cost or duty, would always enjoy the best marketing opportunity. The Japanese and the US markets with relatively low import volume will continue to satisfy their supplies from nearby countries. The other growing Asian markets like Singapore, Hongkong, South Korea, etc. are hardly in a position to accept large volumes. One also needs to consider the emerging new markets in Middle East countries and Russia. The best option however, would be for all the major flower growing nations in the region to concentrate on developing their own domestic market. With increasing consumer wealth, population concentration in major urban areas and improved living standards, India, China, South Korea among others appear to be potentially important flower consuming countries for the future. In these countries, as also in others, most flower consumption occurs around major holidays, like New Year, Valentine's Day, Memorial Day, Mothers' Day, Fathers' Day, Christmas, and in wedding seasons. The governments in these countries are making sincere efforts to develop the philosophy of people with regard to flowers through strong promotional efforts. Foreign investors are being encouraged to help develop their floriculture industry through joint venture projects. All these efforts will hopefully lead to making available quality produce for the consumers. The governments will need to invest more on developing the required infrastructure, particularly for marketing like auction centres etc. (already being developed in Korea and India), and increase access to new and improved technologies and quality inputs. The efforts to develop the competitiveness in the export markets, through providing subsidies for capital investments, air freight etc. should also continue alongside.

With the growth of the flower industry the world over and increased exports, there is need to identify varieties that are most tolerant to the stresses of long shipping and storage. Consumers are seeking products which offer diversity in flower and leaf colour, good texture and form, as well as better performance in interior environments.

Efforts to develop techniques which will reduce the costs of the ultimate product for the consumer will need to be strengthened, with identification of low cost production technologies in protected conditions and refinement and reorientation of imported technology to suit our requirements.

Public concern for the environment will require Asian growers to re-examine and change their irrigation, fertilization and insect control practices. They will also have to reduce the use of pesticides and other chemicals, through effective IPM programmes and increased use of biological control methods. This is important due to the fact that, apart from environmental concerns, developing new pesticides is becoming expensive and governments in several countries are enforcing strict registration requirements for new pesticides, besides removing several pesticides from the market.

Floriculture Worldwide

They can be divided into cut flowers & foliage, potted plants, garden plants, nursery stock (trees), flowering leafy, annuals & perennials and bulbs & tubers. Parental material like plant parts from tissue culture, cuttings is used as a start in production systems to grow the end products for the market.

The production of floricultural products has grown quite consistently over the last 20 years. Based on Dutch and American articles the production was estimated 11 billion dollars in 1985 growing to 24 billion in 1990, via 31 billion in 1996 to 44 billion dollars in 2000.

Our figures are not yet complete, but we estimate that the production value was about 60 billion dollars in 2003. This means an average yearly growth of 6 to 9 %. The highest growth has developed in Asia and the USA. And the future is promising.

Production areas

When it comes to the main production areas of ornamentals we see the following:

- In Europe the most ornamentals are produced. The Netherlands is known for cut flowers and potted plants, as well as bulbs, for annuals and perennials. Germany has a name in nursery stock and garden plants. In Italy lots of flowers and potted plants are produced. While Denmark is famous of their potted plants. France is a broad player when it comes to the different types of products. UK, Belgium en Spain are small players in this field.

  The total production value is about 10 billion dollars, and has stabilized. Countries with the largest share in cut flower production are Germany (11%), Italy (18%) and the Netherlands (35%).
- In North and South America ornamental production mainly consist of flowers and cuttings. In North America (USA and Canada), where 80% of the flowers and potted plants are grown, California and Florida are the most important production regions. In South America, Colombia (6%) and Mexico (3%) together with Costa Rica and Ecuador have developed rapidly over the last decade as producers of flowers and parental material. Also Brazil (6%), for a long time, has been a producer of parental material and cuttings of potted plants.
- Production in Africa has increased over the last decade, with Kenya in the frontline followed by Tanzania, South Africa and Uganda. Flower production, of roses in particular has made enormous progress. Production in countries like Zimbabwe and Ivory Coast has decreased because of the political situation in the country. For Africa we see rapid growth in a high risk environment. Besides an estimate of 0,2 billion dollar production there are hardly figures available about ornamental production in this continent.

- About Asian production we hardly have any figures from the developing countries. Japan is traditionally a producer of specialities in ornamentals. South Korea, India and Thailand are coming up strongly as producers of ornamentals, followed by Taiwan. The production of cut flowers in Israel has decreased because of the political situation. Local production for local markets seems to be big. Export is starting from countries like China, Vietnam, India, etc.
- In Oceania, Australia and New Zealand produce mainly cut flowers, but are relatively small producers.

Trade Patterns

This paragraph elaborated on the main streams of export of ornamental produce world wide. Also the value of export streams of ornamental between continents will be shown. There is hardly any relation to production values of continents and countries, except that they are smaller a lot. Quite some countries for instance mainly produce for local markets.

Cut Flowers

Based on figures of 2002, Europe is by far the largest market for cut flowers. The countries within together have an export value for Europe of 2660 million. US dollar. Further more Europe imports from main suppliers like South America (228 mln. USD), Africa (196 USD), Asia (174 mln USD) and North America (65 mln USD). So, the export in and outside of Europe totals about 3425 mln USD).

North America is the second largest market. USA and Canada have an export to each other of about 108 mln USD. The continent imports large amounts of cut flowers from the countries in South America (859 mln USD). Europe is with 141 mil USD the second supplier, followed by Asia with 26 mln dollar export value on cut flowers. So, the export in and outside of North America totals up to 1100 mln USD)

From our figures about Asia shows an internal export between countries of about 94mln USD. This is added by export of cut flowers of 53 mln USD out of European countries, 22 mln USD out of Oceania and 15 mln USD out of South America. This adds up to 184 mln. USD total.

From Africa it is not known that they import cut flowers out of other continents or in between its countries.

Potted Plants

When it comes to potted plants, based on figures of 2002, we can see the patterns differ:

Europe has an internal export between its countries with a value of 3013 million US dollar. In addition Europe imports from Asia for about 47 mln USD. This is followed by imports with a value of 42 mln USD from Africa and 39 USD from South America. In perspective of potted plants this mostly concern propagation material and cuttings for production in Europe. This totals 3140 USD value of imported product for Europe. In North America the internal export between USA and Canada is 329 mln USD. This market imports from Europe mostly end product with a value of 54 mln USD. Also South America is a main supplier with 47 mln USD of products; propagation material as well as potted plants for the markets. This is the same with Asia with an export value of 13 mln USD. Together the trade of North America totals an import value of 440 mln. USD.

Asia shows an internal export of 'potted plants' between countries of about 52mln USD. This is added by export of potted plant of 76 mln. USD out of European countries. This adds up to 128 mln. USD total.

From Africa it is not known that they import potted plants out of other continents or in between its countries.

Worlds Largest Exporting Countries

From the figures available (2002) about world wide trade in cut flowers it is established that the Netherlands is by far the biggest exporter with a value of about 2.000 mln USD. It is followed by Colombia with an export value of about 600 mln USD. They are followed by Israel, Kenya, Belgium and Zimbabwe.

Their role is changing over the last 5 years. We see a negative growth of the export of The Netherlands (-2,7%) and Israel (-6%); the latter because of the political situation. The others show growth of export: Colombia (45%), Kenya (182%), Belgium (130%) and Zimbabwe (46%). Of Zimbabwe we assume that this growth has decreased or that export even has decreased because of the political situation in the country since 2002..

Regarding the inter- and intracontinental trade of ornamentals, the following can be concluded:

- Europe shows by far the largest trade-activity with 3425 mln. USD and 3140 USD for cut flowers and potted plants respectively. North America is second with about 1100 mln USD imports of cut flowers and 440 mln USD for potted plants. Asia comes in third place followed by South America.

The difference between the figures of Europe and North America is exaggerating compared to reality. This is because Europe exist out of a lot of countries, while North America is larger and consist out of two countries. The European market is larger that the Northern American market, but not 3 tot 10 fold.

- If the world wide trade of cut flowers is compared with that of potted plants (materials) it becomes visible that streams of cut flowers are much bigger. An import reason is that potted plants are heavier per volume and therefore more expensive to transport.
- In general we see that south – north is a more import trade direction than east – west.
- In export figures for potted or ornamental plants, also quite an amount of propagation material like tissue culture plants and cuttings are involved. We can not establish what exactly the export values of these are. But they will

grow using the advantages of cheap production input, like labour and land in these countries.

Incentive for Growing International Trade

IT, Communication and Distribution

There are a few global developments which will have their effect on international trade the years to come. There are rapid changes in the possibilities of information technology. Think of the possibilities that handheld computers will mean when it comes to management of the production process. Or what about the introduction of RFID-tags and micro sensors and their impact on tracking and tracing of products throughout the chain. Close to this is the development of communication technology. If the hardware starts communicating with web-based applications all kind of stakeholders of the product chain can access the data and information. This also requires thinking about authorisation of the information access. Next to that more and more of the global population start using English, which make communication easier.

There are also developments in distribution. If the new IT-components en their communication aspects are linked to distribution and route planning, the distribution process can be optimised. Other developments are those in packing and transporting. Packaging called Modified Atmosphere can slow down the process of deterioration of products like fruits, vegetables and flowers. The product can be transported during a longer period of time and still have good quality. This process is also simulated with a ship. The whole shipload can be treated in way that fresh produce keeps its quality. This will increase possibilities of sending orders by ship instead of by plane, which makes transportation a lot cheaper.

International Economic Integration

There are different economic trends which will have their effect on the trade in fresh produce.

For instances, countries more and more unite themselves in trade unions. And these unions are getting more powerful when more countries join. Different examples can be given. The European union existed out of 15 countries. But since may first 2004, another 10 Eastern European countries joined. Then there is ASEAN, a same kind of development in Asia. The USA, Canada and Mexico have there NAFTA-treaty. Trade within these unions is getting easier and cheaper by decreased regulations and trade barriers.

Another development we see is that (home) markets saturate. For producers, this saturation asks for the strategy to keep their share of wallet, either to enlarge the export activities to other continents.

Finally we see that the activities of the World Trade Organisation (WTO-rounds) globally breaking down all kind of trade barriers. This means more and more the free market mechanism can do its work. This will have its effect on trade as well.

Shift in Market for Floricultural Products

The world market for floricultural product has been changing constantly. Going back to 1992 we see that, calculated in dollars, 57% of the world market consisted of the Europe demand, 23% was Japanese demand and 20% was the demand from the USA. Over the years we see these shares changing. While the European market is loosing share, due to growth of market in the USA, Japan is keeping up its position. USA has proven to be a growing market the last decade up until 2002. The developments of the last two years, with dollar that is weakening, importing products into USA is getting more expensive and will strengthen the domestic production.

It is known that the German market has become saturated. The Danish market shows quite a growth. The consumption has increased with even more than 50% looking at the developments in The Netherlands and Spain from 1994 to 2002. The market Great Britain has been enlarged enormously. In 8 years time their people have more than doubled their consumption of ornamentals. The British Supermarkets have played a big role in this as a main outlet for quality products. And this will increase. Furthermore big volumes are needed to supply these parties.

In Japan the consumption in 2002 is one and a half time that of 1994, quite an increased. Compared tot that, with 45% the increase of consumption in the USA is just a little less.

We cannot compare the figures in the table directly to the trends of the shift in share of the global market. To do so, the figures in the table would have to be multiplied with the amount of inhabitants per country or continent.

Patterns Behind the Consumption

That the consumption per capita has increased strongly over 8 years after 1994. Because of a growing income people have started to buy more ornamentals for personal use as well as the gifts. Also the institutional use has risen; companies, events and recreation has started to use more flowers and plants for attractive decoration. In Europe and USA we see more small households occur of one or two persons, with a higher than average salary. So the amount of homes to be decorated per capita is rising. Another reason can be found in a trend called 'cocooning'. People in Europe and USA especially have started to spend more time at home, to get a feeling of peace and safety. Doing so they started to make the home cosier, with decorations like ornamentals. The trend of individualisation results in that people want to show their own life style. They also use ornamentals to do so. Finally there is a demand for quality products has grown, since there is more than enough produce; i.e. the English markets compete in who is selling the best quality products. In Europe there is also a trend in bringing nature back in the daily life, by using natural products. All these developments have contributed to a growth in the consumption of products per capita.

If we look at the developments in trade from 1994 to 2002, we see the following:

The trade within Europe, which was 6055 million dollar in 2002, has changed with 0% in the eight years before. The extra products, that the growth in consumption per capita required, has obviously come from Africa (+74% up until 242 million dollar) and Latin America (+45% up until 274 dollars). Import from Asia seems to have declined by 2% in that period.

With North America the trade has increased with 94% up until 459 million dollar. To be able to supply the increased consumption per capita, further more the imports from Latin America has increased with 72% until 908 million dollar, from Europe with 33% until 361 million dollar and from Asia with 89%. Also trade within Latin America has grown strongly (137% up until 32 million dollar) in the eight year period. Next to this imports from Europe have increased with 52% until 36 million dollar.

Internal Asian trade has been expanded with 28% over the years until 148 million dollars. This might have resulted in a decrease in imports from Europe (-2% until 250 million dollars) and Oceania (-2% until 32 million dollar).

Looking into the figures the size of countries and amount of inhabitants of continents has to be taken into account. While export from The Netherlands to Belgium adds up to the total, the trade between California and New York is unobserved.

Market Opportunities

Considering the developments in trade over the years 1994 until 2002, the following observations can be made when it comes to market opportunities. First the existing markets. Europe offers opportunities for slight growth in either cheap product or high quality product with added value. Also the Japanese market will show a slight growth. Besides very high quality products a niche market for a reasonable price/ quality ratio is developing. To utilize this opportunity producers and traders have to focus on high quality product. Of the existing markets USA offers the most possibilities when it comes to an increase of consumption. The weakened dollar, however makes imported products more expensive.

As possible new markets Africa, Latin America, Eastern Europe and Asia can be considered. Of those Africa offers the least opportunities; there is a lot of poverty and a lot of countries are political unstable. Latin America will offer slight growth, in and about the big cities mainly. Eastern Europe the prosperity has been improving slowly the last decades. Now that 10 countries have joined the EU, an acceleration of this improvement can be expected. Because the level of prosperity still is low, market opportunities in general can be found in relatively cheap product. Because developments in economic growth in South East Asia a big expansion of the market can be expected for instance in China, Hong Kong, Malaysia and Shanghai.

Conclusions about Production, Consumption and Trade Patterns

If we overview the facts about the development of production, trade, and markets from the we can conclude the following:

- Worldwide the production has expanded. We see this in these countries where entrepreneurs see opportunities to enter markets combined with good growing conditions;

- Worldwide consumption has grown in these countries where use of ornamentals could increase and in countries where the level of prosperity has risen;
- With rising income consumption per capita will go up. A higher level of prosperity in countries will increase consumption of ornamentals personal use (life style), gifts and institutional segment (decorations);
- The international trade of ornamental products has increased, following the development in consumption of ornamentals in the different continents and countries.
- Regional supply will be leading. The transportation costs will be the lowest and the product will have the highest possible physiological quality;
- Trade liberalisation will be (or is) followed by non-trade barriers. When the regulations about quota and tariffs are decrease, other regulations increase i.e. on quality levels and phytosanitary condition.

Lessons Learned about production, consumption and trade patterns;

- With higher income in countries and continents, the demand will rise;
- High volume 'bulky' products will be between North-South pattern or interregional because low product cost are essential;
- There are niche markets for high quality product all over the world; when product is special of exceptional, product price is not so much an issue. But constant innovation is essential to be able to 'surprise' them with new products
- Strategies for supplying niche markets should depend on a supply chain concept of 'Quality'; from 'seed to vase'; the supply of product for this market should not be incidental. Structure and year-round supply are prerequisites.
- There is a demand for co-operation for all actors in international chains.

Market Developments

People from all over the world use floricultural products for culture and traditions. In the more prosperous countries they are also used to express themselves in line with their life style. Furthermore businesses and organisations also use these products for decorations to create a certain atmosphere contributing to their image.

To a consumers and institutions a floricultural product is more than a plant with foliage and flowers (core product). Plant go together with a pot, it has labels with treatment information and attributes i.e. for Valentines day, etc. (tangible product). Then the flowers or plants can be combined with others i.e. in a bouquet, with certain packaging information and services i.e. home delivery (extended product). All these features belong to a product concept of a certain brand of product, with a certain exclusiveness (or not), with certain guarantees i.e. on vase life. Consumers buy products from concepts of 'basic' with low price until 'extend' with all kinds of added values. When they repeatedly buy a product from a certain concept they expect it to be of a constant quality and price; these are prerequisites to the suppliers. Consumers and institutions buy their product at different kind of outlets. In Europe, USA and in the big cities all over the world there are local florists or even artist selling ornamentals and decorations. Adapting to developments sometimes they form chains, i.e. in Holland a Business Florist Association was established for the business segment. Then floricultural products can be bought at local garden centres and warehouses. But more and more chains of these, with a franchise element, occur. Furthermore in Europe more and more floricultural products are sold in supermarkets. In the U.K. for instance, the main part of flowers and potted plants are sold thru i.e. Tesco, Sainsbury's, Waitrose supermarkets, etc.

These different outlets get their products from different types of suppliers. Local florists and garden centres may buy their product at the auction directly or at the cash and carry service. In Germany, France and England product can also be acquired at wholesale markets. The chains of florists, garden centres, warehouse and supermarkets, more and more have their (preferred) supplier. It concerns large quantities.

More and more these supermarket chains have, lead by the choices consumers make, a clear market position in either the discount segment or service retailing with added value product. Because of fast growth of these chains, their economies of scale increase and every now and than price battles are the result. The supply chain therefore is forced to further reduce costs. We also see the growth of private label. Ornamentals in supermarkets i.e. in the U.K, France Germany etc. occur under private label. Supermarkets see that to make money, fresh product requires professional approach. The trend is to outsource category management to partners in the chain. For image building the supermarkets use all kind of quality assurance schemes, which have to be met by the supply chain.

These developments are the reason that wholesalers/traders specialise, professionalize and grow rapidly, to be of sufficient scale for their clients. More and more they operate as a logistical service provider to their clients. To be able to meet the demand they start working more and more close with growers instead of solely buying at the auction. This, to be sure, their demand is met. The marketing departments of these wholesalers develop new product concepts with their chain partners.

Supply Chain Management

While product, from all over the world solely used to be sold at the auction to all kind of buyers, more and more product is sold directly to supermarkets and specialised wholesalers. Dependent on their market focus and desired image they have product specifications on quality, diversity and quality assurances schemes to be met. The trend we see is that supermarkets, garden centres and warehouses want to be sure the products are produced in a sustainable way. This is also called production in the triple P-concept; a way of producing combining making money (Profit), with care for the environment (Planet) and a social responsible way (People). The word Profit contains items like assets, return on investment, partnerships, mergers etc. Planet refers to entities as soil, air, water and the effects of the use of fertilizer, chemicals for pest control, use of energy and the amount of waste. ' People' requires that attention is paid to human rights, poverty, workers, health and security. These are schemes like, Eurep-GAP with demands on producing in a environmental and social responsible way. A scheme like this can also be covered with the MPS-label for environmental inputs, the MPS-scheme for socially qualified holdings, and product specs from a label called Florimark, combined in MPS-GAP. The supply chain will have to combine commercial quality (branding) of the product, together with environmental quality (labels/certificates) and social quality (labels/certificates) in their product proposition towards their customers. Additional of managing the logistics of product to the outlets, from the chain partners this requires registrations, certificates, quality audits, information supply in the chain.

How can Producers Benefit?

Requirements Outlets and Wholesalers

The developments in the market, at the outlets and at wholesaler level have implications for producers as supplier. To be able to supply the chains of supermarkets, garden centres, ware houses, producers have to be able to supply enough quantity. They need economies of scale and they have to specialise in supplying this chain to beat the competition. Investing in ICT-infrastructure can be a part of that for communicating with the right information at the right time. Their focus should be on decreasing supply chain cost and on decreasing lead times to get a fresher product in the store. They have to be able to meet the quality assurance schemes of the supermarkets, and assure they produce in a sustainable way (triple P). All these developments require knowledge of the supply chain, marketing and product development to be able to surprise the customer with new products. These are quite some requirements to be met.

Questions Producers Will have to Answer

This requires that producers make quite some choices. What kind of product can I grow with what kind of specifications? Can I deliver added value to my product or is standard be good enough. Which kind of markets and chains am I able to supply? Which kind of customers fit my business? Can I do business with supermarkets or will it be better to have a specialised wholesaler as a customer? Will I be able to meet quality assurances schemes? Can I produce in a sustainable way? Will I be able to develop all this by myself, or would it be better to join a co-operative. And what influence will my choices have on cost, turnover, margins and profit? The more clearly the choices can be made (out of a strategy) the higher the chance of being successful.

Important Aspects of the Flower Supply Chain

In this article it is established that the world wide trade of ornamentals has increased over the last decade. This also means there is quite a bit of competition in supplying the markets. The markets and different type of outlets set the standard. Further more, their hosting countries and continents have their regulations. So if exporting product to other countries and continents is considered, the following has to be taken in to account:

- Disease free propagation material; to deliver floricultural products free of disease, for Asian producers it is important to start with healthy propagation material. Because southern countries have good climate conditions for producing propagation materials like seeds, cuttings it is interesting for European and USA growers to import this material. Because of phytosanitary measures in their countries and also their own growing results, it is important that this material is guaranteed disease free.
- High level crop management. To be able to meet the demand and the arrangements made about amount of product, product quality, moment of delivery, it is important that crop management is of high level to assure this.
- Grading according to international standards is important to make the delivery accepted.
- Independent quality inspection. When propagation material or end products are inspected at borders or at the customers', the process will take shorter when it can be shown that it has been inspected by a independent quality inspection organisation in the exporting country. If these are organisation working world wide, i.e. S.G.S., this will build trust. This kind of inspection also will assure the exporting party the product won't returned or destroyed.
- Flexible marketing system. Exporters will have to try to find different (types of) outlets for their produce. Each customer will have a different demand and different product specifications. The more the exporting company is able to play the customers with more different type of specifications, the higher the chance of being able to sell the produce.
- Specialised outlets. Outlets try to build an image and to play target groups by marketing. They fill this in with their outlet buildings and locations, their product assortment and price setting contribute to this. When potential suppliers are able to strengthen this formula with their product and services, they will be welcome as a supplier. Next to that they can distinguish themselves from other suppliers to invest in this relation
- Climate controlled storage and transport. 'Due diligence' or 'do as you say' is extremely important when it comes to meeting the quality level agreed on. During the transport, the quality of the product will decrease. With a climate control focussed on the physiological needs of the product, the process of deterioration will be slowed down. This will rise the chances that the product will arrive with the quality of the right level.

Expectations Nearby Future

Looking at the facts presented about the developments in production, trade and consumption in the last 10 to 15 years, combined with the actual trends and lessons learned, we come to the following expectations for the nearby future:

- Higher demand. The demand will rise because of higher income and because certain markets have not saturated yet. Rising income and prosperity will increase the demand in the Asian, Latin American and Eastern Europe;
- Higher quality products. The rise of the average income, the competition in production, market strategies of professional supermarkets will lead to a demand for higher quality product;
- Volume/Logistics. For their competitive edge for supermarket chains it will be more and more important to be delivering year-round, to cut in the amount of suppliers and to constantly have the story full with product. Suppliers will have to be able to help solve these problems by being able to supply big volumes, and to have the logistic well arranged to be able to meet the arrangements made;
- Professionals/providers (category management/vendor management inventory). Because supermarkets sell a whole range of products they could use help in the treatment of the fresh products category. Sometimes this category has problems i.e. with a lot of waste. They have to improve this because of the competition. If suppliers are able to offer category management for the supermarkets in a way that there is more turn over and less waste, it will benefit both parties;
- Long term commitment. This offers possibilities to invest in the relation and to optimise the fitting of the primary process of the enterprises in the supply chain. This also will benefit the parties in the chain;
- International alliances. Continents, countries and unions have their own regulations, and business culture. For exporting producers it is important to understand and cope with this. International alliances will make market access easier.

Market Information Needs to be Fed

Watching trends in production, trade and consumption on markets world wide makes it possible to draw ones conclusion about their own position. In order to do so, there is a (constant) need for good statistics. With the help of a network of specialists and local sources throughout the world this could be achieved.

Cut Flower Production in China - Yang Xiaohan, Liu Guangshu and Zhu Lu

China's cut flower industry began in Beijing, Shanghai and Guangdong in 1984. During the next 12 years, cut flower production grew steadily, and developed faster after 1990. The main reasons for this growth are: a) the consumption of cut flowers which has increased due to an increase in the standard of living and the rapid development of the tourist industry; and b) growers can gain relatively high profits from growing cut flowers. For example, the annual profit is US$18,750-65,625 per ha from cut flower cultivation, as compared to US$11,250 per ha from vegetable cultivation. Currently, cut flowers are grown almost all over the country, from Hainan in the South to Heilongjiang in the North, and from Shandong in the East to Xinjiang in the West. From a small start in 1987, commercial cut flower production in Yunnan Province grew at a higher speed than in any other area in China, moving this Province into number one position. There are more than 30 cut flower species which are commercially grown in China, including *Alstroemeria aurantiaca*, Anemone, Antirrhinum, Asparagus, Anthurium, *Calendula officinalis*, Calla, *Callistephus chinensis*, Carnation, *Centaurea cyanus*, *Chimonanthus praecox*, Chrysanthemum, *Dianthus barbatus*, Freesia, Gladiolus, Gypsophila, Gerbera, Heliconia, Hippeastrum, Lilium, Lisianthus, Limonium, Moluccella, Orchids, Peony, *Pulsatilla chinensis*, *Prunus persica*, *Prunus mume*, Ranunculus, Rose, *Salix leucopithecia*, Solidago, Sunflower, *Tagetes patula*, and Tulip. The growers can be divided into 5 categories: a) state farms or companies; b) collective farms; c) private farms; d) sino-foreign country joint ventures; e) foreign companies. Due to the nature of constantly expanding activities, it is difficult to obtain official statistical figures regarding China's cut flower industry. The data shown in this paper only partially represent the real situation.

Present Situation of Cut Flower Production

In 1996, the production area under cut flowers was about 3,000 ha and the yield was 1.09 billion stems. The major areas where cut flowers are grown commercially are Yunnan, Shanghai, Sichuan, Zhejiang, Guangdong, and Beijing. The production-acreage in Yunnan, Shanghai, Zhejiang and Beijing accounts for about 50% of the total acreage in China. The annual production of cut flowers in these four regions accounts for about 79% of the total production in the country.

The top ten cut flower crops in China are rose, carnation, chrysanthemum, gladiolus, calla, gypsophila, anthurium, gerbera, lily and limonium. Roses are mainly grown in Beijing, Guangdong, Shanghai and Yunnan. In 1994 about 120 million stems of rose cut flowers were produced, accounting for about 17% of all cut flowers. In 1995 the production-acreage of rose cut flowers was 300 ha, of which 100 ha were unprotected, and the annual production increased up to 200 million stems, accounting for about 18% of all the cut flowers produced in the country. The major cultivars of rose cut flowers include Samantha, Red Success, Madelon, Kardinal, Carl Red, Dallas, Baccara, Rouge Meilland, Gabriella, Golden Emblem, Gold Medaillon, Golden Times, Gold Medale, Cocktail 80, Blami, Diplomat, Flamingo, Prima Donna, Sonia Meilland, Leading Lady, Bridal Pink, White Success, Tineke, Athena, Bridal White, Carte Blanche, Marina and Mercedes.

Carnation prevalently of the standard type, accounts for about 25% of all the cut flowers sold in the Beijing and Kunming wholesale markets. The production areas center in Yunnan and Shanghai. The major cultivars of carnation are from Israel, the Netherlands and Germany, such as Tasman, Francesco, Omaggio, Red Rimon, Presto, Ondina, Dona, Red Corso, Hermes, Pink Francesco, Pallas, Napoleon, Sugar Lee, Rimo and Camba. The spray carnations are not popular in China because of consumer's preference for standard carnations.

Chrysanthemum accounts for about 20% of all the cut flowers sold in the Beijing and Kunming wholesale markets. The production areas are concentrated in Guangdong,

Shanghai and Beijing. The major cultivars of chrysanthemum include late-spring-flowering, summer-flowering, autumn-flowering and early-winter-flowering cultivars. Most of them are standard chrysanthemums. At present the spray chrysanthemums are not popular in China because of consumer's preference for standard chrysanthemums. Among various cultivars, yellow is the most appreciated colour, followed by white and red.

Gypsophila accounts for about 20% of all the cut flowers sold in the Kunming wholesale market. The production areas are concentrated in Yunnan and Shanghai. There are semi-double and double cultivars with white flowers.

Gladiolus is mainly grown in Liaoning. There are 28 cultivars from Jining Province, 11 from Gansu Province, 11 from Hubei Province, 9 from Liaoning Province, and 26 from the Netherlands.

Lily is mainly grown in Gansu. The major cultivars of lily cut flowers include Navona, Acapulco, Elite, Lorina, Solemio, Pollyanna, Adelina, Akita, Her Grace, Jessica, Maremma, Amanda, Ankra, Apropas, Merostar, Wisdom, Snow Queen and White Satin. Gerbera is mainly grown in Yunnan and Shanghai. The major cultivars of Gerbera include Laurentius, Terranero, Michelle, Feugo, Shanghai, Estelle, Terramor, Clementine, Terraparva and Terratuba.

Anthurium is mainly grown in Shenzhen and Beijing. The major cultivars of Anthurium include Alexia, Anneke, Candia, Gino, Gloria, Lydia, Margaretha, Mauricia, Nette and Rosetta. Calla is mainly grown in Shenzhen and Beijing. There are only white flower cultivars. Limonium is mainly grown in Yunnan, Sichuan, Guangdong and Beijing.

Cut flower production is year-round in various areas, and is usually under protected conditions except Hainan, Guangdong, Guangxi and Fujian where there are more cut flowers produced in open fields. The acreage of cut flower production under protected conditions is larger than that in open fields, but accurate data about the ratio of production between protected conditions and open fields are not available. Generally, during fall-spring, cut flowers are produced in greenhouses; and during spring-summer, they are produced out doors. Tunnels covered with plastic are widely used in growing cut flowers throughout the country. Modern greenhouses are also used, especially in Northern areas. In the last few years, new greenhouses and modern equipment have been introduced into China from Israel, the Netherlands and France.

In recent years, the growers in Beijing, Shanghai and Kunming have started using soiless culture and drip irrigation system to grow cut flowers. Controlled-release fertilizers have been used in Beijing for cut flower production. In the summer season, shade is widely used to reduce light to optimum levels for cut flower production. Artificial light has been applied by a limited number of growers in some regions such as Beijing and Shanghai to promote flowering of gladiolus, gypsophila, and lily during the winter, or to regulate flowering of chrysanthemum for year-round production. The technique of forcing bulbs at optimum temperatures is also practiced.

Production of Planting Material

Most of the cuttings, tissue culture plants and grafted planting material are produced in Shanghai, Yunnan, Beijing and Sichuan. Some of the propagation materials are virus-free. Although the companies in Liaoning, Jining, Beijing, Shanghai, Shaanxi and Shandong have produced some bulbs, most of the bulbs necessary for growing cut flowers in China are from the Netherlands.

The Beijing Flower Seedling Company which owns 5,000 square meters of glasshouse and 400 square meters of modern factory with tissue culture facilities, can produce 3 million plantlets every year, including rose, carnation, eustoma, chrysanthemum, lily, gerbera, etc. The North Star Garden Plants Company in Beijing can produce 1 million plants of anthurium, gerbera, *Phalaenopsis amabilis* and lily per year by means of tissue culture under automatically-controlled greenhouses (temperature and mist). Fifty thousand bulbs of lily are produced in Changping, Beijing, annually.

Each year, the Shanghai Floriculture Experimental Farm supplies 2.5 million carnation and gypsophila cuttings, gerbera plantlets and grafted rose plants, and 1.5 million bulbs of lily, gladiolus, iris and freesia. The Shanghai Sino-dutch Flower Co. Ltd. produces 1 million planting materials of gerbera and carnation annually.

Recently a newly-built company equipped with computer-controlled facilities has been established in Yunnan that can produce 30 million planting materials annually. Another newly-built company in Kunming has a production capacity of 10 million carnation cuttings. A company in Liaoning Province supplies propagules year-round, including 11 million bulbs of gladiolus and 0.2 million bulbs of lily. A company in Sichuan Province has the capacity to produce 1.5 million bulbs and 1.2 million plantlets annually.

Marketing

More than 90% of the cut flowers produced in China are consumed within China. The cut flowers are sold through whole sale and retail distribution channels. At present, there are 670 cut flower markets and 7,000 flower shops across the country, and 7 regional whole sale markets which have been established in Beijing, Shanghai, Kunming, Guangzhou, Fujian, Chengdu and Liaoning. The major consumption areas are Shanghai, Beijing, Zhejiang and Guangdong. The majority of the flower shops are located in these areas. In 1996, 260 million stems of cut flowers were consumed with about 20 stems per capita in Shanghai, and 100 million stems were consumed with about 10 stems per capita in Beijing.

It is inferred from the analysis of the whole-sale prices of rose, carnation, chrysanthemum and gladiolus in the Beijing and Kunming markets during the period from January 1996 to March 1997 that these cut flowers were cheaper during the period from May to October than during the period from November to April. It is also obvious that the average wholesale price in Beijing is higher than that in Kunming. Yunnan supplies fresh cut flowers to more than 20 other regions such as Beijing, Guangzhou and Shanghai.

Table : Flower Shop Distribution in China

Regions	*Number of Flower Shops*	*% Total*
China	7000	100
Zhejiang	2000	29
Shanghai	1300	19
Shenzhen	500	7
Beijing	500	7
Guangzhou	400	6
Hebei	200	3
Dalian	150	2
Hefei	107	1
Shenyang	100	1
Hainan	30	0.4
Guiyang	30	0.4
Jiangxi	20	0.3
Other regions	1663	23.9

The export of cut flowers to other countries mainly Hongkong, Japan, Singapore and Thailand, is very limited, less than 10% of all the production in China. In 1993, China exported cut flowers of about US$1 million. In 1994, Yunnan exported cut flowers of US$4 million. In 1996, Beijing exported cut flowers worth US$750,000 (about 3 million stems through the Beijing Flower Service Cooperation), and Kunming exported more than 18 million stems.

Potential for Cut Flower Production Development

China has an abundance of wild and garden plants which can be developed into cut flowers. For example, Lilac and Forsytia are good early-spring cut flowers, but they are used only as garden plants. There are rich plant germplasms (15 species) of peony in China. Recently, China and Australia have done joint research to develop cut flower cultivars and the related cultivation and postharvest techniques.

There exist three types of regions suited to growing cut flowers in China: a) Dalian in Liaoning Province, and Weihai and Qingdao in Shandong Province, where the climatic conditions are similar to those in the Netherlands, with a very short day and freezing weather in winter and a very long day and warm weather in summer. With the appropriate technologies, it is possible to produce in Dalian, Weihai and Qingdao all the cut flower crops which are grown in the Netherlands; b) the eternal spring region like Kunming in Yunnan Province and Xichang in Sichuan Province, where the climatic conditions are similar to those in the Bogota plain in Colombia, with about the same length of day and night, and cool nights and warm days almost without change throughout the year; with the appropriate technologies, it is possible to produce in the Kunming and Xichang region all the cut flowers which are grown in the Bogota plain; c) the tropical region like Hainan Province and Xishuangbana in Yunnan Province, where the climatic conditions are similar to those in the Caribbean Islands and Thailand, with about equal length of day and night throughout the year; with the appropriate technologies, it is possible to produce in the Hainan and Xishuangbana regions all the cut flowers which are grown in the Caribbean Islands and Thailand. These three types of regions give China a unique potential to produce commercially all the known cut flower crops throughout the year.

China has a growing strength of labour because of the agriculture modernization and the increase in population. Currently, there are about 100 million extra labourers from the rural areas which can be used for growing cut flowers. The cost of labour is very low compared to the developed countries. Presently China produces 1.09 billion stems of cut flowers annually. This represents a consumption of about 1 cut flower per capita per year, a very low level compared to other countries. More than 1.2 billion people are living in China. An increase of one cut flower in per capita consumption means an increase of more than 1.2 billion stems in the China's cut flower market. The people in China have a natural love for flowers, which is expressed by the care of gardens and public parks. We believe that with the improvements in the economic situation, the cut flower industry will expand rapidly.

Both the Yunnan and Guangdong Provincial Governments have evolved a series of policies to promote the development of cut flower industry. For example, Yunnan Provincial Government and local governments invested US$5 million in the cut flower industry annually in the last few years. They also reduced the cost of air freight and charged a low tax, only 2% of the gate value.

Constraints in Cut Flower Production Development

Production areas are concentrated in the rural environments, but the consumption is concentrated in the cities. There is little direct communication between growers and consumers.

Presently, the yield of cut flowers per square meter in only 50-80 stems in China, a very low level compared to other countries. The quality of most cut flowers produced in China is only equivalent to that of the third grade in the international markets. The main reasons for the low yield and poor quality are:

a) lack of specific knowledge and technology; for example, most growers who originally grew vegetables still treat flower crops as vegetables, using the local soil as growing media and traditional methods of irrigation and fertilization; in most growing areas, the local soil is heavy, unsuitable for continuous cultivation under the regime of heavy fertigation which tends to destroy the soil structure; the mist and drip irrigation systems are seldom used.

b) lack of financial source to modernize the cultivation facilities.

c) lack of disease-free planting materials; in order to save money, many growers use recycled materials for propagation instead of buying commercial planting materials.

d) lack of cool chain from grower to retailer; treatments by chemicals for increasing vase life is not practiced extensively; the quality of cut flowers produced in Yunnan is reduced after they are transported to other areas such as Beijing and Guangzhou; until now there has been no quality standards for plant materials and fresh cut flowers.

The supply of cut flowers is centered around spring and autumn seasons, but during the Chinese lunar calendar new year Spring Festival when the consumption is the highest in the year, the supply is the smallest.

China has not joined the international organization UPOV yet. Therefore, it is difficult for the Chinese cut flower industry to obtain the newest and best cultivars bred by the breeders in other countries. There is an urgent need to pass the Plant Breeders' rights Law in China. This process has been started and China will have this law in the near future. This law is expected to promote investments in breeding efforts within China.

Conclusions

Cut flower production in China is being industrialized. The Chinese government has made "The Ninth Five-Year Plan from 1996-2000 and the Long-Term Objective Programme for 2010 for the improvement of cut flower production" as follows:

a) The annual cut flower production will reach 2 and 4 billion stems by the year 2000 and 2010, respectively, in the country.

b) The average per capita consumption in the urban areas will reach 10 stems by the year 2000.

c) Cut flower production will be concentrated in Beijing, Shanghai, Guangdong, Yunnan, Sichuan and Hebei.

d) The production of planting materials will be concentrated in Sichuan, Yunnan, Shanghai, Liaoning, Shaanxi and Gansu.

e) Modern cut flower trade centres will be built in Beijing, Shanghai, Guangzhou, Kunming, Chengdu and Shenzhen.

f) Optimum-scale cut flower whole-sale markets will be built in metropolitan centres across the country.

Cut Flower Production in India

India has a long tradition of floriculture. References to flowers and gardens are found in ancient Sanskrit classics like the Rig Veda (C 3000-2000 BC), Ramayana (C 1200-1300 BC), Mahabharata (prior to 4th Century BC), Shudraka (100 BC), Ashvagodha (C 100 AD), Kalidasa (C 400 AD) and Sarangdhara (C 1200 AD). The social and economic aspects of flower growing were, however, recognized much later. The offering and exchange of flowers on all social occasions, in places of worship and their use for adornment of hair by women and for home decoration have become an integral part of human living. With changing life styles and increased urban affluence, floriculture has assumed a definite commercial status in recent times and during the past 2-3 decades particularly. Appreciation of the potential of commercial floriculture has resulted in the blossoming of this field into a viable agri-business option. Availability of natural resources like diverse agro-climatic conditions permit production of a wide range of temperate and tropical flowers, almost all through the year in some part of the country or other. Improved communication facilities have increased their availability in every part of the country. The commercial activity of production and marketing of floriculture products is also a source of gainful and quality employment to scores of people.

Present Situation of Cut Flower Production

Inspite of the long and close association with floriculture, the records of commercial activity in the field are very few. The information on the area under floriculture and the production generated is highly inadequate. As commercial floriculture is an activity which has assumed importance only in recent times, there are not many large farms engaged in organised floriculture. In most part of the country flower growing is carried out on small holdings, mainly as a part of the regular agriculture systems.

Production Areas

The estimated area under flower growing in the country is about 65,000 hectares. The major flower growing states are Karnataka, Tamil Nadu and Andhra Pradesh in the South, West Bengal in the East, Maharashtra in the West and Rajasthan, Delhi and Haryana in the North. It must, however, be mentioned that it is extremely difficult to compute the statistics of area in view of the very small sizes of holdings, which very often go unreported. This perhaps would be the reason for unrealistically small areas reported for floriculturally active states like Maharashtra, Uttar Pradesh and Madhya Pradesh.

Table : Area Under Flower Production in India

State	*Area (ha.)*
Karnataka	19,161
Tamil Nadu	14,194
West Bengal	12,285
Andhra Pradesh	5,933
Maharashtra	3,356
Rajasthan	1,985
Delhi	1,878
Haryana	1,540
Madhya Pradesh	1,270
Uttar Pradesh	1,000
Others	2,166
Total	64,768

More than two thirds of this large area is devoted for production of traditional flowers, which are marketed loose e.g. marigold, jasmine, chrysanthemum, aster, crossandra, tuberose etc. The area under cut flower crops (with stems) used for bouquets, arrangements etc. has grown in recent years, with growing affluence and people's interest in using flowers as gifts. The major flowers in this category are rose, gladiolus, tuberose, carnation, orchids and more recently liliums, gerbera, chrysanthemum, gypsophila etc.

The production of flowers is estimated to be nearly 300,000 metric tonnes of loose flowers and over 500 million cut flowers with stem. In the case of production also, the estimates could be at variance from the actual figures as some of the flowers like rose, chrysanthemum, and tuberose are used both as loose flowers and with stem.

It may be mentioned that almost all of the area reported here is under open field cultivation of flowers. Protected cultivation of flowers has been taken up only in recent years for production of cut flowers for exports. The estimated area in production is about 200 hectares, which is likely to increase to over 500 hectares by the year 2000.

Recognising the potential for low cost production for export, in view of cheap land, labour and other resources, several export oriented units are being set up in the country. These projects, located in clusters around Pune (Maharashtra) in the West, Bangalore (Karnataka) and Hyderabad (Andhra Pradesh) in the South, and Delhi in the North, are coming up in technical collaboration with expertise mainly from Holland and Israel. More than 90 percent of these units are for rose production, on an average size of 3-hectare farm, while some projects for orchid, anthurium, gladiolus and carnation are also being set up. Nearly one third of over 200 proposed projects, have already commenced production and export.

Major Cut Flower Crops

Rose is the principal cut flower grown all over the country, even though in terms of total area, it may not be so. The larger percentage of the area in many states is used for growing scented rose, usually local varieties akin to the Gruss en Tepelitz, the old favourite to be sold as loose flowers. These are used for offerings at places of worship, for the extraction of essential oils and also used in garlands. For cut flower use, the old rose varieties like Queen Elizabeth, Super Star, Montezuma, Papa Meilland, Christian Dior, Eiffel Tower, Kiss of Fire, Golden Giant, Garde Henkel, First Prize etc. are still popular. In recent times, with production for export gaining ground in the country, the latest varieties like First Red, Grand Gala, Konfitti, Ravel, Tineke, Sacha, Prophyta, Pareo, Noblesse. Virsilia, Vivaldi etc. are also being grown commercially.

Gladiolus is the next most important cut flower crop in the country. Earlier it was considered a crop for temperate regions and its growing was restricted to the hilly areas, particularly in the north eastern region, which still continues to supply the planting material to most parts of the country. However, with improved agronomic techniques and better management, the northern plains of Delhi, Haryana, Punjab, Uttar Pradesh, as well as Maharashtra and Karnataka have emerged as the major areas for production of gladiolus. Tuberose, a very popular cut flower crop in India is grown mainly in the eastern part of the country i.e. West Bengal, and also in northern plains and parts of south. Both single and double flower varieties are equally popular. Tuberose flowers are also sold loose in some areas for preparing garlands and wreaths.

The other main cut flower item is orchid. Its production is restricted mainly in the north-eastern hill regions, besides parts of the southern states of Kerala and Karnataka. The main species grown are Dendrobiums, Vanda, Paphiopedilums, Oncidiums, Phalaenopsis and Cymbidiums.

Among the traditional crops grown for loose flowers, the largest area is under marigold, grown all over the country. In most parts of the country only local varieties are grown for generations. African marigolds occupy more area as compared to the small flowered French types. Jasmine flowers in view of its scent are also very popular as loose flowers and for use in garlands and *Veni* (ornament for decoration of hair by women). The major areas under this crop are in Tamil Nadu, Karnataka in South and West Bengal in East. The varieties are mainly improved clones of *Jasminum grandiflorum*, J. *auriculatum* and J. *sambac*. The chrysanthemum, particularly the white varieties are much in demand as loose flowers during the autumn period of October-December when other flowers like jasmine, tuberose are not available for use in garlands etc. Among other traditional flowers grown in large areas are crossandra in southern states of Tamil Nadu, Karnataka and Andhra Pradesh and aster in Maharashtra.

Research Support

Research work on floriculture is being carried out at several research institutions under the Indian Council of Agricultural Research and Council of Scientific and Industrial Research, in the horticulture/floriculture departments of State Agricultural Universities and under the All India Coordinated Floriculture Improvement Project with a network of about twenty (20) centres. The crops which have received larger attention include rose, gladiolus, chrysanthemum, orchid, jasmine, tuberose, aster, marigold etc. The thrust till recently had been on crop improvement, standardization of agro-techniques including improved propagation methods, plant protection and post harvest management. In view of the fact that most of the cut flower production is being done under open field conditions, the research efforts generally relate to open cultivation. In recent years, however, technologies for protected cultivation and tissue culture for mass propagation have also received attention. A large number of varieties suitable for cut flower use, as well as garden display have been developed. Production technology, particularly the agronomic requirements and control methods for important diseases and insect pests have also been developed. Contribution by the private sector in research activities in floriculture is negligible.

Planting Material

The requirement of planting material to cater to the large area under flower crops, is largely met from domestic production. Since efforts to set up large commercial farms generally suffered due to lack of quality planting material in sufficient quantities, this aspect has received greater attention in recent years in the breeding centres, which are producing sufficient quantity of planting material. Most of the nurseries propagating planting material are in the private sector. In the absence of any mechanism to register nurseries, it is very difficult to ascertain their exact number, but at a very conservative estimate there are more than 100,000 nurseries, spread out all over the country, producing seeds and other planting materials for flower growers. The states with larger numbers of nurseries include Maharashtra, West Bengal, Karnataka and Tamil Nadu. Most of the nurseries are small, with little or no improved facilities like mist propagation unit, green houses/net houses etc. For meeting the demand of flower seeds, several large seed companies have production units in Punjab, Himachal Pradesh and Jammu & Kashmir in the North, Karnataka in the South and West Bengal in the East. A few of the leading multinational seed companies have tied up with local seed companies or producers for custom production of seeds of their varieties. In the case of bulbous plants, most of the planting material is produced in the north eastern hilly regions of West Bengal (Kalimpong) and Sikkim, though for some crops, it is also produced in hilly regions of northern India. The introduction of a revised seed policy by the government of India in 1989 has enabled unrestricted introduction of many new and superior varieties into the country, increasing the variety in the floral basket.

Tissue culture has, in recent years, been recognized as an important tool in agriculture development. With its diverse climatic zones and qualified manpower, India is well placed to exploit the benefit of tissue culture based applications to floriculture crops. Most popular application of tissue culture has been micropropagation using *in vitro* technique for mass multiplication of planting material. Tissue culture plants of ornamentals have found ready acceptance by the commercial growers and their production increased significantly from 130 million plants in 1985-86 to 680 million in 1994-95. At present 30 commercial tissue culture units with annual capacities of 0.5 to 15 million plants each are in operation, resulting in total capacity of about 110 million plants. While most of it is exported, a small percentage of cut flower crops like carnation and gerbera are finding good market within the country.

Marketing

Marketing of cut flowers in India is very unorganised at present. In most metropolitan cities, with large market potential, flowers are brought to wholesale markets, which mostly operate in open yards. A few large flower merchants generally buy most of the produce and distribute them to local retail outlets after significant mark up. The retail florist shops also usually operate in the open on-road sides, with different flowers arranged in large buckets. In the metros, however, there are some good florist show rooms, where flowers are kept in controlled temperature conditions, with considerable attention to value added service. The government is now investing in setting up of auction platforms, as well as organized florist shops with better storage facilities to prolong shelf life.

The packaging and transportation of flowers from the production centres to the wholesale markets at present is very unscientific. The flowers, depending on the kind, are packed in old gunny bags, bamboo baskets, simple cartons or just wrapped in old newspapers and transported to markets by road, rail or by air. The mode of transportation depends on the distance to the markets and the volume. Mostly, flowers are harvested in the evening time and transported to nearby cities by overnight trains or buses. In recent years, the government has provided some assistance for buying refrigerated carriage vans. A large number of export oriented units have built up excellent facilities of pre-cooling chambers, cold stores and reefer vans and their produce coming for domestic market sales are thus of very good quality and have longer vase life and command higher price. The government programmes for floriculture development include creating common facilities of cool chain in large production areas to be shared on cooperative basis. Formation of growers' cooperatives/associations are being encouraged.

In view of the unorganized set up, it is difficult to estimate the size of flower trade, both in terms of volume and value. A study conducted in 1989 estimated the trade to be worth Rs. 2050 million. It is in the period of the last five years or so that this business has really boomed in India, which is reflected in the number of new florist outlets in all cities and increase in the public's purchase of flowers as gifts. This would put the current trade at several times the earlier estimate. A recent study of Delhi market alone put the value of flowers traded on wholesale as Rs. 500 million.

The loose flowers (traditional crops like marigold, jasmine etc.) are usually traded by weight. The average price of different flowers in major markets varies considerably depending on the period of availability.

Table : Average Market Price for Major Flower Crops

Flowers	*Unit*	*Price (US$1 = Rs.40) Rs./kg or doz or each stem*
Marigold	kg.	3-60
Jasmine	kg.	15-150
Crossandra	kg.	20-120
Chrysanthemum	kg.	5-25
Tuberose	kg.	5-30
Rose	kg.	6-60
Gladiolus	doz.	20-75
Carnation	doz.	30-75
Gerbera	doz.	36-75
Orchids	each stem	10-45
Liliums	each stem	10-45
Anthuriums	each stem	15-45

The net returns to the growers depend on the packaging and transportation costs. The cut flowers with stem have a limited overall market in terms of volume. The share of cut flowers has almost doubled from 30 to 60% in the last decade.

The value of cut flower export from India has increased twenty five fold during the last five years. With more export oriented units coming into operation, exports are likely to grow further in the coming years. The major share of the export trade is for roses, in addition to orchids, gladiolus etc. The major markets are Europe (Holland, Germany and U.K.) and Japan. The exports of roses to Japan, have really picked up in the three years from Rs. 360 million in 1993-94 to Rs. 6090 million in 1995-96. As per the estimates for 1996-97, India has been the largest supplier of roses to Japan (volume wise).

Table : Export of Floriculture Products from India (Rupees in Million)

Item	*1991-92*	*1992-93*	*1993-94*	*1994-95*	*1995-96*
Cut Flower Fresh	3.99	10.90	9.96	29.98	100.33
Dried Flowers	64.35	74.31	109.21	195.78	364.56
Live Plants	40.15	30.56	30.52	60.43	81.48
Dried Plants	23.99	25.45	23.72	10.35	35.83
Bulbs, Tubers etc.	12.06	7.83	14.95	11.83	19.21
Total	144.54	149.05	188.36	308.37	601.41

Potential for Cut Flower Production Development

The availability of natural resources like favourable and diverse climatic conditions permit production and availability of a large variety of flower crops round the year. Cheap labour leads to reduction in production costs, increasing access of the consumer to good quality flowers at affordable prices, besides increasing our competitiveness in the export markets. Being a new concept in the agri-business, it took some time for scientific commercial flower production to take roots, but with the appreciation of its potential as an economically viable diversification option, its growth is slowly stabilising. The government also has, during the last few years, recognized floriculture as an important segment for developmental initiatives. Model Floriculture Centres being set up in 11 major production zones, to serve as focal units for development in the region, have a mandate of making available quality planting material, new/ improved production technologies and also to provide training in production and post harvest management. There are also special government programmes for area expansion in floriculture with state assistance. The National Horticulture Board, a major developmental agency for horticulture, also makes available finances as soft loan for setting up integrated projects for production and marketing. As mentioned earlier, the government is investing in improving the infrastructure for marketing in the domestic sector.

Production of cut flowers for exports is also a thrust area for support. The Agricultural and Processed Food Products Export Development Authority (APEDA), the nodal organization for promotion of agri-exports including flowers, has introduced several schemes for promoting floriculture exports from the country. These relate to development of infrastructure, packaging, market development, air freight subsidy etc. The 100% Export Oriented Units are also given benefits like duty free imports of capital goods.

All these efforts indicate the government's commitment for improving the sector and creating a positive environment for entrepreneurship development in the field.

Constraints in Cut Flower Production Development

Being a new concept, the requirements of scientific and commercial floriculture is not properly understood in the country. The developmental initiatives of the government have to keep in mind the low knowledge base, small land holdings, unorganized marketing and poor infrastructural support.

While long experience of flower growing in the open field conditions enable sufficient flower production for domestic markets, the quality of the produce, in view of its exposure to various kinds of biotic and abiotic stresses, is not suitable for the ever growing export market. The production technology for flowers under protected environment of green houses needs to be standardized. There is hardly any post harvest management of flowers for the domestic market. Availability of surplus flowers from exports for sale in the domestic market, has increased the appreciation of quality produce and the demand for good quality flowers is increasing. With the introduction of new varieties of crops in the country, facilities for generating their planting material for large scale production need strengthening. Special attention needs to be paid to strengthen the marketing infrastructure like organised marketing yards, auction platforms, controlled condition storage chambers etc.

Greater research efforts are also needed for integrated pest management, development of location specific package of practices for traditional flowers, value addition to traditional flowers etc.

The initial cost and availability of finance is a critical matter in the development of large commercial projects requiring heavy investments. More options for developmental finance, such as the soft loan scheme of the National Horticulture Board need to be identified. In the initial years of commercial floriculture development, the governmental support in terms of subsidies etc. needs special attention.

The potential for growth of export market is always linked to the strength of domestic market its capacity to absorb surplus and over production, and quality consciousness of consumers. Though we have a large domestic market, the marketing system and facilities need to be modernized.

The production for exports at present has suffered due to a few constraints. While our growers have been successful in producing world class quality at low cost, high air freight rates, low cargo capacity available, imposition of import duties, inadequate export infrastructure etc. have reduced their competitiveness. There is also a shortage of trained manpower to handle commercial floriculture activity. The demands of the growing export oriented industry would require adequate attention to be paid for human resource development, particularly at the supervisory level.

Conclusions

India has a long floriculture history and flower growing is an age old enterprise. What it has lacked is its commercialization. The growing demands of flowers in the domestic as well as the export market will require a concerted effort on the part of the government as well as the private entrepreneurs to develop floriculture on scientific lines. Paying attention to the input needs, better resource management and making various policies entrepreneur friendly would lead to a balanced growth of the industry.

Table : Cut Flower Exports from India

Country	*1993-94*	*1994-95*	*1995-96*
Japan	322.50	8255.58	35932.56
Netherlands	1004.61	9102.49	24799.90
U.S.A.	1175.38	2495.21	17652.50
Germany	957.61	2538.63	9256.00
U.K.	1420.93	1113.78	3345.86
U.A.E.	2120.40	3388.19	2459.29
Italy	210.28	164.96	2200.49
Hongkong	730.02	903.15	1504.02
Singapore	78.45	437.37	1190.54
Nepal	11.86	36.09	292.64
Kuwait	24.22	5.00	274.13
Saudi Arabia	413.98	169.52	272.77
Switzerland	136.96	258.20	242.83
Hungary	-	286.69	181.82
Thailand	-	86.49	177.82
Australia	-	-	132.89
Russia	368.46	20.96	119.85
Others	988.68	719.39	9293.82
Total	9964.34	29981.60	109329.73

Cut Flower Production in Indonesia

Indonesia is the world's largest archipelago of more than 13,000 islands with a total coastline of 81,000 kilometers. With a population of 190 million, Indonesia is the world's fourth most populous nation after China, India and the United States.

The country is in the process of accelerating the development of its economy through industrialization on the basis of self reliance, including rural economy. This will create employment opportunities and improve the welfare of millions of people living in the rural areas, as well as increase production quantitatively and qualitatively to meet the growing demand of the rapidly expanding population. This will offer strong investment opportunities especially for floriculture and other horticultural crops.

Ornamental horticulture, especially cut flower production has sprouted from cultural and hobby based activities into a lively and prospective business. Current government efforts to monitor the domestic and global demand and production has provided sufficient evidence that the cut flower industry could contribute substantially to the GDP of Indonesia. A productive cut flower industry could provide additional economic strength into the process of industrialization in the 21st century.

The government policy on horticulture development is directed to:

- stimulate investment and create opportunity in the horticultural sector, based on agribusiness orientation;
- reduce the sharp price fluctuation in order to maintain economic stability;
- reduce import and increase export, to raise foreign exchange earnings;
- create job opportunities and increase community income;
- fulfill the demand for beauty, harmony and natural environment.

In the sixth Five Year Development Programme (Pelita VI), expected contribution from horticultural commodities to Product Domestic Bruto is 6.1 percent and growth of horticultural production is 5 percent per year (Repelita VI Pertanian, 1992). Approaches in the Second Long Term Development Plan for the horticultural sector will be focused on agri-business orientation which is modern, economically efficient and sustainable.

Present Situation of Cut Flower Production

Production centres of cut flower and ornamental plants have mostly developed on the basis of climatological and soil conditions and their distance from larger cities. Main production centres at present are:

a) Jabotabek:	Jakarta, Bogor, Tanggerang and Bekasi
b) North Sumatera:	Brastagi, Kabanjahe, Tanjung, Morawa and Medan
c) Riau:	Riau Islands
d) West Java:	Cipanas, Cianjur, Bogor, Cisarua/Lembang, Sukabumi, Tasik Malaya
e) Central Java:	Bandungan, Yogyakarta, Magelang, Tegal, Pemalang and Purbalingga
f) East Java:	Tretes, Batu, Pujon, Pasuruan and Madura

Due to the lack of statistical information, it is difficult to obtain figures about total area cultivated, production and productivity. For development purposes it is necessary to keep track of these data as is done for other horticultural crops.

The most important cities with regard to existing flower and ornamental sales as well as growing of orchids and ornamental plants are:

Java:	Jakarta, Bandong, Semarang, Malang and Surabaya
Sumatera:	Medan
Bali:	Denpasar
Sulawesi:	Ujung Pandang

The main features of the production centres for flowers and ornamental plants in Indonesia are described here under:

Jakarta (Jabotabek)

In general, every large city has its own orchid gardens which supply its own consumers. In Jakarta however, due to a scarcity of land and high price, orchid growers make investments outside Jakarta, for example in Tanggerang, Bogor and Bekasi.

Estimates of a 1993 census by DKI Jakarta reveal that a total of 102 hectares were cultivated under orchids and ornamental plants. Of this, 73 percent or 75 ha were in use for orchids. Orchid genera mostly grown by Jakarta growers are Dendrobium, Vanda, Arachnis, Oncidium, Phalaenopsis and Cattleya.

Major production areas of orchids in Jakarta are:

West Jakarta:	Kebon Jeruk, Grogol and Cengkareng
South Jakarta:	Kebayoran Lama, Cilandak, Lebak Bulus and Pasar Minggu
East Jakarta:	Jatinegara, Kramat Jati, Pulo Gadung, Pasar Rebo and Pondok Gede

Although orchid growing was primarily done in small shade net base gardens, presently some large scale orchid farms are established in Jakarta and its surroundings. Orchid growing is capital intensive compared to other cut flowers. An orchid farm of 1000 square meters will require a working capital of around US$ 12,000 to 15,000.

North Sumatera (Berastagi, Kabanjahe, Tanjung Morawa) and Riau Island

Production areas in Kabanjahe are scattered over 4 districts: Kabanjahe, Barusjahe, Tigapanah and Simpang Empat. Tanjung Morawa is located 17 km from Medan and is known as the center of production of ornamental plants in North Sumatera. Production of cut flowers include gladiolus, chrysanthemum, aster, gerbera, dahlia, anthurium and sedap malam (*Polyanthes tuberosa*). A few varieties of lilies and rose are also found. Farm sizes for flower growing in North Sumatera varies from less than 1000 square meters to one hectare. Most of the flowers are grown in combination with vegetables. Compared to Java less workers are used. Dry fish waste is used as fertilizer.

Gladiolus is planted in a phased manner so that harvest can be done continuously. In some fields, old plants are left for bulb production; generally, yellow, pink, red and dark red varieties are popular. Chrysanthemum is grown in open fields; white chrysanthemum is grown and pinched to yield one big flower, whereas in yellow and red varieties up to three flowering buds are left; rejuvenation is done after 18 months. Aster, gerbera and dahlia are more widely grown in North Sumatera than chrysanthemum. Some flowers are also sold without stem.

Most farms at Tanjung Morawa concentrate on more or less the same kind of ornamental plants, such as palms, pines, ixora, dieffenbachia, dracaena, cycas and croton. Plants are transplanted from the nursery into black plastic bags.

The wholesale market is located at Berastagi, especially on Tuesdays and Fridays, where flowers are traded and distributed to different markets and flower shops in Medan.

Cut flower growers in Riau island produce Heliconia, one of the most prospective and commercial tropical flowers.

West Java (Cipanas, Bogor Cisarua/Lembang and Sukabumi)

Cipanas has been well known as a center for flower and ornamental production for a long time. Due to its favourable location and climate many inhabitants of Jakarta visit Cipanas on the weekends. Ornamental plants are sold directly to consumers alongside the roads. Major cut flowers produced are chrysanthemum, rose, gladiolus, gerbera etc.

Chrysanthemum from Cipanas is well known for its high quality. In rose, mostly local varieties are grown. Cut flowers are supplied to Jakarta using bamboo boxes with banana leaves and are transported to hotels, florist shops and Rawa Belong market in Jakarta. Ornamental plants produced at Cipanas are pines, adiantum, azaleas, dieffenbachia, dracaena, bougainvillea, hibiscus etc. which are produced in small pots by villagers and sold alongside roads.

Bogor is known for its heavy rainfall with an estimated 320 days of rainfall in a year. Bogor has a very good connection with Jakarta and can be reached in half an hour by road. Cut flowers such as carnation, chrysanthemum, gerbera and calalily are produced in plastic greenhouses at Megamendung and Ciawi. Ornamental plants such as palms, pines and adiantum are produced at Ciapus Bogor.

In Lembang cut flowers are grown on a small scale and are less important than ornamental plants. At Cisarua, a desa (village) next to Lembang, the villagers are producing ornamental plants in almost every home garden. It is estimated that there are around 100 growers with an average farm size of 500 to 1000 square meters. Plants are grown in plastic bags in a medium of rice hulls mixed with stable manure. They mainly consist of flowering pot plants such as baby roses, asters, mini carnation etc. Young plants of pines, cycas, azaleas, dracaena and palms are also produced.

Cut flowers such as gladiolus, anthurium and gerbera are mostly produced at Selabintana and Sukabumi. Sedap Malam (*Polyanthes tuberosa*) is produced at Selabintana/Sukabumi, Mayak/Cianjur and Indihiang/Tasik Malaya. Area under tuberose in Sukabumi is 6 hectares, in Cianjur 39.15 hectares and in Tasikmalaya 9 hectares. From the wholesale market the flowers are transported to Jakarta.

Central Java (Bandungan, Tegal, Pemalang, Purbalingga and Magelang)

Bandungan is the main production area for cut flowers in Central Java. Most of the flowers at Bandungan are produced on small farms by a large number of farmers in a mixed cropping system with vegetables. From an agricultural census conducted in 1986 it appeared that there were 1530 farmers, who produced both flowers and vegetables. Thirty six of them had an area larger than 0.5 hectare. Bandungan produces cut flowers of aster, roses, tagetes, gerbera, dahlia, gladiolus, chrysanthemum, carnation, lily and amaryllis. Quality of flowers varies a lot and grading is done by the trader.

Tegal and Pemalang produces jasmine (*Jasminum sambac*). It is used as an ingredient for the taste of green tea. Banjarnegara has the largest jasmine plantation area of 345 hectares while Banyumas has 45 hectares.

Sedap malam (*Polyanthes tuberosa*) is produced by growers in Central Java at Desa Citrosono of Magelang and Bandungan. Harvested area under tuberose at Citrosono-Megelang is around 14.5 hectares and at Bandungan-Semarang 12.3 hectares.

East Java (Batu, Pujon, Tretes, Pasuruan and Madura)

Batu is the center of cut flower and ornamental plant production in East Java. Batu and Pujon, are located at high altitudes of 1000 to 1300 meters above sea level. Most of the farmers at Batu and Pujon use mixed cropping patterns of flowers and vegetables with an average of 0.5 hectare per farmer. Cut flowers produced are gladiolus, rose, chrysanthemum, orchids, gerbera, lilies and tagetes.

In Tretes ornamental plant growers have formed an association called Aspeni (Asosiasi Pengusaha dan Petani Flora Indonesia). Ornamental plants produced at the 3 locations above are pines, palm, cycas, ficus, ixora, dracaena, azalea and dieffenbachia.

Pasuruan is the center of sedap malam (tuberose) production in East Java. Total harvested area of tuberose is 75 hectares at Bangil and 69 hectares at Rembang, which are sub-districts of Pasuruan. Growers in Pasuruan produce jasmine beside sedap malam. Harvested area under jasmine in Pasuruan is 15 hectares.

Madura is one of the most famous areas for jasmine production in East Java. Area under jasmine at Bangkalan-Madura is 30 hectares. Most jasmine production in East Java is used for accessories at wedding ceremony and other religious activities.

The local wholesale market of cut flowers is located at Batu. Some of the growers act as traders as well and packing of flowers is done at their home.

Technology Generation

Research in floriculture is carried out by the Research Institute for Ornamental Plant (RIOP) in Jakarta which has had its mandate from the Ministry of Agriculture since 1995. RIOP has two installations at Cipanas and Segunung. Major areas of research are breeding, agronomy, pest and disease control, post-harvest technology and economics. Attention has been directed to cut flowers namely orchids, rose, jasmine and tuberose.

The on-going research programmes within the RIOP are as follows:

- Germplasm collection, characterization, elevation and conservation.
- Quality improvement on cut flower priority and ornamental plants.
- Growth media and plant nutrition.
- Cropping system including hydroponics.
- Pests and disease control of cut flower priority.
- Vaselife prolongation of cut flower.
- Socio-economic studies, to increase grower income and marketing.
- Seed technology and planting material production.

Research on orchids mostly relates to quality improvement, virus resistance, comparison of growth media and plant nutrition from the tissue culture stage through transplanting and cut flower production. Ways of rapid multiplication of orchid plants with tissue culture have been studied as well. Types of orchids include Dendrobium, Phalaenopsis and Vanda. On roses work was done on variety improvement, adaptability testing, planting material production and socio-economic aspects. Prolongation of vaselife of cut flowers and colouring tuberose by testing several formulas have also been studied. A germplasm collection of cut flowers and ornamental plants is maintained at the Research Institute for Ornamental Plant.

Impact of research result is clear through interaction between growers and the research institute. Moreover, dissemination of technology could be further enhanced by seminars and publications. At present there is an institution called BPTP which has the mandate for technology assessment, development and extension work in almost every region.

Production of Planning Material

At present many varieties are being grown in Indonesia both local as well as imported. It needs to be determined which ones are more suitable for local growing conditions and match consumer preferences. With the rapid development of the local market of cut flowers and ornamental plants, larger quantities of high quality planting material will be required. Specialization on growing planting material of a certain kind of cut flowers and growth stage of ornamental plants is promising and looks prospectful.

Multiplication by means of tissue culture of cut flowers and ornamental plants has been found promising because of the rapidity of producing a large number of plantlets in short time, and the uniformity of planting material which is free from diseases.

Propagation of ornamental plants is done in partial greenhouses which provide shading and protection against rain and pests. Simple overhead sprinkler irrigation system provides easy water supply and control of humidity.

Table : Estimation of Cut Flower Planting Material Demand, Import and Production

Crop	Seedling Requirement	Import	Seedling Production
Orchid	66,200,000	1,100,000	65,100,000
Rose	5,600,000	-	5,600,000
Chrysanthemum	76,400,000	100,000	76,300,000
Anthurium	10,800,000	400,000	10,400,000
Carnation	28,800,000	500,000	28,300,000
Lilium	32,400,000	100,000	32,300,000

Source: Directorate of Horticulture 1993.

Planting material production in Indonesia is not yet developed. There is no company breeding for cut flowers at present in Indonesia. The market segment which uses high quality planting material is small but shows a growth of about 10 percent per year.

Marketing

In regard to the market for cut flowers in Indonesia, it must be noted that no integrated domestic market exists. The market consists of a number of confined markets, larger cities and their supply regions. A distinction should be made between Jakarta and other larger cities in Indonesia. In Jakarta giving flowers for all social occasions, such as birthdays, weddings, illness, business attention etc. has become rather common and substitutes for personal visits. As a result of this, Jakarta has become by far the largest market for flowers in Indonesia and consumes approximately 76 percent of the total consumption. In Jakarta more than 900,000 stems of cut flowers are sold per week, accounting for an approximate value of US$ 5.1 million per year, whereas the total amount for all major cities is estimated as US$ 6.8 million per year.

Table : Consumption of Cut Flowers in Jakarta (stems)

Year	Rose	Chrysant.	Tuberose	Gladiolus	Anthurium	Dahlia	Others	Total
1993	8,038,800	7,035,000	6,156,800	7,543,500	614,300	2,473,100	3,078,400	33,939,900
1994	8,823,300	7,718,700	6,753,500	7,180,700	673,800	2,714,300	3,376,700	37,241,000
1995	9,682,300	8,466,700	7,406,000	8,878,100	739,000	2,978,400	3,703,000	40,853,500
1996	10,622,400	9,384,700	8,119,200	8,641,100	810,100	3,267,300	4,059,600	44,804,400
1997*	11,650,800	10,178,700	8,898,200	9,475,400	887,900	3,583,400	4,449,100	49,123,500
1998*	12,775,100	11,155,100	9,748,400	10,387,000	972,700	3,928,900	4,874,200	53,841,400
1999*	14,003,700	12,220,800	10,675,700	11,382,500	1,065,200	4,306,300	5,337,900	58,992,100

* Projection

Potential for Cut Flower Production Development

Tissue culture technology for multiplication of plantlets is rapidly developing and has been accepted on a wide scale for several flowers and ornamental plants. Advantages lie in less time needed, uniformity of produced plantlets and virus free planting material.

Indonesia has ideal climatological conditions for growing tropical plants. Presently an interesting domestic market has developed, where relatively high prices are fetched, based on the scarcity of the product more than the actual cost price.

To be able to achieve the quality standards required in the international market and its effective marketing, cooperation with international growers is a feasible alternative to obtain results in the not too distant future for Indonesian growers. In order to enter the world market to some significant extent Indonesia needs to start selection and breeding of its own varieties. A substitute for soil mixed with rice husk as a growth media for ornamental plants is required in case of export since most countries prohibit soil and rice husk to enter their territories.

Constraints in Cut Flower Production Development

a) Although several commercial tissue culture laboratories are in operation in Indonesia, all have limited operations to serve their own requirement or a small circle of associated farmers. By doing so, the scale of operations remains too small for a profitable undertaking and consequently investments in motherstock of planting material are not made.

b) No specialized nursery for young ornamental plants is in existence. Growers are not able to follow the market trend quickly and multiplication is done in an inefficient way.

c) Indonesian private growers of ornamental plants and flowers were left on their own for variety testing and developing appropriate cultivation methods. This has resulted in high cost and slow spread of adapted varieties and cultivation techniques. Support from RIOP as technology generating institution and BPTP for transfer of technology is needed in every region.

d) Institutional promotion to increase popularity of flowers and ornamental plants in Indonesia is still limited. Similarly, information about flowers and plants which can be supplied from Indonesia is almost lacking in the world market.

e) Most of the cut flower production centres are scattered and located too far from the distribution point or air port.

f) High interest rate on credit for capital on agribusiness which is treated the same as other commercial investments.

g) Limited area/land at reasonable price.

Research on cut flowers and ornamental plants is very important to support and solve the problems faced by growers. The cooperation between private and public institutions needs to be strengthened and developed.

The development of the domestic market is a prerequisite for the development of export which requires high standard quality and large volume as well as competitive price.

The development of a well established wholesale market is needed in Jakarta to have a positive impact on the development of this sector.

The short production cycle of cut flowers and ornamental plants, makes the country an ideal place for multiplication of planting material, which needs to be explored.

Cut Flower Production in Malaysia

The cut flower industry in Malaysia is a relatively recent development compared to other agricultural enterprises. From its rather humble origins as a hobby industry, the cut flower industry in Malaysia has developed into a very viable commercial enterprise with the most marked growth in the mid-Eighties. In fact it has shown such tremendous growth in the last decade that production has increased tenfold and export twelve-fold in response to local and foreign demands. The trend is expected to continue in the future with growing affluence of the local population and that of the developed countries as well as improved market opportunities. The cut flower market consists of 3 important components viz. temperate flowers, orchids and other lowland flowers. In general, the area of cultivation of cut flowers in Malaysia is determined by the climate and topography of the land. For instance, highlands such as the Cameron Highlands are the major growing areas of temperate flowers. Other cut flowers adapt better to the hot humid conditions in the lowlands with orchids constituting the major share of the production.

In the National Agriculture Policy (1992-2010) and the Seventh Malaysia Plan (1996-2000), cut flowers have been identified as a priority group of crops with good potential to meet the growing domestic and international demand and to generate higher income for producers. The main thrusts of the policy are reflected in the following areas:

- Substantial expansion of flower production of both the lowland and highland varieties to meet the demand of an expanding market, in particular world demand.
- Cultivation of lowland varieties on marginal as well as former mining areas.
- Efficient and reliable supply of planting materials.
- Intensification of R & D in the development of better varieties, cost effective production technology and post harvest handling.
- Diversification of exports beyond the traditional markets.
- Encouragement of foreign investment to bring in more advanced technology and provide access to international markets.

It has been estimated that at least 1,000 farmers are involved in the floriculture industry that contributed RM 370 million to the economy in 1995. Orchids contributed

40 percent of total value of production, followed by temperate cut flowers (33 percent) and ornamental plants (27 percent). Although the contribution of cut flowers to the total of Malaysian exports is not significant, the annual growth rate over the last year was 24%. If the forecast of 10% annual growth rate of floriculture production holds true for the 7th Malaysia Plan period (1996-2000), the floriculture industry will enjoy the highest annual growth rate compared to other major commodities including rubber and palm oil over the same period.

Present Status of Cut Flower Production

Major Production Areas

The major production areas for temperate cut flowers are in the Cameron Highlands, Pahang State (638 ha), Gua Musang, Kelantan State (10.3 ha), and Ranau, Sabah (85.1 ha). In the lowlands the main orchid cut flower production areas are located in Johore Bahru, Johore State (162.5 ha), Batang Padang, Perak State (52.4 ha), Kota Tinggi, Johore State (32.89) and Petaling, Selangor State (27.36 ha). Smaller quantities of orchids and other lowland cut flowers are also produced in the states of Melaka, Negeri Sembilan, Kelantan, Perlis and Sarawak.

Area Under Cultivation and Flower Types/Varieties

The total area under cut flowers is estimated to be over 1,218 hectares, out of which 580 hectares are under orchids and 638 hectares are under temperate cut flowers. In terms of number of flower stalks, temperate flowers contribute 71.46% towards the total cut flower production while the remaining 28.54% is from orchids.

The most popular orchid types cultivated are Dendrobium (11.17%), Aranda (8.0%), Oncidium (5.01%) and Mokara (3.5%). Together these 4 orchid genera produced 97% of the country's orchid cut flowers in 1994. The most important temperate flower types are rose (33.48%), chrysanthemum (22.62%) and carnation (9.02%). These three flower types contributed 91.1% of total temperate cut flower production.

Open Field and Protected Production Systems

The type of production system used for different species of cut flowers in Malaysia is dependent on weather conditions, cultural requirements and quality maintenance, as described below:

Temperate Flowers

The climate of the Cameron Highlands, where the mean day and night temperatures are 14°C and 22°C, respectively, is very conducive to the cultivation of a wide range of temperate cut flower species. However, the total annual rainfall of 2700 mm that falls mainly in April-May and October-December affects growth and vigour of the plants, increases disease incidence and pest infestation, reduces yield and quality and causes erosion of top soils. Thus, almost all temperate cut flowers such as rose, carnation, chrysanthemum, peacock and lilium are grown under simple plastic shelters. These are simple open structures with height of 3 m and each consists of a wooden or galvanised iron frame with a polyethylene roof to protect the cut flowers from rain. Anthuriums, on the other hand are planted in shadenet houses with shading level of 75-80%. Irrigation is generally provided by overhead sprinkler daily during the first month of growth and gradually reduced during later phases. Surface and drip irrigation is used when the flower starts to open. Processed organic fertilizers, compound fertilizers and foliar fertilizers are used at different stage of growth. Chrysanthemum is different from the other flower types because there is a need for disbudding, application of plant growth regulator (to regulate height, strengthen stems and for quality blooms) and the use of artificial lighting and light interruption for vegetative development and floral initiation, respectively.

Table : Main Orchid Cut Flower Types and Varieties and Percentage Composition of Total Cut Flower Production (1994).

Main Flower Types	*Main Varieties*	*% of Total Cut Flowers Produced*
Orchids		
1. Dendrobium	D. Sonia, D. Tomie Pink, D. Shavin White, D. Chanel, D. Sharifah Fatimah, D. Tuang Pink	11.17
2. Aranda	A. Wan Chark Kuan, A. Noorah Blue, Aranda Kooi Choo, A. Happy Beauty, A. Anne Black	8.00
3. Oncidium	O. Golden Shower, O. Gower Ramsay, O. Taka	5.01
4. Mokara	M. Chark Kuan (Pink, Orange, Red), M. Dinah Shore, M. Khaw Phaik Suan	3.50
5. Arachnis	A. Maggie Ooi	0.56
6. Aranthera	A. Anne Black, A. James Storei	0.13
7. Vanda	MAS Los Angeles	0.09
8. Holttumara	H. Loke Tuck Yip, H. Chee Kiong	0.06
9. Phalaenopsis	P. Sun Prince Pink, P. Hawaiian Clouds x P. Carmela Pink	0.01
10. Kagawara	K. Christine Low	0.01
11. Cymbidium		0.00
Subtotal for Orchids		28.54

Table : Main Temperate Cut Flower Types and Varieties and Percentage Composition of Total Cut Flower Production (1994).

Advances in Production Technology

Varietal Improvement

Three programmes for varietal improvement of orchids are currently being carried out, namely conventional breeding, genetic engineering and mutation breeding. The germplasm collection has increased to 500 species representing 100 genera. Another 196 accessions are being maintained as "studs" for hybridization work. Conventional breeding has resulted in the release of at least 7 hybrids consisting of 4 Dendrobiums, 1 Vanda, 1 Cattleya and 1 Kagawara. Research in genetic engineering has just been initiated to induce colour change in established varieties, for instance changing yellow Oncidium (O. Golden Shower and O. Gower Ramsay) to white. Meanwhile studies on *in vitro* mutagenesis include the use of mutagens and gamma ray irradiations on Aranda, Mokara, and Vanda orchids. Induced mutation research on Dendrobium has indicated variations in pigmentation and flower size in response to irradiation.

For temperate cut flowers, varietal evaluation trials have identified promising varieties of chrysanthemum, rose, carnation, lilium, anthurium and gerbera for highland cultivation. Induced mutation studies on 3 chrysanthemum varieties have also indicated that the recommended optimum dose for the mutagenesis of these varieties is 30 Gray.

Cultural Practices

In orchids, the application of fertilizers at night and in the morning helps to increase inflorescence yield. Oil palm bunch compost when mixed in charcoal medium improved seedling growth by three times in Oncidium, twice in Dendrobium and one and a half times in Aranda seedlings. The quality of chrysanthemum cut flower can be improved significantly through the application of calcium up to 150 mg/1. Cyclic light interruption as opposed to continuous night light interruption reduces energy cost and thus lowers the cost of chrysanthemum cut flower production.

Disease Control

In orchids, necrotic specks in Dendrobium blooms are common in plants grown under plastic shade. The spraying of 1 g/1 Ronilan every 10 days can effect complete control of this problem. The black nose disease of orchids has been confirmed to be caused by the association of Fusarium monoliforme with thrips. This disease can be controlled by alternate sprayings of Mesurol and Dithane M 45. Chemical control measures have also been developed for Sclerotium rot and pseudobulb rot, Cercospora leaf spots and alga.

The white rust problem on chrysanthemum that has restricted its import by countries such as Australia and Japan has been overcome with rust resistant varieties such as Tigerrane, Tiger, Target, Naru and Puma Yellow. Effective chemical control is also obtained with fungicides such as bromuconazole, bitertanol, flutolanil, myclobutanil, tridemenol and propiconazole. Powdery mildew in roses can also be controlled by chemicals such as sulphur and cupric oxide.

Pest Management

Leaf miners are a major pest problem of cut flowers in the Cameron Highlands. Studies on biological control have indicted that two eulophid parasitoid species (*Hemiptarsenus varicornis* and *Chrysocharis sp.*) parasitize both species of leaf miners (*Liriomyza sp.* and *Chromatomyia horticola*) found in a wide range of cut flowers.

Meanwhile neem extracts have been found to be effective against leaf miner in potted chrysanthemum.

Studies have identified the most serious nematode pests as *Eadopholus simillis* on chrysanthemum and *Pratylenchus coffeae* on chrysanthemum and anthurium. These pests can be controlled by the application of nematicide such as fenamiphos, carbofuran, dazonmet and isazophos and adoption of preventive phytosanitary measures. Effective chemical control measures have also been developed for the beetle *Oulema pectrolis* and snail.

Post Harvest Handling

Studies indicated that the telescopic box type is suitable for packing temperate cut flowers while treatment of rose and carnation flowers at the right stage with certain flower preservatives has reduced bent-neck problem and extended its shelf-life. In addition, preservation techniques have been developed for rose, carnation, peacock, golden rod, anthurium, statice and cosmos for use as dried and pressed flowers. Although orchids are less perishable than temperate cut flowers, storage temperature does affect its quality. Dendrobiums can be stored at temperatures between 10-15°C without any sign of scorching.

All mites and thrips on roses, carnations, chrysanthemums and orchids are killed by methyl bromide fumigation without affecting the shelf life of the flowers. Evaluation of several chemicals as fumigants has yet to come out with an effective alternative to methyl bromide.

Production of Planting Material

A survey of the cut flower industry conducted in 1995 indicated that 107 orchid growers produced their own planting materials, 241 producers obtained their planting materials from local nurseries while 122 growers imported theirs from other countries. The corresponding figures for non-orchid cut flowers are 107 (own production), 294 (local) and 134 (import). Based on the application of import permits, import of planting materials during the first 6 months of 1996 consisted of 3.43 million cuttings/plants, 1.01 million bulbs/corms, 156 kg of flower seeds, 51,595 flask of tissue cultured orchid materials, 4,000 community pots of orchids, and 330,450 orchid seedlings.

Although many temperate cut flower growers produce planting materials, a large number purchase these from local nurseries or importers. In the case of chrysanthemum, the most widely grown temperate flower, two companies namely Van Der Kemp and Fides, produce most of the rooted and unrooted cuttings. Van Der Kemp has a 15-acre nursery that produces cuttings in imported Dutch greenhouses under drip and mist irrigation systems. Fides operates a smaller 10-acre nursery with cuttings produced under plastic rain shelters that are equipped with similar irrigation facilities. These two nurseries together produce an estimated 50 million chrysanthemum cuttings. About 80% of these materials are supplied to growers in the Cameron Highlands, while the rest is exported to nearby ASEAN countries. It is estimated that a total of 300 million chrysanthemum cuttings is required by growers in the Cameron

Highlands and both these companies can only produce between 10-13% of the local needs. There are more than a dozen commercial tissue culture units producing orchid materials. These are mostly small operations producing either meristem culture or seed culture. Some major orchid growers also contract out the production of planting materials of orchid hybrids to large tissue culture laboratories in Thailand and Singapore.

Marketing

Marketing of cut flowers in Malaysia is less organized than that in Western countries with different degrees of sophistication depending on flower varieties traded, farm size and destinations of the produce. The multi-tiered marketing and distribution system is rather complicated and according to growers does not pay premiums for good quality. Malaysia is studying the possibility of establishing an auction market. However, growers still play an important and direct role in the distribution to retailers in the local market as well as to importers in the foreign market. The grower can either deliver directly (large farms) or through an export agent (small farms) to the importer. The flowers are then sold directly to wholesalers or in auction markets (foreign countries) before distribution to the retail markets and from there to the consumers. There are several associations of cut flower producers but these are formed to provide service and supplies to members only. The more significant ones are the Cameron Highlands Floriculturist Association and the Commercial Orchid Growers' Association of Malaysia. Although these associations do not collectively market their produce there is growing realization of the benefits of such marketing arrangement. Several temperate flower producers have already grouped together to form a trading company to deal directly with importers in foreign countries. Some companies also specialize only in export and buy directly from growers who have production contract with them.

The exports of floricultural products increased from RM 6.5 million in 1985 to RM 54 million in 1994 showing an eight-fold increase over a ten-year period. However, conservative official records show a decline in the export value of cut flowers from a high of RM 40.63 million in 1992 to a low of RM 18.31 million in 1995, although there was an increase in cultivated production area. This two conflicting trends appear misleading but may be explained by the low market prices that plague the industry in 1995, under declaration of import volume and value, movement to lower-price markets and increased domestic consumption. The main export markets are Japan, Singapore and Hongkong. Japan is the leading export market for orchids, Singapore for ornamentals and Hongkong for temperate flowers.

Import of cut flowers has shown a fourfold increase from RM 0.73 million in 1992 to RM 2.77 million in first eleven months of 1995. The main countries exporting cut flowers to Malaysia are the Netherlands, Singapore and Thailand.

Potential for Cut Flower Production Development

The prospects for the cut flower industry in Malaysia are very bright due to the growth in domestic and export markets. The domestic per capita consumption has increased from RM 0.71 in 1990 to its current level of RM 2.50. This trend is expected to increase further with greater consumer affluence and growing appreciation for fresh cut flowers. Other factors that contributed to this forecast is the rapid economic growth and the fast rate of urbanization in the country.

There is also an increasing demand for cut flowers in the international market that is expected to continue at a growth rate of 6% per annum. This will result in a corresponding increase in value that is expected to reach RM 20 billion and RM 36 billion in the years 2000 and 2010, respectively.

There is tremendous scope for expansion of the export potential of Malaysian cut flowers, as major consuming countries are facing rising production cost and will increasingly depend on lower-cost producers to supply their requirements. Malaysia has several advantages including a wider product range, longer shelf-life for Malaysian orchids, potential to exploit its biodiversity for product development and ability to export flowers year round. New markets especially for Malaysian orchids are emerging in the Middle East, Northern Europe and East Asia. Existing markets such as Belgium, Finland and Ireland are expected to expand further.

Constraints in Cut Flower Production Development

Although the cut flower industry in Malaysia has shown impressive growth in the last decade, it has not been able to realize its full potential because of several constraints. Among the moro important ones are the following:

Table : Exports of Cut Flowers According to types from 1992-1996 (RM '000)

Type/Year	*1992*	*1993*	*1994*	*1995*	*1996**
Orchids	6,259.60	4,268.40	9,034.49	6,828.24	6,069.18
Temperate Flowers	33,305.84	20,450.93	23,283.37	9,632.21	9,140.85
Other Cut Flowers & Flower Buds, Other than Fresh	1,063.82	753.08	1,394.76	1,853.90	1,239.78
Total Export	40,629.26	25,472.41	33,712.62	18,314.35	16,449.81

Source: Statistics Department, Malaysia

* Data based on Jan-Nov. 1996

Table : Import of Cut Flowers According to Types from 1992-1996 (RM '000)

Type/Year	*1992*	*1993*	*1994*	*1995*	*1996**
Orchids	114.17	14.28	39.03	74.88	977.10
Temperate Flowers	287.46	171.69	434.87	433.27	701.20
Other Cut Flowers & Flower Buds, Other than Fresh	326.03	392.55	461.15	747.47	1,093.30
Total Import	727.66	578.52	935.05	1,255.62	2,771.60

Source: Statistics Department, Malaysia

* Data based on Jan-Nov. 1996

Economic

The intense competition for land resources with more lucrative economic activities, high development cost, short land tenure and environmental considerations (highland production) impede the expansion of the cut flower industry both in the highlands and lowlands. This is further compounded by the high initial investment for planting materials, irrigation systems, shade and other infrastructure facilities. Financial institutions are also unwilling to provide credit for floriculture projects due to perceived high risks.

Technology

The development of new varieties has not been able to keep pace with changing consumer preferences. Moreover, inconsistency in the supply of uniform and high quality planting materials and the slow rate of technological advancement in critical areas such as post harvest handling, storage and transportation have resulted in export of lower quality produce with shorter shelf-life.

Infrastructure Support

Existing marketing practices do not ensure premium prices for good quality. Prices are also not normally fixed at the point of sale in the open account payment system. This often results in non payment, high arbitrary discounts and commissions especially by buyers from Singapore and Hongkong.

Insufficient facilities for refrigerated transportation and storage have led to high post harvest losses and export of poor quality produce. Exports are also confined to traditional markets due to insufficient air cargo space and handling facilities at airports.

Institutional Issues

The high tariff barriers imposed by Taiwan, Thailand and Europe and the strict phytosanitary requirements enforced in Japan, the United States of America, Australia and New Zealand, are other impediments to further development of the industry. These problems are further compounded by the lack of a standard domestic grading and inspection system.

Human Resources

There is a general lack of skilled human resources in important areas such as production, post harvest handling, product handling, product development, biotechnology and auctioning. This can become a serious handicap in the highly competitive cut flower industry, where innovation and responsiveness to changes can give local growers the edge over their rivals.

The prospect for cut flowers is very bright taking into consideration the growing demand in the domestic and export markets and the fact that flowers are relatively recession-proof. The Malaysian Government has recognised this fact and has accorded priority to floriculture in the National Agriculture Policy, providing various tax and financial incentives to attract investments. Malaysia should make full use of its inherent advantages and attempt to rise above current constraints and provide an adequate and consistent supply of quality cut flowers during periods of maximum demand. Current production and marketing activities including development of desired varieties, and the search for new markets should be upgraded with the full cooperation of all parties concerned. These efforts and other future development plans are expected to enable Malaysia to make the quantum leap and become one of the leading producers of tropical cut flowers in the world. A successful development plan will have to pay particular attention to the following areas:

- Expansion of domestic production through the efficient utilisation of domestic resources to exploit the opportunities that exist.
- Strengthening market and market capabilities through better cooperation between private sector and Government agencies in order to expand existing markets and penetrate new ones.
- Strengthening the institutional support and support services for continuous innovation and development of capital-intensive production technologies to improve product quality and to meet changing consumer demand.
- Exploitation and enhancement of processing and downstream capabilities.

Cut Flower Production in the Philippines

The Philippine Ornamental Horticulture Industry has come a long way since the 70's when cut flowers were considered its major component and the growers were hobbyists or plant enthusiasts. In the mid 80's the industry became a huge income source and potential for a foreign revenue earner. Interest in the commercialization of horticultural products grew to a point where the government recognized its importance and contribution to the local economy. The number and size of farms had continued to increase adopting modern technologies where appropriate and in some cases foreign technology was sourced to boost production and quality. Promotion on the use of Ornamental Horticulture products increased people's awareness of the benefits that can be derived from the industry.

The floriculture industry in the Philippines is almost synonymous with the ornamental horticulture industry; the former connotes flowers; the latter, includes flowers and other ornamentals. The country offers the following horticultural products for both domestic and external trade:

- Cut flowers fresh flowers and flower buds that have been cut from the plant suitable for bouquets, wreaths, corsage and special flower arrangements.
- Foliage and other plant parts fresh leaves and branches of trees, shrubs, bushes and other plants, grasses, mosses, lichens and the like suitable for ornamental purposes.
- Dried ornamentals - dried, dyed, bleached plant materials such as grasses, statice, eucalyptus.
- Other ornamentals - trees, shrubs, bushes, roots, cuttings and slips used indoor or for landscape purposes.

The phenomenal growth of the Ornamental Horticulture Industry in the last few years has led the Department of Agriculture to include the cut flower in its Key Commercial Crop Development Program (KCCDP) for 1996 - 2000. The Department of Trade and Industry adopted the ornamental crops as its "Export Winner" and the Department of Science & Technology recognized it in its Science & Technology for National Development (STAND 2000). Young as it is, the industry has shown its vital role in the local economy, and slowly it is penetrating the export market and emerging as a true global competitor.

Present Situation of Cut Flower Production

Cut flowers are grown throughout the Philippines but the industry was originally confined only to a few, small growers. In the last few years, an increased awareness and recognition of high return on investments, rapid population growth, higher standard of living, more hotels and restaurants, influx of tourists has led to more demanding and choosy clients. An increased demand triggered more production but despite the larger area devoted to cut flower, there is still a short fall in the supply. The demand for the domestic market is so big that the country has no option but to import some cut flowers, mainly chrysanthemum and orchids from other countries. This is strongly evident during Valentine's Day (Feb. 14), All Saint's Day (Nov. 1), School Graduation (March and April), May Flower Festivals and Yuletide Seasons (December). The major commercially grown cut flower varieties in the Philippines and their area of cultivation.

In general, gladioli, aster, sampaguita, chrysanthemum and heliconias are grown in the open field. Depending on the cultural and light requirements of the cut flowers grown, greenhouses and net houses are usually constructed both in the highlands like Benguet and areas of low elevations. Semi-terete vandas are grown in the open, usually mounted on driftwoods; strap leaf vandas and dendrobiums on commercial basis are usually protected by 2 - 3 layers of fish nets and raised in pots/baskets. Anthurium production generally uses a more advanced technology in greenhouse complete with computerized fertigation facilities. In some areas, net houses provide the necessary light requirements of the cut flowers. Roses are grown either in the open or in greenhouses, the same is true with the gerbera. Greenhouse structures protect the cut flowers from the intense heat of the sun and heavy rains. Cultural and management practices are also easier to control inside and can be scheduled regularly regardless of the weather outside. The flowers of *Jasminum sambac* or sampaguita are usually used as *leis* or garlands usually with religious connotations. The plants are grown near houses or along roadsides for easy access and convenience on the part of the grower.

Flowering is controlled in chrysanthemum and aster production, usually in the form of supplementary lights. Sometimes growth regulators are applied in chrysanthemums if they are intended as potted flowering plants. Cold storage to break the dormancy of corms and cormels of gladiolus is available. Post harvest facilities for cut flowers are also available at the University of the Philippines, Los Banos. Big growers have their own storage facilities.

Production of Planting Materials

Depending on the kind of cut flowers grown, planting materials are produced through cuttings, bulbs, tubers, corms, tissue culture, embryo culture, etc. For orchids, community pots, flasks/culture bottles, top cuts and overgrown seedlings are purchased either from abroad (Malaysia, Thailand, Hawaii) or from local nurseries. From the initial purchase, the cut flower grower can produce some of his planting materials for the farm.

Chrysanthemums are propagated mainly by cuttings, roses by budding, marcotting and cutting, anthurium by division or production of suckers, gladiolus through corms and cormels, heliconias by division. For big farms, the initial planting materials are imported, as in anthurium and chrysanthemum. Mist beds for rooting the cuttings and laboratories for embryo and tissue culture of orchids are available. Government agencies like universities and state colleges, regional offices of the Department of Agriculture and Department of Science and Technology usually offer services for embryo culture of hybrids produced by the grower for a small fee. In this manner, the grower can have his own planting materials. These agencies also sell their own produce to the public. Several growers specially the orchid breeders have their own laboratories and offer their hybrids for sale.

Marketing

The key players in the marketing of the cut flowers in the Philippines are the breeders/growers, input suppliers, traders, cooperatives, transporters, brokers, exporters, importers, institutional buyers and walk-in consumers.

New and improved varieties of cut flowers form the backbone of the industry. Breeders, therefore, play an important role by developing the varieties and hybrids and help in the production of planting materials for the commercial growers. Input suppliers include the distributors and dealers of farm chemicals like fertilizers, pesticides, growth regulators and the like.

The traders can be individuals or groups of individuals who distribute cut flowers from the producers to the demand site. The traders could be assembler-wholesalers, assembler-wholesalers-retailers, commission agents, wholesalers, florists, retailers or vendors. Alternatively, members of cooperatives grow cut flowers and the organization itself may handle the selling and distribution of its produce to buyers.

Transport of cut flowers is performed both by land and air e.g. planes, buses, vans, and jeepneys. Processing of papers of imported cut flowers and other supplies are performed by the brokers. Cut flowers are exported by corporations, organizations and individual entrepreneurs. Institutional buyers include hotels, restaurants, banks, offices, hospitals and churches.

The quantity purchased by walk-in customers is minimal, depending on the occasion. Employees, students and housewives are the most common walk-in customers.

There are no auction centres in the Philippines. There are only a few cooperatives that market their produce. Attempts have been made by some people to channel their cut flowers through their cooperatives but somehow, it is easier and more convenient to sell their produce individually.

Export of cut flowers is on a limited scale, with gladiolus topping the kind of cut flower exported. Almost all of it, which is about 46% of the total export, goes to Korea. The other cut flowers exported to Japan, Brunei, Hongkong and Italy include anthuriums, liliums, and roses. Chrysanthemums and carnations constitute the bulk of imported flowers from Australia, Holland and Malaysia. Almost half of the flowers are imported from Holland.

Table : Cut Flowers Exported, 1996

Cut Flower	*Country*	*% Share*
Gladiolus	Korea	46.10
Anthurium	Brunei, Japan	43.49
Liliums	Japan	8.93
Roses	Hongkong, Japan	1.33
Orchids	Italy, Japan	0.41

Table : Export Markets of Philippines Cut Flowers for 1996

Country	*Cut Flower*	*Quantity (Kg)*	*% Share*
Japan	Anthurium	9,358.62	53.48
	Liliums	1,933.92	
	Roses	279.50	
	Orchid	6.00	
Korea	Gladiolus	10,000.00	46.19
Brunei	Anthurium	55.62	0.25
Hongkong	Roses	10.00	0.046
Italy	Orchids	3.00	0.013

Source: Bureau of Plant Industry Quarantine Division, Department of Agriculture.

Table : Philippines Cut Flower Importation (1996)

Cut Flower	*Country*	*Quantity (Kg)*	*% of Total Importation*
Chrysanthemum	Australia	102	36.98
	Holland	8,389	
	Malaysia	13,566	
	Singapore	586	
Carnation	Australia	638	22.05
	Holland	7,691	
	Malaysia	5,097	
	New Zealand	80	

Pompom	Holland	176	7.29
	Malaysia	4,292	
Roses	Australia	55	6.81
	Holland	3,069	
	Malaysia	513	
	New Zealand	145	
	Thailand	60	
	USA	332	
Amaranthus	Australia	990	5.35
	Holland	2,226	
	New Zealand	60	
Alstroemeria	Australia	990	5.35
	Holland	2.226	
	New Zealand	60	
Lilium	Australia	301	4.19
	Holland	1,984	
	Malaysia	141	
	New Zealand	141	
Gerbera	Australia	78	3.79
	Holland	1,976	
	Malaysia	257	
	New Zealand	15	
Anthurium	Australia	428	3.58
	Holland	1,545	
	Malaysia	202	
	New Zealand	10	
Tulip	Australia	57	2.95
	Holland	1,629	
	New Zealand	71	
	USA	50	
Misty Blue	Australia	72	1.60
	Holland	61	
	Malaysia	852	
Total			61,224

Source: Bureau of Plant Industry, Quarantine Division, Department of Agriculture.

Potential for Cut Flower Production Development

The country has its strengths for cut flower production. It has an ideal climate for year round production with the capability to grow both tropical and subtropical varieties.

Considering the wide availability of land waiting for tillage and production of agricultural crops, it would not be difficult to produce cut flowers since suitable production techniques are already available. The richness and diversity of Philippine flora allows the development of varieties of cut flowers with endemic species as parents. A very important factor that contributes to the cut flower production in the Philippines is the existence of a very active horticulture association, of which commercial growing of cut flowers is one of the main objectives.

The Federation of Cut flower and Ornamental Plant Growers, a non-stock, non-profit private organization was formed in 1990 with the support of the Department of Agriculture to assist the government develop the cut flower industry. Its main thrust is service in the different fields where technical manpower support is lacking. The services are on consultation, skills and training, involvement in the marketing of the products of the growers, establish linkages with non-government organizations like USAID, JICA and Dutch-Flamingo International (FMD). The Federation is involved in advocating new policies for the improvement of the industry and its members.

There is still a substantial shortage of supply of cut flowers in the Philippines. Importation of cut flowers from other countries is quite substantial. Hence, by stepping up the cut flower production within the country, a substantial amount of foreign exchange can be saved.

Constraints in Cut Flower Production Development

The high cost of structures like greenhouse, irrigation and postharvest facilities are some of the major constraints to the producers of cut flowers. The availability of quality planting materials most of which come from abroad is sporadic. For new varieties/types of cut flowers like lilium, gerbera and calla, production technology is still insufficient. With new production technology comes the introduction of new agricultural chemicals which are not readily available locally and work out very expensive when imported. This is coupled with high import duties on other inputs too.

Access to credit sources is also another major problem confronting the growers. The interest rates on loans offered by institutions are high. Growers of roses, chrysanthemums, gladiolus, orchids and anthuriums are in need of financial support during the initial stages of production. Collateral requirements demanded by lending agencies are rarely met and the grower faces the dilemma of whether to continue cut flower growing or not.

On the part of the exporters, there is a shortage of cut flowers volume wise and quality wise. The small growers can not compete with big growers. It will be easier to meet the quality, but the volume needs more attention. The cut flower industry is still in its juvenile stage and not much information on the feasibility and prospect of the business is available. Hence, investors are slow in getting into the business since lending agencies or banks always demand feasibility studies. Big growers are quite hesitant to provide information because of inherent competition.

Cut Flower Production in Sri Lanka

Cultivation of flowers for various religious and cultural festivals has existed in Sri Lanka for ages. It was only after the British rule that floriculture really came into practice not as an industry but mainly as a hobby for pleasure. Later, with many new introductions of tropical and sub tropical plants the trend was gradually passed down to other levels of society.

Floriculture in Sri Lanka started as an industry in 1970. It has grown substantially during the last few years to become one of Sri Lanka's major foreign exchange generating ventures.

Present Situation of Cut Flower Production

Western, North Western and Central Provinces in Sri Lanka are the major areas where cut flowers are grown commercially. Cut flowers grown in the country can be divided into two main categories based on their temperature requirements i.e. Temperate Cut Flowers and Tropical Cut Flowers.

Temperate Cut Flowers: Temperate cut flowers include carnation, rose, statice, gypsophyla, alstroemeria, chrysanthemum, lilies and irises. Among them, carnations and roses are grown mainly for export, in the highlands of the Central Province of Sri Lanka. Carnations are produced entirely from imported planting materials and are graded according to internationally accepted specifications for export. The other species are commercially grown mainly for the local market.

American and Mediterranean carnation cultivars are quite famous in the world market. Pink, white, red, yellow and salmon colours are much popular. Novelties such as striped and frosted types are also becoming increasingly popular. American cultivars grown are silvery pink, karina, barbara, red barron, elsy, royalette, bagatalle, bianca and adelfie. Mediterranean cultivars grown are nora barlo, shainah, lena, castellaro, scania, tanga, roma, pallas and charmeur. Carnations are grown in poly tunnels, covering more than 10 hectares, under fully protected environments.

Roses are second to carnations and production is limited to about 40,000 blossoms per annum. This is mainly due to the highly expert conditions required for cultivation and production. Approximately 2-3 hectares of area have so far been used for cultivation of roses. Roses for the export market are grown under controlled environments in poly tunnels. Popular colours are highly variable. They are generally required in a mix and an acceptable mix would consist of red-50%, Pink-30%, yellow-10% and others 10%. Roses are quite popular in the local market also. Most of them are supplied by small scale growers.

***Tropical Cut Flowers*:** Anthuriums and Orchids are the most popular tropical cut flowers which are being grown commercially for exports as well as for the local market. Anthuriums can be grown at elevations up to 1500 metres above sea level, with texture and the flossiness remarkably enhanced with increase in elevation. Annual production of anthuriums is around 3 million flowers, the majority of which are sold at the local market. The exports of Anthuriums at present are not very significant. A land area of approximately 10 hectares is under Anthurium cultivation at present and the industry is expanding steadily at village level. Almost all plantations are either under poly tunnels or structures with shade netting. However, locally available materials are also used under certain conditions to provide required shade levels. The standard trade types in the world market are the "Avo" lines i.e. Avo Nette, Avo Ingrid, Avo Anneke. The improved standard types e.g. Germa, Cuba, Fuego, Favoriet etc. are also popular. The Royal Botanic Gardens, Peradeniya, being the pioneer institution for the development of cut flower industry in Sri Lanka has produced a few promising Anthurium varieties with export potential. Some of these varieties are being closely studied and mass propagation has just been started. A few of the slected varieties are: RBG - Green Tip, RBG - Soft Sheen, RBG - kandy queen, RBG - Gardens Pride, RBG - Royal Flag, RBG - Lak Isuru, RBG - Wild Beauty, RBG - Krishnas Red.

Tropical Orchids can be grown under warm humid conditions up to 500 metres above sea level. Climatic conditions in the Western Province are quite favourable for cultivation of Dendrobiums, Vandas and Phalaenopsis types which are quite popular in the local and foreign markets. Approximately 3-4 hectares of land are under Orchid cultivation at present and the industry is gradually developing to cater to export markets. Almost all cultivations are under shade netting and the majority of growers have developed mist irrigation techniques. 'Madam Pompadour' (Pink and White), and 'Rena Vapahoo' (Pink and White) hybrids are some of the popular hybrid Dendrobiums in the market. Hybrids of Arachnis, Oncidiums, Phalaenopsis and Dendrobiums are being exported in smaller quantities and the income earned is not very significant.

Gerberas are becoming popular among growers due to the availability of a wide range of long lasting cultivars produced by many modern breeding methods and tissue culture. These cultivars can be broadly categorised into three classes, namely Singles (such as 'Fleur' and 'Apple Blossom'), Doubles (such as 'Marleen' and 'Hildegard'), and Black Centres (such as 'Fabio' and 'Rosseta'). Pink, salmon, orange, red and yellow are the popular colours in the market. Trials have already been started under controlled environments by private entrepreneurs to grow gerberas for the export market. About 2 hectares of land is so far being used for gerberas.

Grower Categories and Specialities

Generally, large scale commercial establishments also have their cultivations under green house conditions, poly tunnels or netting. The medium and small scale growers either have their cultivations under natural shading (under trees) or use locally available materials such as coir fibre mats, cajans (dried and woven coconut palm leaves) or ropes to provide shade requirements of a particular crop.

Sri Lanka's floriculture industry consists of three categories of producers or growers a) large commercial ventures for export; b) middle level growers catering to the local market; and, c) village level producers who may sell their products to either of the two categories mentioned above.

The total land area under floriculture is around 500 hectares at present and the majority of lands are in the Western Province. There are 10 ha under carnations, 3 ha under roses, 2 ha under gerberas, 10 ha under anthuriums, 3 ha under orchids and 472 ha under foliage plants. These figures clearly show that the floriculture trend so far in Sri Lanka has been in favour of foliage plants.

The majority of large scale commercial growers produce plants in collaboration with foreign partners. Technology is shared by these partners and highly advanced methods of production are followed. However, most of these ventures are involved in the production of ornamental foliage plants and very few are involved in the production of cut flowers. Middle level and village level growers usually go in for low cost cultivation with minimum advanced techniques, sticking to conventional methods. The majority of these growers are the ones who produce cut flowers. Therefore, most of the cut flowers produced are used to satisfy the local needs and only the surplus is exported occasionally in small amounts. The exception is the production of carnations which are grown mainly for the export market. However, export oriented cut flower ventures are on the increase.

Production of Planting Material

Flower seeds of many flowering annuals such as asters, petunias, impatiens, phlox and verbenas are produced mainly for export under fully controlled green house conditions. A small part of this production is released to the local market at regular intervals. The bulk of floriculture planting materials exported are in the form of stem cuttings rooted or unrooted. A very small amount of bulbs, corms and tubers are exported to specific customers abroad.

Large tissue culture units are few in number. However, at present, six commercial tissue culture units are operational. One of these laboratories specialises in the production of cut flower anthurium plants through tissue culture. This venture is of recent origin and produces large quantities of anthurium plants now popular in the world market. Four of these laboratories are attached to large commercial, export oriented nurseries involved mainly in ornamental foliage plants and to a lesser extent cut flowers. The sixth of these laboratories is exclusively for the production of orchids for export. There are a few middle level orchid growers who own mini-laboratories for the production of orchid seedlings and other ornamental plants for village growers.

Marketing

Thus far the cut flower market in Sri Lanka has not been able to create Auction Centres as many other countries have done. Retail outlets scattered through the production areas are the popular centres where cut flowers are sold. There are few growers who have created cooperative systems to sell their products.

Exporting of cut flowers is done by a few societies which have a selected group of partners and farmers. In many instances the agents are sent to villages to collect flowers from farmers directly. Exports of floriculture products from Sri Lanka. Main export markets are Europe (72%), and Far East and Middle East (28%).

Table : Export of Floriculture Products from Sri Lanka (1990-1995)

Product	*Value (Million Rupees)*					
	1990	*1991*	*1992*	*1993*	*1994*	*1995*
Bulbs, Corms, Tubers	0.5	0.71	21.2	21.6	10.03	6.1
Live Plants	99.5	123.1	132.4	120.3	162.4	179.0
Cut Flowers	31.0	29.7	21.7	34.5	45.3	54.8
Cut Foliage	60.3	77.74	106.9	111.0	189.1	190.2
Total Rs. Million	191.3	231.25	282.2	287.4	406.83	430.1

Potential for Cut Flower Production Development

Sri Lanka's stable and varied tropical climatic conditions and the geographic terrains from sea level up to 2200 metres of humid mounts have created magnificent macro and micro environments to house many thousands of local as well as sub-tropical and temperate plants. The rich native flora contains many potential ornamental plants which could be developed to satisfy the demands of the flourishing industry.

Sri Lanka has a favourable location to serve different markets in the world. Availability of land and the high literacy rate of the average person would be an added benefit to those who wish to invest in the industry. In addition, the tax benefits and BOI incentives granted by the government would help bring in more investors to the country and facilitate further development of the existing industry.

Furthermore, the new policy framework prepared by the Ministry of Agriculture and Lands has clearly identified the need to initiate a Floriculture Research and Development Programme. This programme will cover many areas such as agronomy, pathology, entomology, mutation, breeding, post harvest and mass propagation to support the industry.

Constraints in Cut Flower Production Development

The following are the major constraints being faced by the majority of growers:

a) *Inadequacy and high cost of air cargo*: Air Lanka, the national carrier has always given priority to perishable cargo. However, the available capacity is not sufficient.

b) *Lack of facilities for research and development*: So far only the Royal Botanic Garden, Peradeniya has been involved in providing assistance by and large to the middle and village level growers, but this is negligible when compared to the ever increasing demands of the industry.

c) *Lack of trained personnel*: The floriculture industry requires trained personnel at each level of production. Education programmes from schools up to University level and training institutes to conduct courses on high-tech practical skills in floriculture are essential for the development of the industry.

d) *Big initial investment on farms*: Duty free facilities for import of vital items not produced in Sri Lanka such as shade nets, uv stabilised polythene, irrigation and fertilising systems etc. would help to promote the industry.

e) *Lack of improved systems of marketing*: Lack of proper or organised systems for marketing and inadequacy of current international market information on prices, trends, volumes and data on competitive countries etc., severely affect the development of the industry.

f) *The uneconomical size of floriculture industry*: Difficulty in acquiring suitable land and lack of infrastructure facilities also adversely affect the industry.

g) *Lack of information on pesticides*: Current information on pesticides and their acceptance in various countries is essential.

h) *Phytosanitary clearance*: Phytosanitary inspection just prior to shipment are inconvenient and expensive, besides leaving no time to rectify any problems.

Floriculture has a great potential in Sri Lanka even though the existing market share is less than 0.2% of the world market at present. Interest in the cut flower industry is encouraging and many investors, both local and foreign, have started their nurseries to cater to the expanding market. A comparison of the world trend and the production figures of Sri Lanka clearly show that we are not in line with the world trend. However, this situation will soon change with the new policy changes taking place in the country. The Ministry of Agriculture and Lands, Export Development Board and various Floricultural Associations of the country have got together to discuss various issues in this regard and progress so far is very encouraging. Steps should, however, be taken to encourage floriculture research to cater to the special needs of the existing industry and to facilitate long term research programmes to explore the rich flora of the country to develop novelties to satisfy the growing market.

The Government should take immediate steps to alleviate constraints mentioned above and to create an Institute or a Centre for Floriculture Research and Development. This could be a semi-governmental Centre or an Institute partially supported from the growers annual income.

Cut Flower Production in Thailand

Administratively, Thailand is divided into 76 provinces and 5 regions (North, Northeast, Centre, East and South). The country has abundant water supply and suitable agro-climatic conditions for the production of various kinds of orchids and other cut flowers. Depending on the location, tropical, subtropical and temperate cut flower species can be grown. Bangkok and the nearby provinces are good for orchids and other tropical flowers, whereas the subtropical and temperate cut flower species are best grown in the north and northeastern parts of the country, especially in the winter season. The country was the world's fourth largest exporter of cut flowers in 1993 and 1994. Exports consisted almost exclusively of a wide range of cut orchids and amounted to 748.65 million baht (US$ 65.37 million) in 1993 and to 782.45 million baht (US$ 68.20 million) in 1994. The volume of cut orchids exported in 1994 was 11,897 tons. In 1994, Thailand exported orchid cut flowers to Japan, Italy, U.S.A., Germany, Taiwan and the Netherlands. During the same year the country also exported garlands made from orchids to U.S.A., Germany and Taiwan. Besides cut orchids, Thailand exported other cut flowers which were mainly tropical flowers, though the volume was very small. The flowers exported included Jatrophia, Lotus, Ananas, Heliconia, Jasmine, Rose and Marigold.

Although Thailand exported a lot of orchids, it also imported small quantities of cut orchids from Netherlands, India, Malaysia and Singapore, in addition to other cut flowers, mainly temperate, which were imported in larger volumes. The total value of imports in 1994 was 63.58 million baht.

Present Situation of Cut Flower Production

Categories of Cut Flowers

The cut flower production can be divided into two categories i.e. cut orchids and other cut flowers.

Cut Orchids

The production area of orchids was 13,000 rai (6.25 rai = 1 ha) in 1988 and increased to 14,412 rai in 1994. The production was 18,750 tons in 1988 and increased to 25,900 tons in 1994. Main production areas were Samut Sakhon, Nakhon Pathom, Bangkok, Ratchaburi, Pathum Thani and Ayuthaya.

Other Cut Flowers

***Jasmine*:** The area had increased from 3,600 rai in 1988 to 5,326 rai by 1994. The annual production was 6,718 tons in 1988 and increased to 9,447 tons in 1994. Main production areas were Nakhon Pathom, Nakhon Sawan, Bangkok, Samut Sakhon. The production cost per rai was 75,116 baht or 40.26 baht per kilogram. Average income was 104,365 baht per rai. The profit was 29,249 baht per rai.

***Rose*:** During the years 1988-1993 the production area decreased from 3,500 rai to 2,231 rai. In 1994 it increased to 3,911 rai. The annual production had decreased from 11,257 tons in 1988 to 9,457 tons by 1994. Main production areas were Nakhon Pathom (1,377 rai), Samut Sakhon (905 rai), Nonthaburi (610 rai), Bangkok (235 rai), Ratchaburi (150 rai), and Chiang Mai (106 rai). The average cost of production was 87,883 baht per rai, or 0.76 baht per stem. The profit was 98,050 baht per rai.

***Gerbera*:** The production area of double-strain gerbera had decreased rapidly from 2,000 rai in 1988 to 368 rai by 1994, mainly because the growers changed from the old-fashion double-strain gerbera to European strain with longer vase life. The European strain which is single is more suitable to grow on the highland and is in demand in the local market. Major production areas were Ratchaburi (100 rai),

Nonthaburi (64 rai), Phattalung (45 rai), Ranong (25 rai) and Chiang Mai (23 rai). Cost of production was 81,347 baht per rai or 1.09 baht per stem. Average income per rai was 154,859 baht, and the profit was 73,511 baht per rai.

***Chrysanthemum*:** The total area was 580 rai in 1988 and had increased to 998 rai by 1994. The annual production was 1,204 tons in 1988, and had increased to 2,886 tons by 1994. Main production areas were Nonthaburi (432 rai), Chiang Mai (194 rai), Yala (53 rai), Nakhon Phanom (42 rai), Phat Thalung (40 rai). The average cost of production was 59,940 baht per rai or 1.64 baht/stem. The income was 110,634 baht per rai. The profit was 50,694 baht per rai.

***Aster Peacock*:** The production area of 200 rai in 1988 had increased to 407 rai by 1994, and the annual production from 256 tons in 1988 to 529 tons by 1994. Main production areas were Chiang Mai (213 rai), Nong Khai (57 rai), Nonthaburi (35 rai) and Songkhla (31 rai). Production cost per rai was 35,921 baht or 41.21 baht per kilogram. Average income was 48,633 baht per rai. The profit was 12,712 baht per rai.

Production Area

The total area of cut flowers in the year 1994 was 39,893 rai. The largest growing area was Bangkok with 8,853 rai, followed by Samut Sakhon, Nakhon Pathom and Nonthaburi with 7,097 rai, 6,780 rai and 5,735 rai respectively.

Table : Cut Flower Production Area by Province in Thailand (1994)

Rank	*Province*	*Area (Rai)*	*Percent*
1	Bangkok	8,853	22.19
2	Samutsakorn	7,097	17.79
3	Nakhonpathom	6,780	17.00
4	Nonthaburi	5,735	14.38
5	Phayao	1,397	3.50
6	Nakhonsawan	1,300	3.26
7	Ratchaburi	1,033	2.59
8	Lampang	982	2.46
9	Chiangmai	869	2.18
10	Suphanburi	743	1.86
11	Kanchanaburi	385	0.96
12	Kamphaengphet	282	0.71
13	Ranong	277	0.69
14	Songkhla	268	0.67
15	Udonthani	244	0.61
16	Other Provinces	3,648	9.14
	Total Area	39,893	100

Source: Department of Agriculture Extension

The major kinds were orchids - 14,412 rai, jasmine - 5,326 rai, lotus - 4,371 rai, marigold - 4,072 rai, rose - 3,911 rai, giant milk weed - 2,054 rai, champaca - 1,725 rai and chrysanthemum - 998 rai; the other cut flowers included golden rod, aster peacock, gerbera, torch ginger, gladiolus, heliconia, globe amaranth, anthurium, curcuma, tuberose, red ginger and jatrophia. The main production area of orchids and other tropical cut flowers were Bangkok and the nearby provinces; while the temperate cut flowers namely lily, freesia, tulip, calla lily, alstroemeria, bird of paradise, statice, gypsophilla and carnation were grown in the Northern part (Chiang Mai and Chiang Rai) and the Northeastern Part (Nakhon Ratchasima, Ubon Ratchani, Udon Thani, Kohon Kaen, Nong Khai and Nakhon Phanom).

Cultivation Conditions

Cut Flowers Grown in the Open Field: Cut flowers which are grown in the open field are: aster peacock, champaca, curcuma, giant milk weed, gladiolus, globe amaranth, heliconia, jasmine, lotus (in water), marigold, tuberose.

Cut Flowers Grown under Protected Conditions: Cut flowers which are grown under protected conditions are:

Shading house (saran) 50% shading:	Orchids
60% shading:	Red ginger, Torch ginger
70% shading:	Anthurium
Rain-proof house (Vinyl house):	Carnation, Chrysanthemum, Gerbera, Lily, Rose
Insect-prevent house (Net house):	Chrysanthemum

Table : Cut Flower Production Area by Kind of Flowers Grown in Thailand (1994)

Rank	*Flowers*	*Area (Rai)*	*Percent*
1	Orchid	14,412	36.16
2	Jasmine	5,326	13.36
3	Lotus	4,371	10.97
4	Marigold	4,072	10.22
5	Rose	3,911	9.81
6	Giant Milk Weed	2,054	5.15
7	Champaca	1,725	4.33
8	Chrysanthemum	998	2.50
9	Golden Rod	644	1.61
10	Aster (Peacock)	407	1.02
11	Gerbera	368	0.92
12	Torch Ginger	191	0.48
13	Gladiolus	104	0.26
14	Heliconia	74	0.19

15	Globe Amaranth	63	0.16
16	Anthurium	41	0.10
17	Curcuma	32	0.08
18	Tuberose	24	0.06
19	Red Ginger	10	0.02
20	Others	1,025	2.57
	Total Area	39,893 rai	

Source: Department of Agriculture Extension
Remark: 6.25 rai = 1 hectare

Advances in Production Technology

Breeding and Selection: Many Universities, the Ministry of Agriculture, the Royal Projects and many companies in the private sector have done good work in improving, introducing and selecting good cultivars of cut flowers for the market. The methods employed for different crops included:

Hybridization:	Orchid, Gerbera, Lily, Anthurium, Curcuma
Somaclonal variation:	Orchid
Induced mutation:	Orchid, Carnation, Chrysanthemum
Induced polyploidy:	Curcuma and Lily
Protoplast fusion and genetic engineering:	Orchid

Micropropagation: Mass rapid clonal propagation using tissue culture techniques has been employed for commercial multiplication of plant material for the growers in orchid, gerbera, carnation, chrysanthemum, anthurium, curcuma, red ginger, torch ginger, lily, calla lily. There are around 50 commercial tissue culture units in Thailand.

Irrigation Technology: Advanced irrigation methods like sprinkler (for orchid and anthurium) and drip irrigation (for rose, gerbera, carnation, chrysanthemum, lily) are used for automatic watering.

Production Cycle Management: Technology for off-season and year round production for various flower crops were standardised using photoperiodic control (chrysanthemum, aster peacock, curcuma), temperature control (lily, gladiolus) and pinching technique (orchid).

Post Harvest Technology: Research efforts were directed to standardization of post harvest technologies related to improving vase life of cut flowers through using appropriate holding solutions or controlling atmosphere/temperature in cold storage.

Production of Planting Material

The requirement of planting material for cut flower production was mainly met through introduction of seed (marigold, statice, bird of paradise), bulbs (lily, gladiolus, calla lily), cuttings (carnation, chrysanthemum, jasmine), divisions (gerbera, aster peacock, red ginger, heliconia, torch ginger, alstroemeria, tuberose) and budding and grafting (rose). In addition, as mentioned earlier, micropropagation (tissue culture) was used for mass production of planting material for various species.

Marketing

Most cut flowers are for domestic consumption, except orchids 50% of which are exported. During the period 1988-1994, the demand for domestic consumption of orchids and other cut flowers increased from 33,696 tons to 46,330 tons.

The price of cut flowers depends on grade and quality, the season and demand during special occasions, like Buddhism days and Graduation days. During 1994-95 the average price for major cut flowers was: for orchids 2.49 baht/stem; rose 1.61 baht/stem; chrysanthemum 3.02 baht/stem; gerbera 2.07 baht/flower; aster peacock 1.61 baht/stem; jasmine flower 55.93 baht/kilogram.

Constraints in Cut Flower Production Development

Although Thailand has the fourth rank in world export of cut flowers, there are several constraints to be considered for future development of this industry. These are as follows:

(i) The multiplication rate of the selected cultivars by conventional methods is slow; therefore, tissue culture micropropagation is used.

(ii) High initial investment, especially the cultivars that have to be imported or grown under controlled greenhouse conditions.

(iii) Increasing labour cost.

(iv) Short vase life.

(v) Production depends on the season, and not always on the market demand. Sometimes there are not enough flowers, and sometimes there are too many flowers, which results in unstable prices.

(vi) Improper application of chemicals for controlling diseases and pests.

(vii) Post-harvest technology and packaging suitable for Thailand are not available.

(viii) No cool storage at the airport.

(ix) Insufficient air cargo space during the peak season for export.

(x) Lack of efficient systems in communication and information, resulting in difficulty to plan production according to the market demands.

(xi) The strict phytosanitary requirements enforced in the importing countries.

Potential for Cut Flower Production Development

Thailand has the capability to develop a sound cut flower industry for export due to favourable climatic and infrastructural conditions, cheap land, low labour cost, relatively low capital investment and high value addition. Future potential for the cut flower industry in Thailand is very good, but it is necessary to ensure that the flowers are of high quality, there is a reliable and continuous supply of flowers, sufficient quantities of selected cultivars and competitive prices.

The Eighth Economic Development Plan for cut flower development in Thailand in the next five years (1997-2001) indicates that the production area should increase by 2.67%, the production per rai by 5.63% and the export value by 4%. The production area of cut flowers was 45,000 rai in 1997 and is expected to increase to 50,000 rai in 2001. The total annual production volume is 62,550 tons, and is expected to increase to 86,850 tons. At present, Thailand has the fourth rank in the world cut flower exports market with orchid as the main crop. Orchid cut flowers form only a relatively small fraction of the world cut flower market. The top-seller cut flowers in the world market like carnation, chrysanthemum, rose, gypsophyla, gladiolus, asiatic lily, tulip, iris, oriental lily, daffodil and cape flower, can also be grown in Thailand. Research and development efforts have to be strongly supported for the improvement of the cut flower industry in the country.

Cut Flower Production in Vietnam

Vietnam is located in South East Asia between latitudes 80 and 230 North. The total land area is 33,099,093 hectares (330,990 km^2). Its population is approximately 76 million (1996). The country can be divided on the basis of its topography, soil pattern and climate into nine agro-ecological zones.

Table : Agricultural Land Use in Vietnam (1993)

Item	*Area (ha)*	*Share (%)*
I. Annual Species	5,523,899	75.00
a. Rice	4,252,107	57.80
b. Other Cereals	1,124,542	15.30
c. Vegetables	15,158	0.20
d. Flowers	1,585	0.02
e. Other Annual Species	130,081	1.73
II. Perennial Species	1,247,161	16.90
a. Industrial Crops	525,342	7.10
b. Fruit Trees	187,309	2.50
c. Other Perennial Species	532,767	7.20
d. Nurseries	1,743	0.02
III. Grazing Land	304,274	4.10
IV. Water Surface for Agriculture Use	273,115	4.00
Total	7,348,449	100.00

The country has a tropical monsoon climate with a pronounced dry winter season. The Northern delta has a cool winter from November to April with average temperature around 16°C, and frequent light drizzle from February onwards. Summer in Northern provinces is very hot, rainy and subject to typhoon. The Southern delta has a more equable monsoon with an average temperature of 25 - 30°C. Rainfall in most areas varies from 1,600 to 2,500 mm per year.

Soils are formed from parent rocks in the NE and NW, alluvium in the deltas, chronic buvisols in the coastal lowlands and fertile soils (basalt) in the central highlands. The soil and climatic conditions in Vietnam are very suitable for flower cultivation round the year, though presently it occupies a very small share in agriculture. The flower production area covers only 0.02 per cent of the total cultivated area, even though flower production is a traditional activity in Vietnam. Flowers are mostly cultivated in urban areas of Haiphong, Hanoi, Ho Chi Minh, Dalat and provincial towns. In the past decades, flower and ornamental production has been the business of small holders and could not be established into a viable industry, largely owing to a limited market. To make it a profitable industry, flower producers must be oriented to accessions of festivals, while the profit from flower production should attain the level of other crops.

Since 1986, with the economic renovation process, the situation of the country has changed as the life of the majority of the population has improved significantly. This is reflected on flower production through a significant increase in the area under flower cultivation, with new villages/regions becoming involved in flower production; also with product diversification with orchids, chrysanthemums, lilies etc. being introduced for commercial production; furthermore with research on flowers being started for variety improvement and standardisation of improved production technologies.

The flower industry in Vietnam is about to be established and form a new economic area involving thousands of farmers. This process will be accompanied by the economic growth of the country and raised living standards of the people. Foreign investments can contribute in this process and Vietnam flower producers could have a share in the international market.

Present Situation of Flower Production

Production Area

The area of cut flower production in Vietnam at present is rather small. It is estimated that only 1,585 hectares are under cut flower production, which cover about 0.02 per cent of the total agricultural land. The major production areas are concentrated in big cities such as Hanoi, Ho Chi Minh City, Hai Phong and Dalat. The area under cut flowers has increased significantly during recent years and is likely to increase further in coming years.

However, still big specialised cut flower farms do not exist. Cut flowers are mainly produced by small holders in their gardens. To meet the increased demand of flowers, the area under cut flower production needs to be increased through involvement of new producers, and the existing producers must increase their production area by renting or buying new land, formerly under other crops. It is expected that the production area will increase substantially if Vietnamese cut flower producers can enter the export market. The explosion of population, urbanization and industrialization also increases the demand and market for cut flowers.

Table : Area Under Flower Cultivation in Vietnam

Province	*Area (ha)*
Hanoi	500
Haiphong	320
Ho Chi Minh	200
Dalat	75
Others	490
Total	1,585

Flowers are cultivated in Ngocha, Quang an, Nhat tan, Tay tuu, and Vinch tuy village of Hanoi; also, at Dang hai commune, An hai suburb in Haiphong, Go vap and Hoc mon districts in Ho Chi Minh City and Districts 11 and 12 in Dalat. These areas cover 70-80 per cent of total land area under flowers. Flower production gives high economic return which is 10-20 times more in comparison to rice. Hence, farmers in these areas have better living standards than others.

Major Flower Crops

Table : Major Cut Flowers of Vietnam

Crop	*Area (%)*
Rose	40
Chrysanthemum	25
Gladiolus	15
Orchid	10
Others (Carnation, Dahlia, Lilium, Gerbera, Violet, Tuberose, Arum, Peach, Lotus, Camellia, Forget Me Not etc.)	10

Rose is the most popular flower in Vietnam. It can be planted and harvested round the year in the Red River Delta, Dalat, Ho Chi Minh City and other provinces. The popular varieties are the white and red rose in Quang an; light yellow, light pink, dark yellow and dark red rose in Dalat; and black red and creeping rose in other places. Chrysanthemum is also popular in Vietnam, particularly in the Red River Delta, Dalat, Ho Chi Minh City and other provinces. Popular varieties/species include *Chrysanthemum morifolium, C. indicum, C. leacanthemum*, Taiwan yellow, Japan yellow, Xuxi chrysanthemum etc. Popular varieties of gladiolus are white, violet, red, light pink, yellow, and short pink. Most of these flowers are grown in Vietnam in the open fields and only in certain areas are grown in plastic greenhouses, nurseries and experimental gardens. Orchids and Camellia are grown in shadenets for protection from the sun in the summer. But these are not protected against rain, storm and floods, and hence their quality and harvesting are affected.

The country has two different climatic regions i.e. tropical in the South and sub-tropical in the North. Besides, in both parts of the country, there are regions with a particular temperate climate, where temperate flowers can be grown such as Sapa and Tamdao in the North and Dalat in the South. Due to this fact, large number of flower varieties could be included in production. These regions are also important for the introduction of new temperate flower varieties, seed production and storage of planting material during summer time, when temperate flowers are difficult to handle. A tendency of increasing the number of varieties of temperate flowers in production has been observed. These varieties are easily introduced into production because of their commercial value. In recent years a range of varieties of roses, anthurium, chrysanthemum, carnation, orchids etc. were introduced and were quickly established in the local market. As a result, these have become the most important commercial varieties.

Production Systems

As was mentioned earlier, production systems in Vietnam have not yet been developed well; open field is employed for the cut flower industry in most cases. For temperate flowers like rose, chrysanthemum, carnation and lilium, simple shading cover from local material (bamboo) is used; special production system viz. greenhouse is yet to be developed. In rare cases in the South small greenhouses are used for cultivation of orchids. But generally orchid production is not popular because of the high investment to build up controlled production systems.

Irrigation is provided daily by overhead or surface watering in the summer time. In the rainy season a cover for temperate flowers is needed to avoid development of diseases. At different stages of growth different kinds of organic fertilizers are used.

Due to the fact that producers usually have small area of land, they undertake intensive continuous cultivation. Flower producers therefore practice renewal of the top cultivable layer of soil by removing 10 - 15 cm of top soil and adding new soil every few years. This practice proved to be a good and cheap method to avoid built up of soil-borne diseases, nematodes and mutation exhaustion.

Advances in Production Technology

As mentioned above, cut flower production has become important in Vietnam only in recent years. Research on variety improvement was initiated in the Institute of Agricultural Genetics (AGI) in 1994. It has resulted in releasing a variety of chrysanthemum (CN-93) and collections of local orchids and chrysanthemum. The work on characterization and evaluation of these genetic resources is in progress.

Studies are being conducted to assess the influence of light on blooming of chrysanthemum and the influence of NPK fertilizers on quality of flowers. Official program on disease control, pest management and post harvest handling has not yet been initiated. However, specific studies are being carried out by some groups of researchers in different insitutions to resolve particular problems.

Production of Planting Material

Planting material is supplied in many cases in the form of exchange between flower growers. Specialized nurseries on production of planting material do not exist.

Production of planting material via tissue culture is employed in some cases for roses, carnation, chrysanthemum, authurium and orchids. However, planting material is mainly supplied by the importer or multiplied by conventional methods. There is a great demand for planting material of introduced temperate flower varieties of chrysanthemum, anthurium, carnation and orchids, but an effective system of *in vitro* production has not yet been developed. The main reason for this situation is the lack of investment and effective technology of *in vitro* production.

Most of the flower varieties in Vietnam are cultivated from seed (Chrysanthemum, *Dianthus caryophyllus, Gladiolus communis*), branch (Chrysanthemum, *Dianthus caryophyllus*) bud, branch and cuttings (Rose, Orchid) and bulbs (*Gladiolus communis, Lilium longiflorum* Thumb). Propagation from tissue culture is employed for orchids, *Dianthus caryophyllus* and Chrysanthemum. Conventional methods are simple, cheap and easy to apply, but virus and other diseases develop easily, which lowers the quality. Seedlings from *in vitro* methods are strong, but more expensive. Since the market for flowers is limited, farmers are more interested in conventional methods.

Marketing

Marketing of cut flowers in Vietnam is not organized. There is no organization or association for marketing and distribution of cut flowers; hence it traditionaly follows the sequence of producers Õ businessman Õ market. This situation adversely affects cut flower production. For certain types of cut flowers the country has a potential to increase production for export but can not realize it.

In Vietnam, flower production is largely undertaken for local consumption. Previously, Vietnamese flowers were exported to Russia and East Europe (Chrysanthemum, Gladiolus), though in a small quantity. Presently, there is only one company dealing with export import.

Potential for Cut Flower Production Development

Cut flower production has a good potential for further development in Vietnam, and it appears that the cut flower industry is on the way to being well established in the next few years. The agro-climatic conditions are suitable for cut flower production development. The Government is also encouraging this activity.

Constraints in Cut Flower Production Development

The export market has not yet been developed; therefore, there is no motivation to make use of the country's potential. Also, advanced technologies are not available. Moreover, equipment and transport facilities to ensure good quality of flowers are lacking.

Vietnam is a country with good potential for cut flower production development. To promote this, it is necessary for Vietnam to collaborate with other countries in the region in order to be able to apply advanced techniques in production, harvesting and storage of cut flowers.

5

Cut Flower Production

Flower forcing is an operation or treatment to the plant, after it reaches the ripeness-to-flower stage, in order to stimulate it to flower at a specific date (e.g. on New Year's day), or during off-season period. The flowering date/period may be earlier or later than the normal date/period of flowering.

The goals of flower forcing are off-season production and specific-date production.

Cut flowers which are available during the normal season are plentiful, thus fetching a low price. Sometimes the farmers have to sell their produce even at a loss. In some cases, flowers which could not be sold are either left on the plants or are spoiled after being harvested. Thus, it would be beneficial for farmers to produce cut flowers during the off-season period to obtain higher price, although the inputs may be higher.

Similarly, the demand for cut flowers is generally very high during certain occasions such as New Year, Christmas, Mother's Day, Memorial Day, Valentine Day, Graduation Exercise Day, etc. Thus, it will be to the farmers' advantage if they can produce cut flowers to be available on these specific dates.

The objectives of forcing plants to flower during off-season or at certain specific dates are:

- To avoid surpluses of in-season cut flowers
- To avoid wastage or spoilage of surplus cut flowers
- To avoid danger of epidemics
- To distribute employment throughout the year
- To increase farmers' income
- To reduce imports and trade deficit
- To satisfy customers at the time of their needs.

Physiology of Flowering

The process of flowering follows the following sequence:

Proper stimuli (temperature & photoperiod).

Ripeness-to-flower stage, Vernalin, Florigin, Flower.

Factors affecting flowering are photoperiod, temperature and humidity.

Flowering behaviour of plants is controlled by seasonal changes. There are two types of flowers with respect to the seasonal effect on flowering:

a) Little influence of seasonal changes: e.g. roses, marigold, chrysanthemum, heliconia, etc.

b) Great seasonal influence: e.g. jasmine, dendrobium orchids, etc.

Seasonal factors can be of various types viz;

i) Photoperiodic influence: This includes short day plants, which are temperature and humidity influenced; also long day plants, which are also temperature and humidity influenced.

ii) Temperature influence: This includes low temperature requiring plants, which are photoperiodic and humidity influenced; as also, high temperature requiring plants which are also photoperiodic and humidity influenced.

iii) Humidity influencc: Including low and high humidity requiring plants.

Forcing Operation

Flower forcing can be achieved by adjusting the factors effecting flowering behaviour, viz. photoperiod, temperature and humidity.

Chemical flower forcing can be done through the application of fertilizers, used both for retarding and stimulating flowering; and plant hormones, including gibberellin, growth retardants and growth inhibitors.

Mechanical flower forcing is achieved by operations such as, pruning, leaf trimming, ringing, budding or grafting, smoking, low-temperature storage and breaking dormancy.

Forcing of Some Cut Flowers in Thailand

Dendrobium Orchids

Flowers are available all the year round, but more profuse in rainy season and less profuse in winter (which happens to be the time when the demand is high).

Pinching-off flower buds in August/September (to save nutrients for later blooming) and application of special fertilizer (high in P & K) during October/November, 3 to 4 times, induces blooming from November onwards for December/January harvest.

Siam Tulip

Seasonal flowers are produced during rainy season (June to August), requiring longday condition. No flower develops after September when shortday condition commences. Above ground parts wither and die down, and rhizome enters dormancy period until next rainy season. Providing additional light breaks dormancy. Most effective is 3 hours of light in the middle of the night. Should be started soon after daylength is shortened (21 September). In this way, plants will continue to produce flowers all the way up to the New Year day, provided enough humidity and nutrients are given.

Marigold

Flowers are available round the year. It is a day neutral plant which takes 60 - 70 days from seeding to harvest.

Timing of flowering can be maintained by fixing the date of seeding 60 to 70 days prior of harvest date. The recommended period is 65 days. For example, if the flowers are to be harvested for New Year, seeding should be done on 27 October, transplant on 6 November and pinching on 22 - 24 November. Flower buds emerge by 5 December, and blooming starts from 25 December to 5 January.

Lotus

Flowers are available all the year round, but the plant needs standing water. Low temperature during winter reduces the amount of blooming. Thus, it is recommended that the level of water be reduced to 50 cm in order to raise water temperature. In this way, blooming will be the same as in summer, both in terms of amount and size of flower.

High temperature during the summer speeds up the growth of lotus plant and the blooming time, but flower size remains small. Thus, water level should be raised up to 75 cm (from original 50 cm level in the winter). This way, water temperature will be reduced and the amount of bloom and size of flower will be maintained.

During the rainy season, due to the addition of rain water the level of water may increase, thus it is recommended that the level of water be maintained at 50 cm in order to accommodate additional rain water. The blooming will be maintained as normal.

Chrysanthemum

It is a short day plant, with critical value of 14.4 hours. Thus, it will bloom all the year round under Thai conditions, having maximum daylength of 13.3 hours in June.

Daylength can be extended by giving artifical light after sunset for about 3 hours during the early stage of growth to keep seedlings in vegetative stage until one month prior of the planned harvest date. For example, if the planned harvest date is New Year, cutting should be made in September and transplanted to the growing plot for rooting to occur. Seedling is kept under light regime of more than 14.5 hrs by providing artifical light (100 w incandescent bulb) until 1 December (seedlings should be at least 30 cm high). These will bloom on 1 January.

However, as chrysanthemum blooms profusely during the period of low temperature which commences in December, it fetches a low price in the market even during the time of Christmas or New Year. Thus, some farmers avoid producing flowers during such period but shift it to the summer. The problem is that the temperature during summer is quite high and not optimum for chrysanthemum growth. The most suitable place of production of chrysanthemum flowers is on the highlands of northern Thailand where the temperature during the summer is around 16-20°C. The same principle of flower forcing is applied, but in this case the daylength in northern Thailand may be lower than the critical value for certain varieties. As practiced in Doi Phu Kha of Nan Province, black cloth is used to completely cover the plant house from 16.00 to 08.00 hrs. for 30 days after the cuttings have been exposed to long day conditions (supplemented with artificial light).

Jasmine

Flowers are available thoughout the year, but very profuse during the rainy season and scarce during the winter. To produce jasmine for winter-season harvest, the following operations are recommended:

(i) One month before the planned date of harvest, stop watering for 2 - 3 days until the plants show sign of wilting.

(ii) Prune the plant to a round shape so that blooming will emerge from mature branches when induced in stage (iii). In this way the blooms will be of large size and healthy as they receive full sunlight.

(iii) Apply balanced fertilizer (e.g. 15-15-15) at the rate of 30 g/plant and water heavily. Keep watering normally everyday. Flower buds will emerge within 10 days and blooming will occur within 25-30 days after pruning.

Amaryllis

Flowers are available all the year round, although depends on the availability of bulbs which are available from the Netherlands in the summer (June onwards). It takes about one month after the bulbs are planted to have full bloom ready for the harvest of cut flowers.

It is recommended to keep the bulbs in a refrigerator (4°C) for at least two months. The bulbs must be planted one month prior to the required date of flowering.

Gladiolus

Flowers can be available all the year round, provided the weather is optimum (requires cool climate). Blooms 90 - 100 days after planting. In cool climate preheat corms before planting for 2 weeks at 27-32°C. This will force such corms to flower early. In warm climate soak in GA3 solution (10 - 25 ppm) before planting. This will accelerate flower by hastening differentiation of floral primordia. Growth retardants (e.g. CCC) promote initiation of floral primordia (by reducing endogenous GA3 level, or counterbalancing its inhibitory effect on floral initiation.

Roses

Flowers are available round the year but more profuse in the cool season.

For blooming during Christmas - New Year, and Valentine Day, the following procedure is suggested:

(i) Cut the branches in November. This will stimulate flowering during Christmas - New Year. Flowering will take place 43 days after this pruning. Cutting the flowers on 23 December will further stimulate flower bud initiation to bloom on 10 February. Flowering will take place 49 days after this cutting.

(ii) Pinch (i.e. cut only the tip off) instead of pruning as in (i) above, on 10 November for 23 December harvest. This operation automatically stimulates flower initiation for 10 February harvest.

The following operations should be followed:

- Do not pinch healthy flower buds after 10 November as blooming will be due in 30 days. Allow them to bloom for 10 December harvest.
- Do not pinch flower buds which are still very small.
- Pinch all other branches, including those flower buds which are not healthy, deformed, etc. Do not pinch too far down to healthy leaves, unless they are water sprouts which should be cut off from the base.

Potential of Commercial Floriculture in Asia: Opportunities for Cut Flower Development

The Asian flower industry has the potential to become the leading flower industry in commercial floriculture worldwide. Looking to the future, the Asian flower industry could rival, if not surpass, the size and scope of the European flower industry which presently dominates global commercial floriculture.

Evolution of the World Cut Flower Market Place

Forty years ago, demand for cut flower by consumers around the world was satisfied by local cut flower production. In Europe, per capita consumption was significant, and consumer culture required a large supply of cut flowers for gifts, occasions, and everyday use. As a result, cut flower production in Europe was sizable. Gradually as transportation systems developed throughout this region, it became possible to distribute cut flowers grown in southern areas of Europe to northern areas of Europe. Consequently, the European flower industry began to extend its boundaries for cut flower production and along with this expansion grew the influence of the European flower industry. This background history could be considered the beginning of commercial floriculture as we know it today.

When the world energy crisis occurred in 1973, the marketing plan for distributing cut flowers grown in different European countries to Holland for sale through the Dutch flower auction and back to markets throughout Europe became a significant production opportunity for southern European cut flower growers. Increasingly larger quantities of cut flowers were grown in southern Europe to meet the demand for cut flower sales through Holland. Flower growers in the southern regions had a price advantage over growers located in northern regions because cut flower production was more expensive for northern growers during the winter season due to increased energy costs required to obtain quality flowers in controlled temperature greenhouses.

Then, competition for southern European cut flower growers intensified when Israeli cut flower growers, who were located further south entered the market with product to be sold through the Dutch flower auction. Israeli growers had the production advantage of being further south where they could produce cut flowers in open fields or plastic tunnels year round, eliminating most of the overhead expenses for greenhouses and heating systems. But in order to develop a potentially lucrative export cut flower industry for themselves, the Israelis needed to address limiting factors to their success. The two main limiting factors were transportation costs to Europe and a water shortage if production were to expand.

Solutions to these limiting factors were found for Israeli growers. In the case of transportation costs which offset growers cost advantage in terms of energy compared to growers in southern Europe, the government provided transportation subsidies which have reduced the costs to the growers to ship their cut flower product to Europe, thereby maintaining a competitive cost advantage over European growers. As for the water shortage, research on irrigation systems that would conserve water usage was applied to production systems for cut flowers.

Through the 1970's, the activities of the European flower industry had begun to influence cut flower production and sales beyond the borders of Europe. Cut flower sales through the Dutch flower auctions had gained a share of the United States market. This was achieved by promotion activities in the USA supported by the Holland Flower Council which encouraged Americans to purchase more cut flowers for gifts, occasions and everyday use, similar to consumer habits in Europe. Most of the flowers sold to the USA through the Dutch flower auctions are shipped to the USA by air through New York.

Simultaneously, Miami, USA, was being developed as a key import distribution base for cut flowers being grown in Columbia, South America and shipped north. This caused considerable competition for local cut flower growers in the USA. Manufacturers and suppliers from the European flower industry were quick to find opportunity in this situation. Not only were South American cut flower growers purchasing varieties from Europe but flower growers from the USA were persuaded to invest in production systems and equipment from Europe in hopes of becoming more efficient producers like the Dutch growers who had once faced competition from southern European growers. As a result, the United States flower industry owes a significant share of its growth in terms of promotion and sales and improved production systems to the influence of the European flower industry.

It is worthwhile to mention that the Israeli flower industry has become a formidable competitor of the European flower industry. Israeli cut flower producers ship significant quantities of product into the USA market via both New York and Miami. This compensates Israeli producers for the reduction in cut flower sales to the European market which is increasingly being supplied by flower growers from regions in Africa, especially Kenya. Also, Israelis have been successful in selling their production equipment and varieties to flower growers in other countries. Continuing to advance in the 1980's, the European flower industry began seeking further opportunity and expansion in Asia by 1985. Japan's bubble economy was starting to inflate and discretionary income spending by the Japanese was rising. European flower imports made headway into the lucrative market in Japan. Within a few years, as economies in Korea, Taiwan, and Hongkong strengthened, the European flower industry moved into these markets with their cut flower exports as well.

Since the early 1990's, the European flower industry, as a worldwide leader in commercial floriculture, has been impacting the rest of Asia with cut flower imports from Holland and sales of flower varieties, production equipment, and technology for new production operations in Asia. Israeli cut flower producers, manufacturers, and suppliers have followed but, one step behind. The main difference between the European flower industry and the Israeli flower industry is that the European flower industry enters their new markets by launching aggressive marketing campaigns which call attention to the quality and image of Dutch flowers. These campaigns stimulate demand by new consumers for their cut flower products. So far, the Israelis have not particularly created an image for end consumers of Israeli flowers. This difference is one of the factors which contributes to why the European flower industry is the world wide leader in commercial floriculture.

Initially, commercial floriculture production in southeast Asia was developed because of increasing need for low cost flowers by the European cut flower market place. European flower traders identified commercial floriculture production in southeast Asian countries as a source of supply. Ironically, Dutch auctions often served to re-distribute this product to the Japanese market. By the mid to late 1980's, Dutch importers/exporters had begun selling floriculture product in Japan. With economies expanding, the "little tigers", i.e. Taiwan, Korea and Hongkong were the next Asian targets with market needs for floriculture products from Europe and potentially from other Asian countries which could produce floriculture products less expensively.

The development of the commercial cut flower industry in Asia has been unlike that of Israel, African countries, south and central American countries. In the latter regions, cut flowers have been a product produced mainly for export with no thought of a potential domestic market. On the other hand, in Asia, whereas cut flowers were initially produced for export, the market potential has rapidly changed to include opportunities for supplying the local market as well. This unique development is on account of the rapid strengthening of economies in the region, high population densities, and the consumer perception which has been promoted heavily by the European flower industry that the use of fresh flowers in one's every day life represents an improved, quality lifestyle.

Today and in the future, the potential for commercial floriculture expansion in Asia, including production for domestic and export sales of cut flowers, is greater than ever before. The elements for success needed to transform the Asian flower industry into the worldwide leader of commercial floriculture have been implied by reviewing the evolution of the world cut flower market place.

Opportunities for Cut Flower Development in Asia - Markets

There is considerable opportunity in both export and domestic markets for cut flowers from southern Asia. In case of export markets, it seems reasonable to set short, middle, and long term goals with respect to entering these markets. Simultaneously, efforts to develop domestic markets within and among countries in southeast Asia should be undertaken.

Goals for Establishing Export Markets from Southeast Asia

Short - Term: Countries in Asia - Japan, Korea, Taiwan, Singapore

Already, cut flowers have been exported from southeast Asia to other countries in Asia, especially Japan, Korea, Taiwan, and Singapore. The advantages in marketing product to these countries is that they are relatively near, shipping requirements in terms of postharvest storage have been minimal, though this situation may not persist, and often contacts to find buyers have been made only through relative or previous business relationships.

In the case of Japan, it is well known that custom regulations and quality requirements are among the strictest in the world, though the price fetched for product is high. This incentive is attractive despite the difficulty to achieve success in marketing cut flowers to Japan.

In terms of distribution to these countries, it is important to examine what are the practical transportation routes to these countries from southeast Asia. It could be that developing a transportation hub in Bangkok or Singapore would be convenient for shipping cut flowers from southeast Asian countries, including cut flowers from Yunnan province in China, to countries in northeast Asia. Numerous commercial flights already are in place regarding this possible shipping pattern for cut flowers from the region. The transportation concept of a shipping hub would imply that initially all cut flower production intended for export to northeast Asian destinations would first be shipped to the hub, e.g. Bangkok, and then re-distributed to specific destinations in Japan, Korea, Taiwan, and Singapore. This routing for northeast Asia could be extended to Australia and New Zealand.

This centralized shipping hub would require expanding postharvest handling and storage facilities at the airport terminal and increasing freight forwarding activities for cut flowers in Bangkok. At the same time, shipping procedures and storage requirements for cut flowers to be transported from southeast Asian countries would be streamlined by the ability to focus on transportation requirements to Thailand rather than an array of destinations. In the short term, this could serve as a way to shorten the time to improve distribution of cut flower product from all southeast Asian countries to these lucrative northeast Asian markets.

Mid-Term: Countries in North America - Canada and USA; Central Europe

At first consideration, it may seem unlikely that cut flowers exported from southeast Asia could penetrate the markets in North America or central Europe as these markets are already well supplied with cut flowers from other regions such as south and central America and southern Europe, Israel, and Africa. However, in the case of North America, the entry points for cut flower imports are mainly New York and Miami, leaving open the opportunity for establishing an import entry point in Vancouver or San Francisco, especially for cut flowers shipped from a Chinese port, i.e. Kunming. Recently, with the return of Hongkong to China, there has been an exodus of Chinese from Hongkong to Canada establishing residence in North America. Already the San Francisco area in the USA is home to many ethnic Chinese.

Due to the high density of Chinese in these two western entry points to North America, it is highly conceivable that Chinese flower importing companies could proliferate if the opportunity were available. This was certainly the case with Miami which already housed a large population from south and central America when flower imports from Columbia, Ecuador, and Costa Rica began to enter the USA market on the east coast. Until now, domestic producers of cut flowers in California on the west coast of the USA have been holding up against imports from south and central America but, would surely retreat if low cost, high quality imports from southeast Asia via China were available. The reason for setting this export goal as "mid-term" is that airline shipping routes from China to these cities are not as well established at this time compared to routes from Bangkok to northeast Asian destinations.

It might be proposed to establish Kunming in China, which is currently being developed as the center for cut flower production in China, as the transportation hub for distributing flowers from southeast Asia to North America.

Another mid-term export goal would be to market flowers from southeast Asia to central Europe. This could be achieved, transportation wise also through Bangkok, which already has airline clearances for many central European cities, including Amsterdam.

Another factor in identifying mid-term export market goals is that both north American and central European markets are already well supplied with an abundance of quality, low cost flowers from other regions. These export markets are also both further away in distance so that shipping costs are not particularly an advantage. Marketing factors which will affect the attainment of these mid-term goals will likely be related to price and customer service. The capability for the Asian flower industry to initially offer these export market customers low prices, a full production mix, convenient and reliable shipping service will increase the chances of gaining market share in a market place which already has supply.

Long-Term: Countries in Eastern Europe

Lastly, there will be opportunities for marketing cut flowers from southeast Asia to Eastern Europe, particularly northeastern Europe. This would be another likely destination for flowers being shipped out of Kunming. At the moment, economies in northeastern Europe need to improve before significant quantities of cut flowers would be in demand, but by that time (5-10 years) it is likely that transportation routes will have expanded greatly from China to its neighbouring territory, northeastern

Europe. As in the past, no doubt winter production of cut flowers in a southern area will surpass cut flower production in the north.

Goals for Establishing Domestic Markets within Southeast Asia

Similar to establishing export markets for cut flowers from southeast Asia, focusing on improving transportation is a key factor to establishing distribution into domestic markets within and among southeast Asian countries. It is proposed that distribution between these countries may be improved if a transportation hub system were enacted. Also, in the case of developing domestic markets, it is likely that efforts to add or improve wholesale markets in major cities would facilitate greater distribution of cut flowers within these countries.

Opportunities for Cut Flower Development in Asia - Production

Opportunities for cut flower development in Asia with respect to production are not location dependent, meaning that any area of southeast Asia should be able to capture significant cut flower production opportunity if pursued. In present and future commercial floriculture markets, production opportunities exist if the goal to reliably produce high quality product in consistent quantities can be attained.

To reliably produce high quality cut flowers in consistent quantities requires optimum production management. Production management strategies for implementation and training of personnel will be different for different types of production operations. Varying types of production operations might be commercial cut flower companies, large or small; government directed agricultural farming communities; or rural development programmes managed by public or private funds. For each of these types of production operations, dissemination of pertinent information needed to manage production may require different strategies. Therefore, a key factor to realizing production opportunities is education. Ideally, educational resources should be developed and supported by public and private sectors.

An example of educational resources which could be critical to the success of implementing optimum production management for cut flower enterprise in Asia would be a training center/demonstration showcase of production technology adapted to local growing conditions for specific cut flower crops. This educational institution or organization could facilitate dissemination of information to management at cut flower operations or provide training of technical personnel involved in cut flower operations.

Enlisting the support of floriculture professionals from outside and within Asia to coordinate activities in research and education for the purpose of furthering floriculture development in Asia would have significant impact on the pace and scope of cut flower development in Asia. Formally organized, it would be possible to solicit funds through international agencies to support an institution set up for this purpose, i.e. International Center for Floriculture in Asia.

Departments related to production and marketing management for cut flower crops would be defined in order to reliably produce high quality cut flowers in consistent quantities regardless of the location in southeast Asia. Such departments could be:

- Cultural production, information and management.
- Postharvest production, information and management.
- Equipment use, design, and engineering.
- Genetic improvement of cut flower varieties.
- Consumer marketing.

If sufficient attention and support were given to attaining the goal of reliably producing high quality cut flower product in consistent quantities among Asian countries, then it would be reasonable to expect that the Asian flower industry would soon overtake any other region's flower industry, including the European flower industry, in production. Moreover, given the rapidly increasing rate of spending among Asian consumers for cut flowers, it is also reasonable to expect that the Asian flower industry will soon surpass consumption rates for cut flowers compared to that of other regions.

Leadership in Commercial Floriculture Worldwide

It is exciting to discuss how and why the Asian flower industry has the potential to not only "take the lead" in volume and consumption of floriculture products but, to "become the leader" of commercial floriculture worldwide. This may suggest the most important marketing and production opportunities for cut flower development in Asia.

Once again, in review of activities of the European flower industry, it can be said that the European flower industry has led the way in floriculture production, research, floriculture education and training programmes, development of floriculture postharvest handling and storage technology; floral marketing strategies; grades and standards for floral product; and transportation systems for distribution of flowers. Countries within the European region and outside of the region depend on product innovation from Europe to yield new flower varieties. Innovative floriculture production systems have been promoted and exported to other regions of the world. Even floral design concepts have been defined by the European flower industry.

The European flower industry exports more than just flowers. The European flower industry exports flowers, equipment and technology to produce flowers, marketing programmes to increase consumer purchases, and with those marketing programmes, cultural traditions and attitudes about flowers which includes floral design, uses of floral products, i.e. decoration, gifts, holidays, and appreciation for flowers.

Celebration of western holidays and the purchases of flowers for these holidays have become a main theme of floral marketing for consumers in Asia. Two examples of this are cut roses for Valentine's Day and cut carnations for Mother's Day. This type comprehensive involvement with floriculture from disseminating the use of production equipment and technology to influencing consumer purchases of flowers by promoting western holidays associated with floral purchases has established the European flower industry as the leader of commercial floriculture worldwide.

Let it be asked, could the establishment of an Asian flower industry spread its influence among countries in Asia and the West? Why not? How would this be done?

There are many opportunities for marketing cut flowers from Asia. In addition to developing a new transportation concept for improving global distribution of cut flowers produced in Asia, there are opportunities in developing marketing strategies which would boost global consumption of flowers produced in Asia. First of all, unifying the image of flowers produced in Asia to create an identity of Asian floriculture would be one step. Then, this could be implemented by introducing an Asian calendar of floral holidays which indicated particular flowers and how they are used in association with these holidays. Even if certain flowers are not presently used in association with a particular holiday or celebration in Asia, there is no reason that floral associations from each country could not invent the concept.

For example, in China, on a certain day in the autumn, the moon and all things associated to be round like the moon are appreciated. Millions of round cakes, labelled "moon cakes" are sold to consumers. Why could not this holiday become a commercial opportunity for the Asian flower industry and pom pom mums become the flower of choice for this day as a gift item along with "moon cakes". If the Japanese celebrate Christmas now with all its commercialism, why could they not celebrate "Moon Day" in the autumn? The same can be said for Americans and Europeans. In Vietnam, flower consumption is auspicious on the 1st and 15th of the month in compliance with Vietnamese traditions. This seems like another convenient way for the Asian flower industry to boost cut flower consumption all over the world twice a month if there were an Asian calendar of floral holidays reminding consumers that it was that time of the month to buy flowers.

Imbedded in this type of floral promotion strategy is the opportunity for the Asian flower industry to develop new varieties of flowers which would characterize these holidays. For example, in case of "Moon Day" not just any type of pom pom mum would qualify, but a particular colour, size, shape, etc. could be developed.

Given the cultural differences among Asian countries, the challenge would be to incorporate as many different cultural traditions representing the different Asian cultures as part of the coordinated Asian flower industry effort. This approach would allow each country to capitalize exclusively on a particular holiday, boosting production and sales for that country according to the holiday celebrated.

Once the mission of developing an Asian flower industry was accepted, it would be certain that marketing ingenuity from within Asian flower associations would take over and demand for cut flower products from Asia would soar.

Opportunities in production of cut flowers in Asia would go beyond just the need to increase the rate of production. New production systems for efficiency or environmental protection could be developed which would be adapted to conditions of flower production in Asia. Research priorities at flower research institutions throughout Asia could be directed towards developing innovative solutions which supported global sustainability. Then, in a technology turnaround, it would be countries in the west which would be sending exchange staff to Asia to study new flower production systems. Novel engineering or design applications in equipment and supplies which would be used by producers of flowers for the Asian flower industry could be patented and exported to flower producers in other countries outside of Asia. Endless opportunities. Endless opportunities.

The establishment of an Asian flower industry could direct the unlimited opportunities for development of cut flower production in Asia and development of worldwide flower markets for cut flowers from Asia.

Conclusion

In conclusion, it is necessary to emphasize that cooperation and commitment, in terms of education, research, funding, and communication from countries in Asia, is requisite in order for the Asian flower industry to succeed in replacing the European flower industry as the leader of commercial floriculture worldwide.

Reputations in Auction Ranking Rose Breeding

A rose is a perennial plant of the genus *Rosa*, within the family Rosaceae. There are over 100 species. They form a group of erect shrubs, and climbing or trailing plants, with stems that are often armed with sharp prickles. Flowers are large and showy, in a number of colours from white through yellows and reds. Most species are native to Asia, with smaller numbers native to Europe, North America, and northwest Africa. Species, cultivars and hybrids are all widely grown for their beauty and fragrance. Rose plants range in size from compact, miniature roses, to climbers that can reach 7 meters in height. Species from different parts of the world easily hybridize, which has given rise to the many types of garden roses.

Fossil records show us the roses we grow in our gardens are the result of an amazing 35 million years of change. Today the flower is still evolving, but mainly due to rose breeding – the crossing of two different parent flowers to make a rose with new characteristics.

This is not just a recent practice – rose cultivation probably began around 5,000 years ago in China and what is now Iraq. But it was the 19th century that saw a big explosion in breeding the flowers when a handful of Chinese roses were imported into Europe. They were repeat bloomers, making them of great interest to hybridizers who no longer had to wait once a year for their roses to bloom. These were crossed with European cultivars to produce the great variety that emerged in the 1800s. For example, In 1724, celebrated gardener the Duchess of Beaufort was only able to bring together 16 types of roses for her collection, but by 1828 there were 2,500 varieties.

Empress Josephine of France also had a great interest in rose-growing and breeding and her gardens were a haven for rosarians (rose cultivators) from 1803-1814, with many new types being bred there.

Old and New

From the introduction of roses from China, experts today tend to divide all roses into two groups. There are "old roses" (those cultivated in Europe before 1800) and "modern roses" (those which began to be cultivated in England and France around the turn of the 19th century).

Until the process of hybridisation was understood in the 1800s, new rose varieties were the results of natural crosses or mutations, which were then nurtured by gardeners.

Breeding roses is a very rewarding activity. From the moment a breeder chooses the parent roses, until a few months later when the seedlings flower for the very first time, there is an exciting anticipation as each of these seedlings will be absolutely unique – the creation of the breeder.

Someone who would know all about this feeling is the acclaimed rose breeder and writer David Austin. Born in 1926, his aim was to create new roses by combining the charm and fragrance of Old Roses with the wide colour range and repeat-flowering qualities of Modern Roses such as Hybrid Teas, Floribundas and Grandifloras.

His first rose, 'Constance Spry', was introduced in 1963. In 1967 and 1968 he introduced 'Chianti' and 'Shropshire Lass'. From these, he developed repeat flowering varieties and called them 'English Roses'.

David Austin started David Austin Roses in 1969 to introduce English Roses, as other rose nurseries were not particularly interested in them at the time. This was short-lived and Austin's roses soon became the most successful group of new roses in the 20th century.

The David Austin nursery in Shropshire now runs one of the largest rose breeding programmes in the world, carrying out more than 150,000 crosses each year, producing about 400,000 seeds. From these about 250,000 seedlings will germinate, from which each year's new introductions (new rose varieties) are selected after nine years of trialing.

The five or six new garden varieties are introduced each May at the Chelsea Flower Show.

Naming Roses

Walk through a big rose garden and you may find yourself asking odd questions, such as whether William Shakespeare smells better than Charles Darwin... New varieties by David Austin have been named in honour of his family, well-known Rosarians, historical events or celebrities. In 2006, ballerina Darcey Bussell had a crimson rose named after her, and the year before Alan Titchmarsh became one of the privileged few to have a rose as a namesake.

Blue Rose

Breeding roses is not always easy – it can take years and hundreds of crosses to reach the desired plant.

The holy grail of horticulture and rose breeding for many years has been the blue rose, and now a chance discovery in a laboratory has meant scientists have found a way to produce one.

Doctors were trying to find out how the human liver breaks down drugs and when they moved a liver enzyme into a bacterium, it turned blue. They then moved this gene into plants. It's expected blue roses will be hugely popular, winning five per cent of the international market for cut roses.

Botany

The leaves are borne alternately on the stem. In most species they are 5–15 centimetres long, pinnate, with (3–) 5–9 (–13) leaflets and basal stipules; the leaflets usually have a serrated margin, and often a few small prickles on the underside of the stem. Most roses are deciduous but a few (particularly from South east Asia) are evergreen or nearly so.

The flowers of most species have five petals, with the exception of *Rosa sericea*, which usually has only four. Each petal is divided into two distinct lobes and is usually white or pink, though in a few species yellow or red. Beneath the petals are five sepals (or in the case of some *Rosa sericea*, four). These may be long enough to be visible when viewed from above and appear as green points alternating with the rounded petals. The ovary is inferior, developing below the petals and sepals. Roses are insect-pollinated in nature.

The aggregate fruit of the rose is a berry-like structure called a rose hip. Many of the domestic cultivars do not produce hips, as the flowers are so tightly petalled that they do not provide access for pollination. The hips of most species are red, but a few (e.g. *Rosa pimpinellifolia*) have dark purple to black hips. Each hip comprises an outer fleshy layer, the hypanthium, which contains 5–160 "seeds" (technically dry single-seeded fruits called achenes) embedded in a matrix of fine, but stiff, hairs. Rose hips of some species, especially the Dog Rose (*Rosa canina*) and Rugosa Rose (*Rosa rugosa*), are very rich in vitamin C, among the richest sources of any plant. The hips are eaten by fruit-eating birds such as thrushes and waxwings, which then disperse the seeds in their droppings. Some birds, particularly finches, also eat the seeds.

While the sharp objects along a rose stem are commonly called "thorns", they are technically prickles — outgrowths of the epidermis (the outer layer of tissue of the stem). (True thorns, as produced by e.g. *Citrus* or *Pyracantha*, are modified stems, which always originate at a node and which have nodes and internodes along the length of the thorn itself.) Rose prickles are typically sickle-shaped hooks, which aid the rose in hanging onto other vegetation when growing over it. Some species such as *Rosa rugosa* and *Rosa pimpinellifolia* have densely packed straight spines, probably an adaptation to reduce browsing by animals, but also possibly an adaptation to trap wind-blown sand and so reduce erosion and protect their roots (both of these species grow naturally on coastal sand dunes). Despite the presence of prickles, roses are frequently browsed by deer. A few species of roses have only vestigial prickles that have no points.

Species

The genus *Rosa* is subdivided into four subgenera:

- *Hulthemia* (formerly *Simplicifoliae*, meaning "with single leaves") containing one or two species from southwest Asia, *R. persica* and *Rosa berberifolia* which are the only roses without compound leaves or stipules.
- *Hesperrhodos* (from the Greek for "western rose") contains *Rosa minutifolia* and *Rosa stellata*, from North America.
- *Platyrhodon* (from the Greek for "flaky rose", referring to flaky bark) with one species from east Asia, *Rosa roxburghii.*
- *Rosa* (the type subgenus) containing all the other roses. This subgenus is subdivided into 11 sections.
  - o *Banksianae* - white and yellow flowered roses from China.
  - o *Bracteatae* - three species, two from China and one from India.
  - o *Caninae* - pink and white flowered species from Asia, Europe and North Africa.
  - o *Carolinae* - white, pink, and bright pink flowered species all from North America.
  - o *Chinensis* - white, pink, yellow, red and mixed-colour roses from China and Burma.
  - o *Gallicanae* - pink to crimson and striped flowered roses from western Asia and Europe.
  - o *Gymnocarpae* - one species in western North America (*Rosa gymnocarpa*), others in east Asia.
  - o *Laevigatae* - a single white flowered species from China
  - o *Pimpinellifoliae* - white, pink, bright yellow, mauve and striped roses from Asia and Europe.
  - o *Rosa* (syn. sect. *Cinnamomeae*) - white, pink, lilac, mulberry and red roses from everywhere but North Africa.
  - o *Synstylae* - white, pink, and crimson flowered roses from all areas.

Uses

Roses are best known as ornamental plants grown for their flowers in the garden and sometimes indoors. They have been also used for commercial perfumery and commercial cut flower crops. Some are used as landscape plants, for hedging and for other utilitarian purposes such as game cover. They also have minor medicinal uses.

Ornamental Plants

The majority of ornamental roses are hybrids that were bred for their flowers. A few, mostly species roses are grown for attractive or scented foliage (such as Rosa glauca and Rosa rubiginosa), ornamental thorns (such as Rosa sericea) or for their showy fruit (such as Rosa moyesii). Ornamental roses have been cultivated for millennia, with the earliest known cultivation known to date from at least 500 BC in Mediterranean countries, Persia, and China. Many thousands of rose hybrids and cultivars have been bred and selected for garden use as flowering plants. Most are double-flowered with many or all of the stamens having mutated into additional petals. In the early 19th century the Empress Josephine of France patronized the development of rose breeding at her gardens at Malmaison. As long ago as 1840 a collection numbering over one thousand different cultivars, varieties and species was possible when a rosarium was planted by Loddiges nursery for Abney Park Cemetery, an early Victorian garden cemetery and arboretum in England.

A few species and hybrids are grown for non-floral ornamental use. Among these are those grown for prominent hips, such as the flagon shaped hips of *Rosa moyesii*. Sometimes even the thorns can be treated as an attraction or curiosity, such as with *Rosa sericea*.

Cut Flowers

Roses are a popular crop for both domestic and commercial cut flowers. Generally they are harvested and cut when in bud, and held in refrigerated conditions until ready for display at their point of sale. In temperate climates, cut roses are often grown in glasshouses, and in warmer countries they may also be grown under cover in order to ensure that the flowers are not damaged by weather and that pests and disease control can be carried out effectively. Significant quantities are grown in some tropical countries, and these are shipped by air to markets across the world.

Perfume

Rose perfumes are made from attar of roses or rose oil, which is a mixture of volatile essential oils obtained by steam distilling the crushed petals of roses. An associated product is rose water which is used for cooking, cosmetics, medicine and in religious practices. The production technique originated in Persia then spread through Arabia and India, but nowadays about 70% to 80% of production is in the Rose Valley near Kazanluk in Bulgaria, with some production in Qamsar in Iran and Germany. The Kaaba in Mecca is annually washed by the Iranian rose water from Qamsar. In Bulgaria, Iran and Germany, damask roses (*Rosa damascena* 'Trigintipetala') are used. In the French rose oil industry *Rosa centifolia* is used. The oil is transparent pale yellow or yellow-grey in colour. 'Rose Absolute' is solvent-extracted with hexane and produces a darker oil, dark yellow to orange in colour. The weight of oil extracted is about one three-thousandth to one six-thousandth of the weight of the flowers; for example, about two thousand flowers are required to produce one gram of oil.

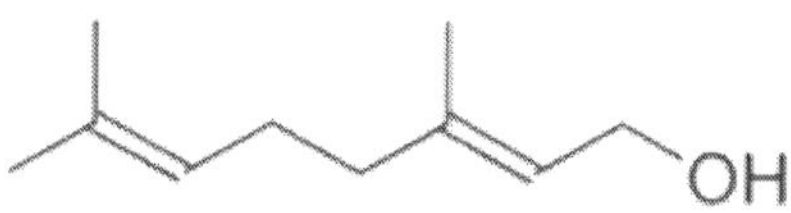

Geraniol $(C_{10}H_{18}O)$

The main constituents of attar of roses are the fragrant alcohols geraniol and l-citronellol; and rose camphor, an odourless paraffin. β-Damascenone is also a significant contributor to the scent.

Rose water, made as a by-product of rose oil production, is widely used in Asian and Middle Eastern cuisine. The French are known for their rose syrup, most commonly made from an extract of rose petals. In the United States, this French rose syrup is used to make rose scones and marshmallows.

Rose Hips

The rose hip, the fruit of some species, is used as a minor source of Vitamin C.

Rose hips are occasionally made into jam, jelly, and marmalade, or are brewed for tea, primarily for their high vitamin C content. They are also pressed and filtered to make rose hip syrup. Rose hips are also used to produce Rose hip seed oil, which is used in skin products and some make-up products.

Medicine

The fruits of many species have significant levels of vitamins and have been used as a food supplement. Many roses have been used in herbal and folk medicines. *Rosa chinensis* has long been used in Chinese traditional medicine. This and other species have been used for stomach problems, and are being investigated for controlling cancer growth.

Culture

Art

Roses are often portrayed by artists. The Luxembourg born Belgian artist Pierre-Joseph Redouté produced some of the most detailed paintings of roses.

Henri Fantin-Latour was also a prolific painter of still life, particularly flowers including roses. The Rose 'Fantin-Latour' was named after the artist.

Other impressionists including Claude Monet, Paul Cézanne and Pierre-Auguste Renoir have paintings of roses among their works.

Symbolism

Roses have been long used as symbols in a number of societies. Roses are ancient symbols of love and beauty. 'Rose' means pink or red in a variety of languages (such as Romance languages, Greek, and Polish). The rose was sacred to a number of goddesses (including Isis and Aphrodite), and is often used as a symbol of the Virgin Mary. The ancient Greeks and Romans identified the rose with their goddesses of love, Aphrodite and Venus. In Rome a wild rose would be placed on the door of a room where secret or confidential matters were discussed. The phrase *sub rosa*, or "under the rose", means to keep a secret — derived from this ancient Roman practice.

Early Christians identified the five petals of the rose with the five wounds of Christ. Despite this interpretation, their leaders were hesitant to adopt it because of its association with Roman excesses and pagan ritual. The red rose was eventually adopted as a symbol of the blood of the Christian martyrs. Roses also later came to be associated with the Virgin Mary.

Rose culture came into its own in Europe in the 19th century with the introduction of perpetual blooming roses from China. There are currently thousands of varieties of roses developed for bloom shape, size, fragrance and even for lack of prickles.

The rose is the national flower of England and the United States, as well as being the symbol of England Rugby, and of the Rugby Football Union. It is also the provincial flower of Yorkshire and Lancashire in England (the white rose and red rose respectively), of Alberta (the wild rose) in Canada, and of Islamabad Capital Territory in Pakistan. It is the state flower of four US states: Iowa and North Dakota (*R. arkansana*), Georgia (*R. laevigata*), and New York (*Rosa* generally). Portland, Oregon counts "City of Roses" among its nicknames, and holds an annual Rose Festival.

Roses are occasionally the basis of design for rose windows, such windows comprising five or ten segments (the five petals and five sepals of a rose) or multiples thereof; however most Gothic rose windows are much more elaborate and were probably based originally on the wheel and other symbolism.

A red rose (often held in a hand) is a symbol of socialism or social democracy: it is used as a symbol by British, Irish, French, Spanish, Portuguese, Norwegian, Danish, Swedish, Finnish, Brazilian, Dutch, Bulgarian and other European labour, socialist or social democratic parties, mostly adopted in the period after World War II.

The White Rose was a World War II non violent resistance group in Germany.

A bouquet of red roses is often used to show love. It is used as a Valentine's Day gift in many countries.

On St George's Day in Catalonia people offer dark red roses as gifts, especially between lovers. The Virolai, a hymn to the Virgin of Montserrat, one of the black Madonnas of Europe, begins with the words: *"Rosa d'abril, Morena de la serra..."* (April rose, dusky lady of the mountain chain...). Therefore this virgin is sometimes known as "Rosa d'abril". The red rose is thus widely accepted as an unofficial symbol of Catalonia.

The rose can symbolize the following:

- The heraldic rose is a common charge in heraldry.
- The rose is one of the state and national symbols in Bulgaria because of the leading traditions in cultivation of Roses for production of Rose oil (especially in the so called Rose Valley near the town of Kazanlak).
- The Republic of Georgia's non-violent bid for freedom during its Rose Revolution;
- The red rose of Lancaster, and the white rose of York, from the Wars of the Roses period;
- The Rose-Cross is a symbol for the Rosicrucian;

- The rose in *The Dark Tower* has symbolic meaning in that series of Stephen King novels
- In some pagan mythologies, no undead or ghostly creatures (particularly vampires) may cross the path of a wild rose. It was thought that to place a wild rose on a coffin of a recently deceased person would prevent them from rising again.
- Since the earliest times, the rose has been an emblem of silence:
  - o In Greek Mythology, Eros presents a rose to the god of silence.
  - o In a Celtic folk legend, a wandering, screaming spirit was silenced by presenting the spirit with a wild rose every new moon.
- Roses were used in very early times as a very potent ingredient in love philters.
- In Rome it was often customary to bless roses on "Rose Sunday".
- In Scotland, if a white rose bloomed in autumn it was a token of an early marriage.
- The red rose, it is believed by some religions, cannot grow over a grave.
- According to some Biblical legends, the original rose growing in the Garden of Eden was white, but turned red as it blushed with shame upon Adam and Eve's fall from grace.
- Rose leaves thrown into a burning flame are said to give good luck.
- If a maiden had more than one lover, it is believed in one mythology, she should take rose leaves and write the names of her lovers upon them before casting them into the wind. The last leaf to reach the ground would bear the name of the lover whom she should marry.
- It is believed that if a rose bush were pruned on St. John's Eve, it would be guaranteed to bloom in the autumn.
- Love
- Social democracy and socialism
- The Virolai, a hymn to the Virgin of Montserrat begins with the words: *"Rosa d'abril, Morena de la serra..."* (April rose, dusky lady of the mountain chain...). Therefore this virgin is sometimes known as "Rosa d'abril".

Rose Cultivation

Roses are the most popular cut-flower world wide. From the early days, many new varieties have been selected aiming at an improved product in both cultivation and cut-flower aspects, like the improving yielding capacity the improved yielding capacity, the improved production, and the improved distribution of harvestable product throughout the year. Another important aspect which has a effect for price development of the product are the improved quality aspects, like flower and stem quality to guarantee an optimal vase life and satisfied customers. Very important in the evaluation of rose quality is the ability to complete the process from bud till ripe flower. The large range in colour, size, and fragrance, allow roses to be selected for each occasion and purpose. Seedlings are the starting point of breeding and the selection of our roses. The seedlings are then grown as cuttings on their own rooting system in an artificial substrate. The use of certain growing techniques, like for example the use of selected rootstock material, may even prove to increase the quality of the selections offered. Another development which already has been given a lot of attention and has been generally accepted is the change in cultivation practices. The concept is to control the plant as growing unit to a maximum. In that respect we try to improve the technical requirements of the system. Since this is done on some physiological characteristics of the rose. There is some information available about a number of the scientific background of the crop cultivation.

Growth Factors

The most important growth factors are light, temperature, and water.

Light

Light is the growth factor which influences both temperature and water, unfortunately it is also the factor which we can control the least. Too much light must be taken away by shading. In areas with a constant high (ir)radiation, shading often is done outside the glasshouse. This has a positive effect on temperature control. In areas with changing conditions, a movable screen inside the glasshouse is recommended. However, this often has negative side effects as temperature increase and through a reduced evaporation, a less active crop is created. As our control over the plant processes improves, shading becomes less necessary. Low light intensities can be improved by applying artificial lighting, which has only a limited effect. It does have some major drawbacks; cost increase, influence on the glasshouse temperature, and the shadow of the installation on the leaves when illumination is not required.

* Primarily light is necessary for assimilation (photosynthesis). Through chlorophyll in the leaves, light is transformed into sugars, a process also requiring CO_2 and water. This energy source is essential for growth and development of flowers, stems, leaves, and roots. The degree to which a plant can use the amount of light available depends on several factors, for example chlorophyll contents of the leaves, CO_2-supply, temperature, and humidity. Each of these factors can prove to be a limiting factor on plant growth.
* Another, more indirect, effect of light is the stimulation of vascular transport through the plant. As light raises the leave temperature, the leaves respond by opening the stomata to allow a higher evaporation in order to cool down. This process also increases the water transport, enabling nutrients, necessary for plant growth, to 3 be distributed through the plant.

Temperature

Temperature influences almost all processes of life. Therefore its relation to radiation should be established. This relation is mainly determined by the production of sugars. Too much light at a low temperature causes an unnecessary surplus of sugars. Even though this is undesired, the other way around (high temperature with low radiation) should definitely be avoided. Due to the high temperature the respiration process works at maximum capacity, while production of energy (photosynthesis) is maintained at a low level. A shortage of energy will be the result.

Temperature is often taken over a 24-hour period, when its relation with light is discussed. The differences in temperature during the day/night (light/dark) period have a major effect on plant growth. It is assumed that through a raise in temperature early in the night period, more energy is put in generative growth (flower development). Higher temperatures during the night period, compared to the day period, will stretch internodes and vice versa. Under poor light conditions this is a possibility to obtain the required 24-hour temperature without creating a weak and elongated crop.

Water

Water plays an important part in the assimilation process, in transport of sugars and nutrients, and in the cooling system of the plant. A plant consists for over 90% of water. The plant is kept upright by regulating water pressure (osmotic systems).

Evaporation is influenced by factors like light and temperature, and to a lesser extent by humidity. Stomata control evaporation, poor water supply will result in a closing of stomata with subsequently no gas exchange and no production of sugars (energy). Therefore it is important to keep a check on water uptake and evaporation. If the evaporation is larger than the uptake (wilting!), the most efficient control measure is light reduction. Water uptake can also be improved by applying directly at the plant (trickle irrigation), and by reducing the salt concentration in the water (E.C. or total nutrient value). When evaporation is low and extra light is not available, try to reduce humidity by increasing the temperature.

Besides a sub-optimal use of light, too little evaporation does not provide an optimal sugar and nutrient transport, resulting in e.g. calcium and magnesium deficiencies. It is important to stimulate evaporation on the total plant canopy. Therefore, vertical temperature differences should be avoided. This can easily be achieved by using a heating system at the base of the crop.

Substrate Cultivation

After the previous paragraphs on growth factors, it may have become clear that a complete control of environmental conditions will result in an optimal growing climate. However, complete control is not possible. Since substrate cultivation allows more control than cultivation in soil, it is often recommended. Soil quality (chemical and organic structure) and the possibility of soil sterilisation should be compared to the additional cost of setting up substrate cultivation before a decision is made on which practice is preferred.

Substrate Choices

Many different types of organic and anorganic substrates are presently used. For growing roses rockwool and coconut dust are most frequently used and preferred.

Since the development of pot cultivation, in stead of coconut dust a mixture of peat and rice chaff is becoming popular.

Requirements for Substrates

- Good air/water balance;
- Good draining and leaching capacity;
- pH neutral and containing none to very few nutrients;
- Stable structure, even in the long run (no or little decomposition);
- Offering sufficient support to the crop.

Rockwool

In choosing rockwool you could take a bread of 7,5 cm by 15 cm by 100 cm (height x width x length). The quality (durance) of the bread is an important characteristic to take into account when purchasing. Since roses are grown for several years on the same piece of rockwool, the quality should be maintained for a longer period. Often special types are available, even though these might be a bit more expensive, the effects of an optimally operating rooting system will become noticeable after some time. A hard rockwool bread has got sufficient air/water capacity. The alkaline nature of rockwool can be corrected by saturating the bread with a nutrient solution at a pH of 5.3-5.5. During cultivation the pH can be adjusted by applying ammonium nitrate in the nutrient mixture.

Coconut Dust

Coconut dust is a waste product from the coconut palm (Cocos nucifera). Presently it is originating from for example Sri Lanka. In fresh material a relatively high E.C.-value is measured (caused by e.g. sodium, potassium, or chloride).

This material is very easily to rinse causing these nutrients to leach. Rinsing often is done at the place of origin, using local rainwater during the monsoon period.

Still a chemical analysis should be completed since high salt concentrations cause serious growth reductions.

Coconut dust can hold both water and air. Remarkable is the ability to take up water even after severe drought.

Stability and decomposition of coconut dust are presently experienced to be good, even after several years no physiological changes have been recorded.

Coconut dust is pH neutral, at a value between 5,5 - 6. At fertilisation one should be aware of potassium present in the coconut dust. Therefore often a dose (1-2 kg/m) of calcium nitrate (CaNO3) is applied at the start of cultivation.

Initially a fertilisation schedule could be used replacing potassium nitrate (KNO3) by calcium nitrate. Please, check the iron (Fe) content regularly, it has been observed that coconut dust may retain iron temporarily. Water supply does not differ from rockwool, perhaps coconut dust requires a little bit more. Drain control is necessary. Coconut dust is presently sold in bags of 15-cm width by 1 metre length. Each bag contains 20 litres of coconut dust. A recent development is pot cultivation with 3 plants per pot of 10 litre.

Cultivation Systems

In the cultivation system recommended by Preesman Roses, the plants are grown 30-40 cm above the ground. This height is necessary to provide space for the bend-out stems and provide enough air circulation through the crop. This improves the evaporation and increases the amount of active foliage. If heating is required, it also allows the tubes to be places underneath the plants. To obtain the elevated flowerbed numerous systems can be developed and several are readily available on the market. When applying drains it also offers sufficient space for installation of gutters. If cultivation in pots is desired, various systems have been developed to allow optimal use of space (frameworks of iron). Since too many systems have been developed we can not deal with all of them specifically, but we like to draw your attention to the fact that each system should have sufficient draining capacity and all irrigation tubing should be level.

Characteristic of trickle irrigation is the frequent application of small amounts of nutrient solution. One should try to have equal amounts of liquid at each outlet. Uneven distributions will cause drought in some areas, and overfertilisation and wet feet in other regions of the glasshouse.

Setting up Cultivation

We recommend cultivation in a so-called 2-row system. Per compartment of 6,40 metre 6 rows of plants are planned.

The distance between the plants in one row varies between 15-20 cm.

This results at 6 to 8 plants per Metre Square (depending on cultivar and cultivation system). In the paths between the plants tubes are planned for heating. These pipes may also be used for transport of carriages. The maximum water temperature should be kept between 40-45 C. The piping of the trickle irrigation system is lying on the ground in between the rows of plants. To provide an optimal CO2-supply, on each flowerbed a tube for CO2 supply is recommended.

Cultivation

Start of Cultivation

At the start of the cultivation a glasshouse temperature of 20 - 22 C is recommended. The relative humidity should be high. Fertilisation is in these initial stages done with an extra 20% calcium nitrate. As soon as the first basal shoots have developed well, it should be returned to normal levels.

The water supply, i.e. the frequency and amount of trickle irrigation, is depending on the time of year and the development rate of the crop. Check the amount drain regularly (minimal 40 - 50%). At trickle irrigation we presume the water gift per drip is at least 50 - 60 cc per supply. When smaller gifts are used the irregularity between drips becomes too obvious. At saturation and the initial weeks of cultivation, the E.C.-

value of the nutrient solution should be lower than the E.C.-value of the rockwool blocks or the jiffy (peat) pots. This stimulates root growth and results in a rapidly established crop.

After this initial period the E.C.-value of the water gift varies between 1,6 - 2 depending on the time of year. The pH of the nutrient solution is maintain preferably at 5,3. If the pH-value in the drain water rises, some extra ammonium nitrate should be applied. If the substrate used is coconut dust, often less ammonium nitrate is required. Besides chemical analysis of the drain water, the coconut dust itself should be analysed itself.

Crop Maintenance

After planting shoots will quickly develop. Only after the flower bud becomes clearly visible the shoots are bend-out.

Since the plants grow about 40 cm above the ground, it is possible to bend down the stems deeply. Be careful not too break the shoots, the plant should remain capable of transporting sugars from these areas to the new developing shoots. The shoots should be bend down so the grafting place or, if a cutting is used, the old top of the cutting will become the top of the plant. The flower buds on these bend-out shoots have to be removed. This system allows the leaves to continue their production of energy, while the local main sincs of the sugars (buds) are taken away. The dominating primary shoot (apical dominance) is removed, causing the plant to respond by developing more basal buds. In the plant hormonal changes take place, which promote shoot development (balance cytokinins/auxins).

After cutting or bending out results in an increased cytokinin level, causing buds to break. The shoots formed are producing auxins, so restoring the hormonal balance in the plant.

Basal Shoots

Depending on the growth potential of a cultivar a number of basal shoots are formed. Per plant 2 to 3 well-formed shoots are allowed to continue growing, if more shoots were formed we recommend bending out of these shoots.

This way a plant has got the use of more active leaf canopy to supply enough energy for development of a heave crop with first quality flowers.

Harvest

At harvest it often was practice to cut back to the first so-called 5-leaf.We recommend cutting back to the just above the original cutting. The length of the remaining stem decides the number of shoots (flower stems) which will grow back. If too much (4-6 cm) stem is left, many shoots are formed of a poor quality. Therefore we advise to cut back to 1 cm. After 1 to 1 year the rose bush is cut back to approximately 10 cm above the original cutting, so creating a new top. Now again only 1-cm stem is left after harvest. Naturally differences do exist between cultivars, in case of doubt do not hesitate to contact us or your local agent.

Crop Maintenance

From all these points it becomes clear that leaves provide energy (source), while the flowers are the main users of this energy (zinc). Therefore it should be common practice save a maximum of inferior. The leaf area is maintained while the flower bud it taken out as soon as possible, since we would like to keep the original top of the cutting, or grafting place, the apex of the plant. The leaves now will be lower than the apex, while care is taken to avoid too many leaves from over lying. The shoots should be bend out carefully to avoid internal damage, this will block the transport of sugars and nutrients and so creates sub-optimal growing conditions. By maintaining differences in developing rate per shoot a crop will be more heterogeneous. This will allow a more equal distribution of energy in time.

Climate

Regulating the glasshouse climate should be done in order to keep the leaf canopy active. Especially when much light is available (sunny conditions), heating under the crop is recommended to prevent large temperature differences in the crop. When less light is available, heating both under as above the crop will be necessary. To remove excessive humidity from the glasshouse, the heating should be combined with ventilation. This depends on the difference between outside and glasshouse temperature and varies with the ventilation possibilities of the glasshouse.

In general with modern glasshouses we recommend ventilation at outside temperatures of 10 C and higher. This also is depending on the wind speed, under windy conditions the ventilation rate will be higher.

High Lines of the Grow System

- Plants are placed approximately 35 cm above the ground;
- Bend-out branches have more space, which improves leaf functions and so create more energy;
- By bending out branches the top of the plant stays at the original place and only basal shoots are formed;
- At harvest, cutting is done just above the original top, giving maximal stem length;
- More uniformity in production between plants gives a higher production per m of a higher quality.

Pests and Diseases

Roses are subject to several diseases. The main fungal diseases affecting the leaves are rose black spot (*Diplocarpon rosae*), rose rust (*Phragmidium mucronatum*), rose powdery mildew (*Sphaerotheca pannosa*) and rose downy mildew (*Peronospora sparsa*). Stems can be affected by several canker diseases, the most commonly seen of which is stem canker (*Leptosphaeria coniothyrium*). Diseases of the root zone include honey fungus (*Armillaria* spp.), verticillium wilt, and various species of phytophthora.

Fungal leaf diseases affecting roses are best prevented by choosing to grow cultivars and species known to be less susceptible to attack, and by using a preventative fungicidal spray program (rather than by trying to cure an infection after it emerges on the plant).

After disease is visible, spread can be minimized through pruning and the use of fungicides, although the actual infection cannot be reversed. Stem cankers are best treated by pruning out infection as soon as it is noticed. Root diseases are not usually possible to treat, once infection has occurred; the most practical line of defence is to ensure that growing conditions maximise plant health and thereby prevent infection. Phytophthora species are waterborne and therefore improving drainage and reducing waterlogging can help reduce infection.

The main pest affecting roses is the aphid (greenfly), which sucks the sap and weakens the plant. (Ladybugs are a predator of aphids and should be encouraged in the rose garden.) In areas where they are endemic Japanese beetles (*Popillia japonica*) take a heavy toll on rose flowers and foliage; rose blooms can also be destroyed by infestations of thrips (*Thysanoptera* spp). Roses are also used as food plants by the larvae of some Lepidoptera (butterfly and moth) species; see list of Lepidoptera that feed on roses. The spraying with insecticide of roses is often recommended but should be done with care to minimize the loss of beneficial insects.

Pests and Diseases of Roses

Roses (*Rosa sp.*) are susceptible to a number of pests, diseases and disorders. A large number of the problems affecting roses are seasonal and climatic. Certain varieties of roses are naturally more resistant or immune than others to certain pests and diseases. Cultivation requirements of individual rose species and cultivars, when observed, often assist in the prevention of certain pests, diseases and disorders.

Pests

Insects that affect roses are often considered pests.

- Aphids: (Greenfly) (Order HemipteraFamily Aphididae) *Macrosiphum rosae* – Likely to be found on new shoots and buds, aphids are soft bodied insects 1-2mm long. Often green but occasionally light-brown, and sometimes with wings, they may cover (in a colony) the complete growing tip of the plant. Aphids are most active in spring and summer and multiply at a prodigious rate feeding on the sap of the plant by piercing the plant cells via a proboscis. In large quantities they may seriously retard the growth of the plant and ruin buds. They are particularly damaging to the new shoots with subsequent damage to the emerging leaves which become malformed with much the same appearance as leaf-curl in peaches.

- Two-spotted mite (Spider-mites or Red spider mite) (Order Acari: Family Tetranychidae) *Tetranychus urticae* – Previously known as red-spider mite these arachnids prefer the underside of leaves and are difficult to see with an unaided eye. Evidence of their presence is silvering of leaves where the mites have destroyed individual leaf cells. Fine webbing and eggs on the undersides of leaves is further evidence of the presence of *Tetranychus urticae*.
- Thrips (Order Thysanoptera) – Thrips are slim-winged insects 1mm in length, resembling fine black slivers of wood. Preferring light-coloured blooms and often appearing in plague numbers flowers are often left looking bruised and lustreless.
- Rose Slugs (Rose Sawflies)- Sawflies are non-stinging wasps (Hymenoptera) in the suborder *Symphyta*, not flies (Diptera). They lay eggs in plant leaves or stems with a saw-like ovipositor. There are three species that commonly cause damage to wild or cultivated roses: The Bristly Roseslug (Cladius difformis) is found in Europe, Siberia, and many areas of North America. The larvae is pale green, up to 16 millimetres (0.6 in) long, and covered with hairlike bristles all over its body. It looks like a catepillar but that term strictly speaking only applies to the larvae of moths and butterflies. It skeletonizes the underside of leaves, with several generations per year. The European Roseslug (Endelomyia aethiops) is found in North America as well as Europe. The larvae is more slug-like (but not slimy), up to 13 millimetres (0.5 in), and skeletonizes the upper surface of leaves with only one generation per year. The Curled Roseslug (Allantus cinctus) larvae is pastel green on the back, marked on the thorax and abdomen with white dots, and up to 19 millimetres (0.7 in) long. It frequently coils up like a snake. After skeletonizing entire leaves except the main veins, it pupates in the pith of canes, with up to two generations per year in North America.
- Caterpillars (Order Lepidoptera)– The tortryx (tortrix) moth *Lozotaenia forsterana* is a prominent pest of roses, although not the sole pest. The caterpillars are green, up to 15mm long, and can be found boring into buds or within curled leaves. When disturbed the caterpillars move swiftly, dropping to the ground on a fine thread. Damage is chewn leaves and flowers and buds with "shot holes".
- Curculio beetles (Family Curculionidae)
- Japanese beetles (*Popillia japonica*) - This species, introduced to North America in 1912, is now an endemic pest in the eastern United States. Adult beetles emerge from the ground in early summer and join into swarms for four to six weeks, devouring blooms and skeletonizing foliage on roses and many other garden plants. Japanese beetles can be partially controlled, albeit slowly, by spreading milky spore bacillus on the lawn areas where the larvae live. The

popular pheromone-baited traps may do more harm than good by attracting beetles from a wide area.

- Scale insects (Order Hemiptera).

  Cottony cushion scale (Order Hemiptera : Family Coccoidea) *Icerya purchasi* – This scale infests twigs and branches. The mature female is oval in shape, reddish-brown with black hairs, 5 mm long. When mature the insect remains stationary and produces an egg sac in grooves, by extrusion, in the body which encases hundreds of red eggs. The insect causes little damage but produces copious honeydew (frass) that can cause damaging sooty mould.

  California red scale (Order Hemiptera : Family Coccoidea) *Aonidiella aurantii*– A hard scale, orange to orange-pink, the female covering being less than 1.5mm across. Often in plague numbers this scale infests upper surfaces of foliage causing yellowing, leaf fall, and twig and branch dieback. Serious infestations can cause plant death.

  Rose scale (Order Hemiptera : Family Coccoidea) *Aulacaspis rosae* – Mainly found on the stems and branches of the plant, lack of control will allow the pest to spread to flower stalks and petioles. At this point the plant would be stunted, spindly and with a white, flaky crust of scales on the bark. Female *Aulacaspis rosae* may live for 1 year and may lay 80 eggs each with several overlapping generations living within milliimetres of the original parent.

  Leaf Cutting Bee (Order Hymenoptera : Family Megachilidae) *Megachile spp.*– Leafcutter bees are 6-16mm long and mostly black with bands of light-coloured hair. They chew pieces from the edges of leaves. The pieces are regular in shape, circular or oval. Damage is not often significant.

  Nematodes (Eelworms)(Order Tylenchida: Family Heteroderidae)

  Root-knot nematode *Meloidogyne spp.*

- Root-knot nematode - symptoms of *Meloidogyne* infestation in roses is stunting, slow-growth, pale green leaves and wilting in mild weather.
- Rose chafer (Order Coleoptera : Family Scarabaeidae) *Macrodactylus subspinosus* – The rose chafer is common to North America, and emerges in late May to mid June. These beetles form aggregations and feed upon foliage, buds, flowers, and fruit of roses and other ornamentals.
- Metallic flea-beetles (Order Coleoptera: Family Chrysomelidae) *Altica* spp. – The small, shiny and metallic *Altica* beetles have thickened hindlegs adapted to jumping, similar to fleas. The insects are 3mm long and chew holes of irregular shapes in young leaves and buds. As the leaves enlarge so do the holes.

Diseases

Fungal, bacterial and viral diseases that affect roses:

Fungal Diseases

- Black Spot (Class Leotiomycetes: Family Helotiales) *Marssonina rosae* syn. *Diplocarpon rosae – Marssonina rosae* causes black spots on leaves. The spots, which may be as much as 12mm across, are generally circular and have an irregular edge often with a yellow halo. Leaves frequently turn yellow and fall early. Sometimes new leaves are produced, and these may also become affected. Continual defoliation will cause weakness, dieback or death of the plant. Some very susceptible species may have stems affected with a considerable reduction in plant vigour.
- Powdery Mildew *Oidium* sp. – *Oidium* produces a very fine, powdery coating on the surface of buds and leaves. Significant cases have stems and particularly thorns, infected. Attacks on young leaves and buds will cause deformity with retardation of growth. Infected buds will fail to open. The disease is likely in hot, humid weather, with fungal spores overwintering on the stems and fallen leaves.
- Downy mildew (Class Oomycetes : Family Peronosporaceae) *Peronospora sparsa – Peronospora* causes purple-red to dark-brown spots on the leaves with irregular margins, however, often angular. Stems, petioles and flower stalks can split and spotted with purple marks. Buds, sepals, petals and calyces can be affected and will present purple spots. New growth affected will be deformed. The disease is spread by wind.
- Rust *Phragmidium mucronatum* – Rose rust appears as yellow patches on the surface of leaves, with orange pustules of spores underneath the leaf. The fungus is spread by wind. Affected leaves fall prior to healthy ones and plants may be defoliated in serious infections.
- Anthracnose *Sphaceloma rosarum* – Spots caused by this fungus originate from a point where leaves are water soaked, usually unnoticeable at first, until they turn black with a very distinct defined edge. As the spots enlarge the centre becomes gray and may fall out resulting in a shot-hole appearance. Defoliation may occur but is often not serious.
- Grey mould (Class Leotiomycetes: Family Sclerotiniceae) *Botrytis cineria* – On roses grey mould is primarily a disease of the flowers and buds, leaves are infrequently attacked. Infected buds rot on the stem and infection may progress down the stem. On petals *botrytis cineria* produces pink rings.
- Verticillium wilt (Class Incertae sedis: Family Verticillium) *Verticillium dahliae*
- Sooty moulds *Alternaria* spp. – Sooty mould appears as black, dry powder on leaves similar to chimney soot. Many sooty moulds grow on the honeydew (frass) produced by sap-sucking insect such as aphids and soft scales. *Alternaria* does no direct damage to plants but surface cover of leaves will reduce the

plants capacity to photosynthesise and may create an unsatisfactory plant appearance.

- Canker *Leptosphaeria coniothyrium* and *Cryptosporella umbrina* – Cankers present as small yellowish or reddish spots on bark slowly increasing in size. *Leptosphaeria coniothyrium* turns brown, increases in size, and may eventually girdle the stem. The tissue within the infection begins to dry out and shrink, presenting a shriveled appearance. If the disease infects only part of the stem, growth above the canker will continue. If it girdles the stem, however, growth will cease and the stem will die.

Viral Diseases

- Rose mosaic – This disease is caused by a complex or viruses and is characterized by yellow patterns on the leaves. The patterns vary considerably, ranging between all-over fine blotches to patterns of lines in waves. The patterns may appear on a few or many leaves. Plants are infected by this virus at propagation using infected plant material.
- Rose Wilt – Rose wilt is a complex of viruses and is referred to as *dieback* in some areas. The disease can be spread by vectors such as aphids. Symptoms are variable and range from stunted growth to curled young leaves. The soft tissue symptoms are more evident in spring and new leaves will reflex towards their own petioles. The affected leaves are brittle and easily fall from the plant. Fully formed leaves will 'wilt' as if the plant were water stressed.
- Rose Rosette Disease - This disease is caused by a phytoplasm (formerly called mycoplasma-like organisms or MLO) that is transmitted by the eriophyid Rose Leaf Curl Mite Phyllocoptes fructiplilus, which inhabits the shoot tips and leaf petal bases of roses, as well as by grafting but not by seed or many other common vectors. Also called Witches' Broom of roses, it is fatal (average lifespan after infection 22 months) in the shrub Rosa multiflora, commonly found wild or as hedges (and considered a noxious weed in some places). It can also infect other rose species, such as garden rose climbers, miniatures, hybrid teas, floribundas, and antique varieties, but the disease is usually less severe. Roses are the only plant known to be susceptible. Symptoms include numerous bright red shoots with stunted, twisted stems and leaves. The leaves may appear red or pale yellow (chlorotic). The lateral shoots growing in different directions are produced 30 to 90 days after infection, and often are much larger in diameter than the canes from which they grew, with higher thorn density. The twisted growth may be mistaken for herbicide damage. There is no treatment for the disease, and control is limited to separating multiflora roses from garden varieties, and destroying infected plants. It is reported that the phytoplasm does not survive in the soil, but can survive in root fragments.

Bacterial Diseases

- Crown gall rot (Class Alpha Proteobacteria: Family Rhizobiaceae) *Agrobacterium rhizogenes* – This disease is characterized by large lumps at the base of the plant stem or on roots. Galls may appear higher on stems as the disease progresses. Galls are soft compared to surrounding plant tissues. The pathogenic bacteria enter the plant via a wound. If the disease affects the plant whilst it is young the plant may be affected to the degree where it will not produce blooms. All affected plants wilt readily and grow poorly.

Environmental Disorders

- Frost will destroy fresh growth causing stems and leaves to wilt, turn black and fall away from the plant. Timing pruning to promote growth after the threat of frost is a means to avoid frost damage.
- Salinity will present in roses as limp and light brown leaves with dry leaf margins. Soil may require testing to determine salinity levels. Symptoms will present if salinity is greater than 1200 parts per million.
- Herbicide Damage – Overspray or soil leaching of herbicidal sprays can present with several symptoms: :Prolonged exposure to overspray of *glyphosate* will cause yellow leaves and new leaves will be small and elongated. Hormone weed sprays (e.g. 24-D & 245-T) may cause grotesque new growth with thin twisted leaves and distorted buds. Plants may die in severe cases. Pre-emergent herbicides contacting the plants' root system via the soil will cause yellowing foliage. Effects of soil borne herbicide may take several years to clear.

6

Potential Cluster Regions: The Case of the Floriculture Industry

The floral industry is one of the major industries in many developing and underdeveloped countries. Floriculture as an industry began in the late 19th century in England, where flowers were grown on a large scale on the vast estates. The present day floral industry is a dynamic, global, fast-growing industry, which has achieved significant growth rates during the past few decades. In the 1950s, the global flower trade was less than US$3 billion. By 1992, it had grown to US$100 billion. In recent years, the floral industry has grown six percent annually, while the global trade volume in 2003 was US$101.84 billion.

The floral industry essentially consists of three major components: the growers, the wholesalers and the retailers whose businesses are quite intermingled. The recent trends are more towards eliminating the intermediaries, the wholesalers between the growers and the retailers, so that the flowers are made available at considerably low prices.

Transportation

Some flowers are sent packed flat in boxes. This enables large amounts of flowers to be packed in small spaces like aircraft holds. Other flowers cannot survive for long periods out of water such as orchids, gerberas (gerber daisies) and water lilies. These are either sent with their own sealed water container (called picks) on each stem end for more expensive or tropical flowers or are transported in buckets of water (This method of transport in water is often referred to as ["Procona]"). The latter method extends the life of flowers and reduces labour time as flowers are ready for sale, but obviously also reduces the amount of flowers that can be transported as they are much heavier than dry-packed flowers and hence air transportation charges are higher.

Flowers take a number of routes to the consumer, depending on where they are grown and how they are to be sold. Some growers cut and pack flowers at their nurseries, sending them directly out to the consumer by mail order. Some flowers are sent to packing companies, who grade the flowers and arrange them in bunches for sale to supermarkets or to deliver by mail order. Some flowers are graded and sleeved by the growers and sold at wholesale flower markets; the wholesalers then sell them on to florists who condition and arrange the flowers for the consumer.

Holland and the History of the Flower Industry

Traditionally, the center of flower production has been near their largest consumers: the developed world, where Japan, Western Europe and North America were both major producers and consumers. The major consumer markets being Germany (22 percent), the United States (15 percent), France (10 percent), the United Kingdom (10 percent), the Netherlands (9 percent), Japan (6 percent), Italy (5 percent), and Switzerland (5 percent).

Holland remains the center of production for the European floral market, as well as a major international supplier to other continents. The flower auction at Aalsmeer is the largest flower market in the world. Since the mid-1970s, the production and distribution of cut flowers in Netherlands has burgeoned. In 1995, Dutch growers produced over 8 billion blooms and the flower auctions collectively traded more than 5.4 billion guilders (about $3.2 billion) in cut flowers and potted plants, contributing over 4 billion guilders annually to the Dutch balance of trade.

New Flower Growing Centres

Experts believe that the production focus has moved from traditional growers to countries where the climates are better and production and labour costs are lower. This has resulted in a paradigm shift in the floral industry. The Netherlands, for instance, has already shifted attention from flower production to flower trading, though it plays an important role still in the development of floricultural genetics.

The new centres of production are typically developing countries like Colombia (second largest exporter in the world), Ecuador, Ethiopia, Kenya, and India. Other players in this global industry are Israel, South Africa, Australia, Thailand and Malaysia. New Zealand, due to its position in the Southern Hemisphere, is a common source for seasonal flowers that are typically unavailable in Europe and North America.

In Africa, Kenya is the largest exporter, supplying a large percentage of Europe's flowers, the industry there is represented by the Kenya Flower Council.

In South America, Colombia is the leading flower producer and exporter accounting for 59% of all flowers imported to The United States in 2006. The United States imports 79% of its flowers. Ecuador has become, in recent years, the leading South American rose producer and is well known throughout the world for its high quality, large headed roses due to the high altitude location of its rose farms INDIA.

Flower Industry Overview

Floriculture Item Sales at All Retail Outlets

2009	$32.5 billion
2008	$33.9 billion
2007	$33.6 billion
2006	$32.5 billion
2005	$31.1 billion
2004	$29.4 billion
2003	$28.2 billion
2002	$27.7 billion
2001	$26.7 billion
2000	$25.3 billion

U.S. Bureau of Economic Analysis, November 2010

Industry Segments (U.S.)

Retail florist shops (2008, latest available):	18,509
Estimated annual florist sales (2009):	$320,000
Floral Wholesalers:	900
Domestic Floriculture Growers (in top 15 states):	6,457
Fresh Flower Growers:	362
Potted Flowering Plant Growers:	1,259
Foliage:	879
Bedding/Garden Plants:	1,806
Fresh Greens:	171
Herbaceous Perennials:	1,416

Where Flowers Come From

Imports account for 68% of fresh flowers sold by dollar volume in the United States.

Top 6 Import Countries (by Value)

Colombia	66%
Ecuador	15%
Netherlands	6%
Costa Rica	3%
Mexico	3%
Canada	3%
Top 6 Fresh Flower Growing States	
California	75%
Washington	7%
Oregon	3%
New Jersey	3%
Hawaii	2%
Florida	2%

An Overview of the BC Floriculture Industry

The floriculture industry in BC is one of the most successful sectors of the horticultural industry. It is highly dynamic and competitive on a worldwide basis. There are no marketing boards, set prices, or quotas to protect local growers against domestic or world trade. Despite strong competition from overseas for the cut flower market, the industry continues to expand to meet the growing consumer demand for fresh cut flowers, potted flowering plants, and bedding plants.

BC floriculture production is technology and capital intensive. In order to compete against countries around the world, it has become essential for BC growers to quickly adopt new technology and the latest flower and plant varieties. The average sized greenhouse is a hectare in size and may cost between $200 to $400 per square meter to build or about $3.5 million in total capital expenditure. Even with state-of-the-art production systems, new operations can be obsolete within five years.

Systems optimization is central to greenhouses of this size. Computer systems are used to monitor and regulate the growing environment to ensure high quality product, rolling tables increase useable production area, supplemental lighting helps to offset our low winter light levels, and mechanization reduces labour costs. Re-using irrigation water is becoming a common practice in many greenhouses. It involves the collection, possible sterilization, and computer monitoring of the return nutrient solution used in plant nutrition. Growers look to re-using irrigation water as a method to reduce possible environmental impacts of greenhouse run-off and as a way to reduce fertilizer costs.

Structure of the Industry

World Floriculture Trade

One of the characteristics of the floriculture industry is its global competitiveness. BC cut flower growers not only compete with the rest of Canada, but with the rest of the world as well. Fast, efficient transportation systems tie together floricultural production on a worldwide basis. Over-production or changes in production in one country can affect markets in other countries. Countries with warmer climates and lower labour costs routinely sell their product to other countries with higher disposable incomes and production costs. A cash-and-carry floral bouquet in your local supermarket could be made of miniature spray carnations from Israel, spray chrysanthemums from Colombia, boxwood from Oregon, and statice from California. However, floriculture trade is bi-directional; BC grown product is shipped across North America and into Europe and the Far East.

Climate and low wages have made some South and Central American countries important producers of floriculture crops. Even when transportation costs are included, the lower production costs of these countries have displaced domestic production of cut carnations, roses, and chrysanthemums in North America and Europe. In fact, Colombia is the second largest cut flower exporter after the Netherlands. Flowers rank forth behind petroleum, coffee, and bananas in export earnings. In the United

States and Canada, growers are moving away from traditional flowers such as carnations and chrysanthemums to different specialty cut flowers such as gerbera, lizianthus, snapdragons, and alstroemeria that are usually more difficult to grow and to ship.

World wide trade in floriculture products was estimated at over US$7.9 billion in 2001. Cut flowers account for 50% of sales, plants were 41%, bulbs made up 9%, and cut foliage accounted for 9%. Seven countries export 73% of the value of the world's floriculture crops: the Netherlands, Columbia, Italy, Belgium, Denmark, the United States, and Ecuador.

The Netherlands continues to dominate the world floricultural industry; it is becoming the epicentre for world flower trading. It was estimated that in 2000 almost 50% of exported floriculture products came from the Netherlands, this figure includes crops that are grown domestically and crops that are imported, brokered, and then resold. Columbia was the second largest exporter at 7.5%, Italy, Belgium, Denmark, the United States, Ecuador, and Germany followed with approximately 3% each of exported products. Kenya, Costa Rica, Israel and Spain produced about 2% each. Major markets are Germany, the United States, Britain, France and the Netherlands. These five countries account for almost 70% of all imports of floriculture products.

The Flower Council of Holland estimated that in 2002, the Netherlands had over 8500 hectares in floriculture crops, including cut flowers, potted plants, propagation material, and bulbs, with a farm gate value of $5.4 billion. Most floricultural production is sold through one of the many floral auctions located throughout the country. In addition to domestic production, cut flowers arrive by the planeload from around the world and are then re-sold to other countries.

The United States is an important trading partner for Canada. The United States Department of Agriculture estimated that in 2002, there were 8,463 hectares in floriculture production with a wholesale value of over US$4.88 billion. California was the leading state with wholesale value of crops at US$962 million. Florida was the next largest with sales of US$877 million. These two states account for 38% of the total. The five largest producing states, California, Florida, Texas, Michigan, and Ohio, had combined sales of $2.62 billion or 54% of the total.

Table : United States Floriculture Sales 2002*

Product Type	***Total Sales (million US $)***
Bedding plants	2,280
Potted flowering plants	822
Potted foliage plants	663
Cut Flowers	410
Cut Foliage	111

* *Source: USDA NASS Floriculture Crops, 2002 Summary*

Bedding plants continue to dominate floriculture sales in the United States. In 2002 bedding plant sales were US$2.28 billion. It represented close to 50% of floriculture sales. Potted geraniums were the largest single bedding plant crop with sales of US$150 million. Impatiens flats were the second largest with sales of US$112 million. In potted flowering plants, poinsettias had sales of US$247 million or 30% of total sales. California was the largest cut flower grower with sales of US$279 million or 68% of the nation's total. The top three cut flowers grown were lilies valued at US$57.7 million, roses at US$56.2 million, and tulips at US$28.3 million. Florida was the largest producer of foliage, accounting for 69% of potted foliage plants and 78% of cut cultivated foliage.

Canada on the World Market

Exports

Canada experienced a positive net balance of trade in floriculture and nursery products for the seventh consecutive year in 2002, reaching a record $165.3 million. Statistics Canada estimated Canadian floriculture and nursery exports in 2002 were $524.3 million, up 80% from 1997. Nearly two-thirds of theses sales were in floriculture crops, mainly potted plants, cut flowers and greens. According to Industry Canada, the main export market for floriculture products is the United States, it generally purchases 89% of Canada's total floriculture exports. Canada supplies about 24% of the floriculture and nursery product imported into the United States. The Netherlands was the next largest Canadian floriculture export market, purchasing 7% of total exports, followed by Germany at 2%. Ontario is responsible for about 63% of export sales, BC ranks second with 22% and New Brunswick third with 11%. Mechanization, crop diversity, and an emphasis on quality enable the two ornamental industries to expand their export markets.

Imports

In 2002 Statistics Canada estimated imports of floriculture and nursery products to be $359 million, mainly in cut flowers and live plants, including cuttings. The largest suppliers were the United States, the Netherlands, Columbia, Ecuador and Mexico. The United States supplied 50% of all floriculture and nursery products imported into Canada. Cut flowers from South America made up 25% of the total imports and the Netherlands supplied 22%, mainly in bulbs.

Floriculture Across Canada

Across Canada, floriculture has enjoyed a steady growth rate over the years. Statistics Canada estimates the farm-gate sales of ornamental plants in 2002, at over $1.4 billion. Ontario accounts for over half of the industry with sales of $745 million. Most of Ontario's industry is located in the Niagara peninsula where growers are within one day's delivery time to over 100 million American consumers. BC is the second largest centre with sales of $312 million and Quebec is the third largest centre with sales of $147 million. Over 1,886 hectares of greenhouses were used in the production of ornamentals and vegetables. Ontario has 958 hectares, over half of the greenhouse acreage in Canada. British Columbia has 24% of the total greenhouse acreage with 461 hectares of greenhouses.

Table : The Canadian Greenhouse Industry, 2002*

Province	*Sales($ million)*	*Size + (hectares, plastic and glass)*
Newfoundland	8.7	6
Price Edward Island	2.8	3
Nova Scotia	29.3	29
New Brunswick	41.1	18
Quebec	147.3	255
Ontario	745.1	958
Manitoba	27.4	25
Saskatchewan	28.7	25
Alberta	78.9	107
British Columbia	312.3	461
Total	1,421.6	1,886

* *Source: Statistics Canada, Catalogue 22-202-XIB*

\+ *The size figures include vegetable and floriculture production.*

Floriculture within BC - Economic Contributions

In 2002, Statistics Canada estimated that ornamental flower and plant sales in BC were $312.3 million. The industry has increased its sales by 59% over the last five years. In the last ten years, it increased by more than 230%. The most significant growth year was in 1996. The years with a small increase in sales reflects the industry concentrating on retrofitting rather than expansion.

Table:. BC Greenhouse Production Input Costs, 2002*

Input	*Number or Cost*
Total Number of Employees	8,490
Gross Yearly Payroll ($)	120,699,000
Total Investment ($)	833,452,000
Total Purchases ($)	78,465,000
Total Fuel Costs ($)	49,891,000

* *Source: Statistics Canada, Catalogue 22-202*

The greenhouse vegetable and floriculture industry directly employs close to 8,500 people and has an annual payroll of almost $120.7 million. The capital investment of over $833 million shows the commitment of growers to the industry. The figure includes land, buildings, equipment and machinery at fair market value. In addition to the capital investments, growers purchased close to $78.5 million of production inputs, such as cuttings, seeds, and bulbs. Fuel costs approached $50 million. Growers have experienced extreme fluctuations in fuel costs since 2000 and volatile prices are expected to continue.

Crops

One of the strengths of the BC floriculture industry is its diverse range of crops. Several thousand plant genera are grown as cut flowers or potted plants. The main cropping strategies for potted plant production on a year-around basis are: a combination of bedding plants and potted flowering and foliage plants, bedding plants and poinsettias, or potted foliage and flowering plants.

Some smaller growers may grow bedding plants on a seasonal basis and only operate a few months of the year. The largest single crop is geraniums with over six million pots grown in 2002.

Table : BC Potted Plant Production (000 pots) *

Crop	*1998*	*1999*	*2000*	*2001*	*2002*
Azalea	1,215	322	-	702	559
Chrysanthemum	867	800	1,297	1,070	1,556
Foliage plants	3,226	3,390	3,732	3,681	4,671
Geraniums	4,918	6,371	8,207	7,529	7,200
Hanging Baskets– Foliage	141	211	172	346	281
Hanging Baskets– Bedding	463	547	-	741	950
Lilies	444	175	242	242	260
Poinsettias	2,156	1,662	1,801	2,285	3,082
Other	13,138	13,210	12,181	21,432	20 550

* *Source: Statistics Canada, Catalogue 22-202*

Other important potted crops not shown on the table are:

Flowering potted crops African violets, potted bulbs, cineraria, cyclamen, exacum, gerbera, gloxinia, hibiscus, hydrangea, kalanchoe, Rieger begonias, gloxinia, mini-rose, and orchids.

Bedding plants Alyssum, begonia, fuchsia, herbs, Impatiens, lobelia, marigold, pansy, petunia, primula, and snapdragons.

Cut flowers include a diverse range of greenhouse and field-grown annuals, perennials, bulbs, ornamental grasses, and woody cut flower crops. Producers grow a combination of these, or specialize in a single crop, either year-around or seasonally. One of the newer areas is perennial production where coolers and/or moving greenhouses are used to extend the natural flowering season of bulbs and perennials. Tulips are the main cut flower grown, with over 21 million stems being produced in 2002. Gerbera production has increased significantly in recent years and is now the second major crop. Cut rose production has decreased in response to a strong increase in South American rose imports.

Table: BC Cut Flower Production (000 stems) *

Crop	*1998*	*1999*	*2000*	*2001*	*2002*
Alstroemeria	5,392	4,097	6,237	11,130	6,569
Chrysanthemum – spray type	1,606	-	112	7,844	-
Daffodil (Narcissus)	4,495	2,830	2,801	2,641	3,087
Freesia	7,893	5,610	5,080	5,149	6,106
Gerbera	-	-	-	7,844	13,018
Iris	3,728	3,437	2,939	2,432	2,463
Lily	3,036	4,293	4,903	6,427	5,382
Rose – Standard	9,267	12,917	12,428	11,559	8,311
Snapdragon (Antirrhinum)	1,555	5,018	1.701	1,316	1,572
Tulip	14,749	11,963	15,745	17,102	21,217
Other	5,537	7,159	6,116	14,976	8,281

* *Source: Statistics Canada, Catalogue 22-202*

There is a large diversity in cut flower production, which is shown by the large numbers of cuts that fall into the "other" category. Some important crops not shown on the table are:

- Greenhouse production - Anemone, lisianthus, matthiola (stocks), orchids.
- Field production - Astilbes, dahlias, delphiniums, gladiolus, liatrus, phlox.

Markets

The market has changed significantly over the past 40 years. In the early 1960's, several large wholesalers dominated the market. In response to the almost "monopoly" powers of the wholesalers, a grower owned and operated co-operative, United Flower Growers' Co-operative Association (UFG) was formed in 1963.

Table : Ornamental Flower and Plant Sales in BC, 2002*

	Sales ($ 000)	*Percent of Total*
Retail Florists	33,928	11%
Domestic Wholesalers	76,561	25%
Mass Market Chain Stores	80,011	26%
Other Greenhouses 1	7,914	6%
Exported	25,462	8%
Direct to Public	27,205	9%
Other Channels	51,212	16%
Total Sales	312,293	

*Source: Statistics Canada, Catalogue 22-202

Over the years, the UFG has gradually become a key part of floriculture production and sales. The importance of corner stores in selling floral product grew along with the UFG. Corner stores developed and capitalized on the spontaneous cash-and-carry market. The next change in the market place was the chain grocery stores selling floral products. Their aim was to make flowers and plants a part of the weekly grocery basket. The most recent change is the role in floral sales by the mass-market retail stores like Wal-Mart and Home Depot. These changes in the floral market have resulted in a significant shift in the market share away from more traditional florists.

Wholesalers' long-time position as the main BC floral marketplace has recently been surpassed by mass market chain stores, including grocery stores and mass market retailers. Sales through the UFG are captured within other channels and accounted for approximately 20% of the product sold in the province. Traditional florists hold 11% of the market. Grocery stores and corner stores account for a large proportion of impulse purchases, while florists continue to dominate the more traditional markets like weddings and funerals. Exports vary with fluctuations in the Canadian dollar. Most exports go to the northwest region of United States.

United Flower Growers' Co-operative Association

The United Flower Growers, or the Auction, plays a significant role in the marketing of BC floral products. It was the first auction established outside of Europe. The next closest auctions are located in Toronto and San Diego. In 2002, it had sales of $55.5 million or about 18% of the market. The UFG has never had a drop in sales since its first year of operation when it had sales of $75,000.

An elected Board of Directors runs the UFG. Currently, it has almost 200 grower members and close to 1,000 customers. It operates out of a 1.8-hectare facility in south Burnaby. Growers pay a sales commission to sell their product through the auction, but they have reduced marketing and promotional costs. A major benefit of selling through the auction is that the UFG pays growers twice a month directly into growers' bank accounts. It eliminates the necessity for growers to grant extended credit to customers and, more importantly, it reduces the risk of bad debts.

Auction sales are done via three electronic, reverse style Dutch clocks. Prices start above the expected selling price and fall as the clock runs until the first buyer using an electronic keypad stops the clock. Unlike traditional auctions, it is the first bidder who gets the product. It is a fast, efficient selling method, with several thousand sales transactions per hour. Prices vary daily depending upon supply and demand. Selling product is a daily gamble since there aren't any guaranteed prices.

Mass-Market Chains

The three main groups of mass-marketers are supermarkets, department stores, and home renovation stores. These large chains attract thousands of customers every day and they have large floral departments. They capture a sizeable proportion of the holiday flowering plant and bedding plant markets. While these stores expose more people to floral materials, the downside is that they often use floral product, especially bedding plants, as loss leaders. Supermarkets and corner grocery stores compete for market share once held exclusively by florists. Most grocery stores have a floral department prominently displayed within the store where they encourage consumers to make flowers a part of their weekly grocery purchases.

Mass-merchandise stores such as Wal-Mart and Home Depot are quickly gaining market share. The large stores deal in large product volumes, so they usually purchase product from larger growers. In addition, they often require growers' product to have the appropriate universal product codes (UPC).

The large volumes and the UPC requirements mean that smaller growers are often not able to fill orders. Another factor is that these large mass-merchandisers operate on forward contact prices with their suppliers. They are not set-up to operate on the spot market such as the UFG. The UFG has responded to this change in the marketplace by creating a Direct Sales Desk that brokers forward contracts between its growers and customers.

Retail Greenhouse Sales

Many cut flower and smaller growers choose to sell the majority of their product through the UFG. It eliminates the need for individual marketing and promotion and allows growers to concentrate on growing a high quality product. However, most large growers and bedding plant growers treat the UFG as a customer, and although they a large proportion of their production to the UFG, they also sell directly to wholesalers, florists, mass-merchandisers, and garden centres. Another option is to sell directly to the public via farm stands or on-farm nurseries. Smaller bedding plant growers often sell direct to the public. It allows these growers to eliminate the middlemen and capture retail prices; however, it involves the extra expense of staffing and running the retail operation. Location plays an important role in the success of on-farm merchandising.

Industry Dynamics

Advantages

Expertise

By having the UFG doing most of the marketing, growers are able to concentrate on growing. Most growers have strong ties to Europe and keep in close touch with research information and production techniques and for variety updates. The high degree of expertise that is needed presents a challenge to new growers, whom not only need a large amount of capital, but also need the skills and knowledge of operating a greenhouse and growing the crops.

Modern Facilities

Lower light levels and high input costs make it essential for growers to have the most up-to-date greenhouses to remain competitive. Growers who don't build or retrofit every ten years lose their competitive edge.

Pacific Rim Trade

BC's close proximity to and the many direct flights to several Asian countries should make them our natural trading partners. In reality, they are difficult markets to penetrate due to volume and quality demands.

Co-operation

Growers frequently meet at the UFG to discuss prices, problems and new crops. Most growers realize that the higher the over-all standards of the industry, the better it is for everyone.

Specialization

By bringing many buyers and growers together at the UFG, growers are able to specialize rather than growing a wide range of crops in an attempt to supply the needs of every customer in the local market. Specialization allows for improved quality and decreased costs of production.

Weather

The Fraser Valley has some of the mildest weather in Canada, allowing lower heating costs and an earlier start to the bedding plant and outdoor cut flower seasons.

Disadvantages

Fewer Pest Controls

The United States and Europe have several new effective pesticides that are not registered in Canada. Canada is a small market and floriculture crops are often high-value. Consequently, large pesticide companies are hesitant to spend the extra time and money needed for Canadian registration. Often these new pesticides are less harmful to the environment and to other biological controls. The decreasing effectiveness and reduced number of registered pesticides places more emphasis on integrated pest management (IPM) practices.

High Costs

Most input costs are rising steadily each year, while floricultural product prices remain the same or decline. Capital costs can run anywhere from $80 to $400 per square metre. Gas and electricity accounts for between one fifth and one tenth of gross sales, however, erratic natural gas prices have seen this ratio fluctuate dramatically during recent years. Labour can also be one third of gross sales. BC land costs are some of the highest in Canada; land can cost up to $40,000 per hectare.

Skilled Labour Shortage

The industry has a shortage of middle management, but it is hoped that educational programs such as the Kwantlen University College's Horticulture program will help to alleviate the shortage. Growers also look to the Netherlands for trained managers.

Lack of Local Research

Research has traditionally focused on food crops rather than ornamentals. The UFG research fund and the Cecil Delworth Foundation, the research arm of Flowers Canada, are in place to stimulate local research.

Weather

Low light levels during the winter require that high quality crops need supplemental lighting.

Issues/Concerns

Environment

Environmental concerns are becoming a major issue for growers. Some of the main areas of concern in production are the re-use of irrigation water, reduced pesticide and fertilizer use, and reduced greenhouse run-off.

Decreasing Margins

Most input costs have risen steadily over the past five years and product prices have not kept up. To remain profitable growers have had to become more efficient in production and management. Growers are forced to adopt changes in technology and new plant varieties or risk falling behind, becoming less efficient and eventually being forced out of the market.

Pest Control

Concerns over pesticide use by the public and growers alike, along with pressures from pests, loss of approved pesticides, the high cost of pesticides, and pest resistance have caused growers to be increasingly receptive to alternative pest control methods.

Integrated pest management (IPM) is playing a larger role in greenhouse pest control. Many growers are now using biological or bio-rational control methods to supplement or replace existing pesticides.

Labour

Labour is an important element in production. Bedding plant and cut flower growers face labour costs of up to one third of gross sales. Unionization is a reality. Growers look to increased mechanization as a solution to their high labour requirements.

Rural-urban Conflicts

Rural-urban conflicts are a fact of life for most of agriculture in the Province. Some municipalities look upon floriculture as more of a factory production industry than agriculture. Growers need to take steps to reduce complaints over lights, noise, greenhouse run-off, pesticide use, and greenhouse appearances to avoid complaints from neighbours.

Regulations

Most municipalities have regulations concerning zoning, noise, lighting and the maximum coverage and set-back limits for greenhouses. There are also numerous environmental and safety requirements at provincial and federal levels that apply to the greenhouse industry in British Columbia.

Capital Costs

Top of the line greenhouse operations can cost in excess of $350 per square meter. This presents a barrier to entry for many potential growers. Field cut flowers and bedding plant production have much lower capital costs, so they are often entry level crops.

To the Future

The future of the BC floriculture industry continues to be optimistic. The "green industries" of floriculture and nursery are the fastest growing sectors of horticulture. Gardening is the number one leisure activity in North America and it has helped to expand bedding plant sales. The demand for floricultural products continues to rise as people realize the importance of plants and flowers in their daily lives.

Similar to other sectors of agriculture, there is a trend towards big, highly mechanized operations. The number of medium sized growers is expected to decline, as large growers become bigger in a move towards providing a more complete range of products for sale to large chain stores or other markets; at the same time, the numbers of smaller growers producing a more specialized product will likely increase.

Growing Pains: The Human Cost of Cut Flowers in British Supermarkets

Flower workers in developing countries like Kenya and Colombia risk their health for unsafe, insecure jobs supplying UK supermarkets. In our new report *Growing Pains* War on Want investigates the human cost of cut flowers in British supermarkets.

Supermarkets sell 70% of all the flowers bought in the UK the highest proportion in Europe. But the workers in Colombia and Kenya supplying those flowers to the supermarkets face low wages, health problems such as repetitive strain injuries and miscarriages through exposure to pesticides. Marks & Spencer, Tesco, Waitrose and Sainsbury's all source from one or both of these countries, and have enormous influence over flower producers and ultimately the health and safety of workers.

Many UK businesses have adopted voluntary standards for their suppliers, but these are still failing to protect the health and safety of workers. War on Want is therefore calling on the Secretary of State for Business, Innovation and Skills, Lord Mandelson, to urge the government to introduce binding legislation to enforce corporate accountability. This must give overseas workers the right of redress in the UK, i.e. the ability to seek compensation for damage to their health and loss of earnings as the result of actions of UK companies or their suppliers.

Flower workers depend on their jobs as vital sources of income. Help us ensure that workers supplying the UK have safe and secure jobs.

Mothers in Colombia growing flowers for sale in British supermarkets for Mothers Day face poverty wages, health problems such as repetitive strain injuries and risk miscarriages through exposure to pesticides. The charity War on Want today revealed these findings in new research that calls on trade secretary Alistair Darling to allow overseas workers hit by UK firms to seek redress in Britain. The call made in the run-up to Mothers Day on Sunday (18 March) has won support from celebrity flower grower Carol Klein, who presents the BBC television programme Gardeners' World.

The World Health Organisation recommends at least 24 hours between the time flowers are sprayed with pesticides to preserve their beauty and when employees re-enter the area. But, as British supermarkets press for completed orders, many workers are told to enter greenhouses to cut the flowers without protective clothing right after fumigation. The report says Colombian women, forced to breathe in toxic chemicals, have above-average rates of miscarriages and children born with birth defects. Exposure to pesticides often results in fainting spells, chronic asthma, eye and breathing troubles, skin complaints, allergies and headaches. And though Colombia's cut flower industry employs less than one in 100 Colombians, flower workers account for one in three cases of carpal tunnel syndrome, which causes pain and muscle weakness in the hand and forearm.

Colombian flower workers complain about their pay, health, working time up to 15 hours a day and firms' anti-union hostility at Florverde (Green Flowers)-certified farms which flower exporters claim ensure ethical treatment for workers and the environment. The research shows Colombian flower workers mostly single mothers earn just over £24 a week, less than half a living wage.

At one Florverde-standard farm, when mother of four Esperanza Botina asked to see a doctor about RSI, the company refused and instead demanded that she worked extra hours. Esperanza declined and was fired without compensation a common fate for flower workers who fall ill. Now, with severe arm pain, she cannot work on a flower farm and struggles to care for her family.

Esperanza says: "I am always short of money. The supervisors were very harsh. If anyone was sick, they would send you a memorandum or a sanction. Right now I feel like a cripple."

Claudia Luca Quevedeo is a pregnant single mother with two small daughters, owed back pay from another Florverde-approved farm whose rich owners suddenly filed for bankruptcy.

Claudia says: "What worries me the most is that I won?t have health insurance by the time the baby is born. And with so many diseases, I don?t know whether he will be healthy."

Yet Marks & Spencer has significantly increased orders from Florverde farms, Tesco buys from them and Asda is also showing increased interest in flowers with the Florverde standard.

The report cites similar poverty wages and unhealthy conditions for Kenyan flower employees. These workers, also mostly single mothers, toil long hours for little more than £5 a week well under half a living wage and cannot meet costs for food, housing, transport, education and health.

War on Want Chief Executive, Louise Richards, said: "Mothers in Colombia and Kenya are paying a dreadful price for growing the flowers many of us will buy for Mothers Day. British supermarkets make huge profits, but fail to ensure decent pay and conditions for these workers. Now the UK government must act to right these terrible wrongs."

Every Week in Britain, 32 Million People Shop in Supermarkets

Inside these cathedrals of modern consumer society, everything is carefully presented: meat trimmed and packaged, potatoes washed, tomatoes uniform in colour and size. Yet there is a darker side to the supermarket revolution, a far cry from the sanitised, neatly-presented world of the retail outlet.

This report is about the supply chains that link the products on supermarket shelves to the people in developing countries who produce them. The structure of supermarket supply chains has changed in recent years, as supermarkets in the increasingly concentrated retail sectors of rich countries cherry pick suppliers from increasingly open developing economies. This situation gives supermarkets more and more power in global markets, which they maximize by, among other things:

- Increasing suppliers' dependence, locking them into exclusive deals or taking up a large percentage of their production.
- Regularly 'delisting' suppliers, or threatening to delist them, to extract more favourable terms.
- Joining together in international buying groups to increase their buying power.

The supermarket squeeze extends as far as luxury items like cashew nuts, where women shellers in India told a familiar story of poverty wages, job insecurity and damaged health.

Pressure from UK supermarkets to reduce producer prices has contributed to an explosion in informal and even illegal processing operations, where a predominantly female workforce has few rights and little opportunity to demand a better deal from employers. Our research in India found workers processing cashew nuts for as little as 30p a day, damaging their health through exposure to corrosive oil during shelling and smoke released in the roasting process.

The growing power of big supermarkets in the UK market is both the product and the driver – a vicious circle – of a way of doing business that is made possible by the exploitation of women workers in developing countries. Organisations like Action Aid have highlighted the problems faced by workers in global supply chains for over a decade. Each time, the reaction from both industry and the government has been that supermarkets must be encouraged to voluntarily clean up their act.

With each year comes a new batch of evidence demonstrating yet more conclusively that this strategy is not working. The Ethical Trading Initiative (ETI), the best and most comprehensive of the voluntary initiatives, has not sufficiently galvanized its member companies into the kind of action needed to stop the rot. Other schemes have been even less successful. While there are isolated examples of good practice, it is increasingly clear that supermarkets will not deliver the widespread improvements that are needed unless they are given more of a push by government.

Trading with Supermarkets is Big Business for Producers in the Developing World

Poor countries earn over £7 million every day, or nearly £3 billion a year, from the food and clothes sold by British supermarkets. By and large, these goods are planted, picked, stitched, packed and cleaned by women, who make up 60% to 90% of the clothing and fresh produce workforce in developing countries.

Always Low Prices

At the beginning of the 1990s, the UK's 'big four' chains – Tesco, Asda, Sainsbury's and Safeway (now owned by Morrisons) – took 47% of British shoppers' spending on food in supermarkets. Since then, a series of mergers and acquisitions, technological innovations, and an explosion in the number and size of stores have raised this figure to 75%.

Tesco in particular has recorded breathtaking growth, doubling its market share from 15% to 31% between 2000 and 2006, and trebling its store count by opening 1,200 new shops during this period. An estimated 32 million people shop in UK supermarkets every week, and over £7 out of every £10 spent on groceries in Britain now goes into supermarket tills.

Indian Floriculture : A Bouquet of Options

In a nutshell, the floriculture industry comprises everything related to flowers – fresh cut and dried flowers or foliage, garlands, bouquets, even floral perfumes, medicines and edible extracts like *gulkand*. More than 53,000 acres of land are being used for flower cultivation – where both traditional flowers as well as ornamental or contemporary varieties are grown.

Unique Floriculture Projects

Maharashtra is currently number one in the production, consumption and export of flowers. The Maharashtra Industrial Development Corporation (MIDC) has come up with a unique project – first of its kind in the world. Land is being offered to individuals keen on starting their own Floriculture business. MIDC also plans to offer a wide range of facilities – from housing for entrepreneurs to professionally managed post-harvest services. This is also a great opportunity for service providers to offer quality facilities through nurseries and post-harvest centres. Since the project land comes under the Agriculture Export Zone declared by the Government of India, the business units will be entitled to applicable subsidies and other benefits. Other states are also realizing the commercial value of their local but exotic flowers like orchids and are focusing on the Floriculture industry.

Initiatives for Boosting Floriculture

The Government of India has urged the European Community to cut down the current import rate on Indian cut flowers to make them more economically attractive in the European markets. It's also considering duty exemption on imported material required for Green House and tissue culture laboratories. Apart from vying for an air freight subsidy for cut flower exports, it's also seeking exemption of EOUs from various custom duty regulations. Import duty on various floriculture-related materials has either been waived or reduced significantly. Floriculture units exporting half of their produce are allowed duty-free imports. Several schemes and training courses on floriculture are also being offered to farmers and entrepreneurs.

Australia Calling !

Australia remains an unexplored market for the fast growing Indian floriculture trade and industry. Its demand pattern, high production costs and the general dislike on the part of its youth to take up farming activities can prove a blessing for the Indian floriculture industry to push up the sales to Australia, where the demand for flowers is growing rapidly, says M B Naqvi after a visit to Melbourne, Sydney and Brisbane.

"Packaging is very important to get a good price.We can grow best quality flowers but if we don't know modern and safe way to deliver them anywhere, the whole effort goes waste. We should be very careful about packaging," emphasised Ali Bhai. It is high time India's expanding floriculture industry and trade, eyeing new export markets, took a good look at Australia. For the hi-tech producers of choice Indian cut flowers, Australia and New Zealand could offer a good potential. A visit to Australia, where annual flower consumption is going up by a good 8-10 per cent, was an eye-opener. It provided an insight into the status of floriculture in that country, its strengths and weaknesses, the trends and preferences of its customers, who have high living standards. The peak demand season there, the high labour and other costs, the general dislike of the younger generation to take to flower growing, or for that matter any farming work, are all favourable factors for India to emerge as an important supplier to the Australian market. Of course, the market there is highly competitive and qualityconscious. India has at present only a marginal presence in the Australian flower market.

As a Sydney-based flower importer, said Ali Bhai, CEO of Roses Galore, "India has vast scope for improving its presence in Australia. It can capitalize on it if it improves its brand image." Ali says he gets small quantities of Indian cut flowers from South India, mainly Bangalore. Initially, the image of Indian flowers was bad because the quality was not up to the mark. The suppliers had no knowledge of consumer trends, which keep changing. The current consumer preference in Australia, for instance, is for pastel shades in roses, with 5 cm bud size. If a supplier can conform to the these standards, his or her flower can fetch a good price, counselled Ali Bhai. Packaging is another aspect where the Indian exporters are lacking, he said. They have experience of importing gerbera from Pune and Anthurium from South India, but they have stopped their import because of packaging problems. "It is a competitive market here, and Indan suppliers could not face its challenges."

Countries exporting flowers to Australia include African nations that mainly supply roses and some traditional flowers, and Malyasia, which sends its colourful orchids via Singapore. Entrants to the Australian market include Columbia, Peru, some European countries and China, which has the advantage of cheap labour and geographical proximity. Because of the growing affluence and changing lifestyles, more and more people in Australia are spending money on flowers and plants. This is a new trend in this country, where in eighties domestic expenditure on cut flowers remained low by world standards.

A noted floriculture writer Helen Moody offers an explanation for it: "Most Australians live in houses surrounded by gardens, and this, combined with the Australian climate and outdoor lifestyle is not conducive to cut flower purchases." But all this is changing now. Australians are spending good amount of money on buying flowers and plants. The per capita flower consumption in the country does not lag much behind that in Europe and USA. The practice of giving flowers as gifts is becoming more popular; hospitals have special kiosks for sale of flowers; flowers are for all occasions marriage or social event, business meet or conference, birthday or funeral. That the domestic market is big and competitive is clear from the pages after pages of ads from florists in the yellow pages of telephone directories as also the on-line and website promotion campaigns. Leading international brands like Teleflora and Interflora have a strong presence in this country.

The supermarkets and prominent flower shops have a well organised delivery system, aided by trained designers and bouquet makers. The networking is strong among the leading florists, whether they are in Sydney, Melbourne, Brisbane, Leeds or Perth. Peak demand for flowers is in June to October, when Australia experiences it winter. This is also the time when its domestic production is low. Here is a situation that should make Indian suppliers sit up and think. June to September is normally a lean season in the Indian domestic market as well as export markets like Europe. Maybe, the Indian suppliers could turn this to their advantage and push up the sales to the Australian market.

Indian florist trade could do much more. For instance, a new trend among the florists in Australia is to get readymade bouquets in bulk. Bouquets have a good demand there and because of manpower problems the florist shops prefer to get them made-up, instead of engaging themselves in the time-consuming task of making them. This probably is one area where the Indian florists' innate skills and artistry can come into play. They can supply well-designed readymade bouquets in bulk to the leading florists in various Australian cities and towns. Bouquets fetch good money 25 dollars (Australian) to 100 dollars a piece, depending on the size. The demand for flowers and bouquets is growing not only in the major cities in this country but also in nearby towns and satellite townships where supplies do not match the demand. The Australian floriculture industry, which of late has started growing, remains focused on exotic or soft flowers for domestic market, says Helen Moody. Most of this production occurs in the eastern states. There are many small farms of less than 5 hectares, but a number of growers of significant scale do achieve "world's best practice." From the beginning of 1990's, says Moody, there has been a sense of enthusiasm and optimism about the Australian cut flower industry. Improved technology, a shift to green house production and greater professionalism contributed to increasing production and raising flower quality. While Australia imports flowers from various countries, it is also an exporter of cut flowers, mainly Australian native, or wild, flowers, to Japan and United States.

Western Australia is home to most spectacular wild flowers and the most sought after Waxflower and Kangaroo Paws and colourful Banksias originate there. According to a Department of Agriculture Bulletin, wild flower exports from Western Australia were worth 20 million Australian dollars in 2006. Queensland has large farms for production of widely popular cut flowers like roses and carnations. Good production centres have come up around Sydney. The growers, with a sprinkling of Asians, are mostly from countries like Italy and Greece. Professional Institutes have come up to train people in floral designing. Women's groups are active, spreading the new concepts of landscaping, garden techniques and plant care. To promote flower sales, special days are organised Daffodils Day, Red Rose Day and White Rose Day, adding to the long list of universally popular occasions like Valentine's Day, Mother's Day, Christmas, New Year and the like. NGOs receive government funds to carry on an awareness campaign on the use of flowers for different occasions. Meetings with florists in Australian cities threw up many new ideas and suggestions. Indian trade should not underestimate the potential of Australian and New Zealand markets. The demand-supply position there would warrant an in-depth study, on the basis of which an action plan can be drawn up. Indian floriculture industry would do well to promote direct interaction with growers, wholesalers, importers, bouquet makers and flower market committees. Australia is a growing market that needs to be tapped by those who can conform to quality specifications and changing tastes.

India has an astonishing floriculture production area of over 103, 000 hectares. Small farms using antiquated growing systems produce tons of marigold, udupi jasmine, ambur jasmine, china aster, chrysanthemum, crossandra, gerbera, gladioli, lilium, carnation, orchid, anthurium, and bird of paradise. The infrastructure is terrible, but many farmers load baskets of fresh flowers on trains or buses to go as far as 650 km to the city street markets, which open only from 6 a.m. to 9 a.m. in the morning. There they crouch on the street pavement to sell yellow and orange marigold garlands, lotus blossoms, a wide variety of greens. Buyers have very little interest in stem length. The flower heads are strung into garlands for temple decorations, weddings celebrations, 'karaga' festival decorations and 'thuroi' the hand-tied round bouquets. The domestic Indian market is growing at 25% per year in the country, 40% per year in New Delhi.

By the measure of purchasing power parity, India now ranks (after the US, China and Japan) as the fourth largest economy in the world, growing at 9.4% per annum in 2006, and according to The Economist. Within 25 years, India's GDP is forecasted to be, after China and the US, the third largest in the world. The potential of the booming Indian domestic market is enormous.

Recognition High

The 2nd International Flora Expo 2006 and an International Conference on Floriculture, entitled 'Advantage India' was held in New Delhi, September 8-10, 2006. It gave recognition to the importance of export floriculture to India. K S Money, Chairman of APEDA, the sponsoring organization, part of the Ministry of Commerce, welcomed thousands of guests. The honorable President of India, Dr. A P J Abdul Kalam, delivered an inspiring opening speech and made sure that one of the most lavish displays at the flower show was from the gardens of the Presidential residence. On the last day of the flower show, the doors were opened to an enthusiastic public.

Foreign delegations to Advantage India came from Europe, Japan and the Mideast and included the commercial manager of the VBA, and one of the directors of the NBV-UGA and the ex-manager of Flora Holland, together with a phalanx of Dutch vendors of equipment, young plants and consultancy within the framework of buyback arrangements.

A decade and a half have passed since Indian growers first exported flowers. Many of the initial rose farms failed due to micro-climate problems, lack of economies of scale, inappropriate technology and the fact that India is still far from being a logistics-driven economy. Conditions have substantially improved. APEDA has made important investments in the cold store, transport and airport facilities in seven major export cities. In India, relative returns on labour and capital are almost opposite to those of Northern Europe. Therefore Indian rose farms are not highly automated. In Pune, rose farms average 22 workers per hectare including grading and transport. This indicates by no means that Indian flower farms are primitive. The Indian farm managers and technicians are world class. They not only guide Indian floriculture with a sure hand, but predominate in floriculture farm management in Kenya and Ethiopia. They have designed excellent greenhouse structures appropriate to their climate and at one third the cost of structures imported from Europe. Authority and responsibility are shared in a collaborative management approach. Workers are very productive. Women figure importantly as flower farm owners and managers. Women labourers are an integral part of Indian agriculture. They work with deftness and alacrity. Their aspirations are high, not for themselves, but for their children, for the next generation, for the sons and daughters they hope to send to college to become titans in IT or some other modern endeavour.

Breeder's Rights Position

Some potential European importing clients decline to deal directly with Indian flower exporters because of Indian ambiguity about breeder's rights. Under pressure from the European breeders, the Indian position has softened. Though more than half of the rose production may still be illegal, an increasing number of the bigger, better farms have made peace with the rose breeders. India is not a signator of UPOV and will continue to hesitate to sign that agreement. Instead it passed PPVAFR in 2001, an act for the Protection of Plant Varieties and Farmers' Rights, which allows farmers to sow, exchange plants, share and sell farm produce without paying royalties as long as they do not label the variety. Dr. Malavika Dadlini of the Indian Agricultural Research Institute, a world recognized expert in seed science, defended this position, but she noted that ornamentals which are asexually propagated, with high value and exported into international commerce should accrue gains to the breeder. She even was categorical about the fact that multiplication of protected varieties was punishable by law. But she also pointed out that bio-piracy of (Indian) natural resources was illegal and unethical, a comment aimed at foreigners in the audience. This is important because much of the world's agricultural development work on new fruit and plant varieties is being done in India. The policy of the Indian government and the academicians is to distribute new types of planting material to domestic growers at cost without royalty protection or patents in an effort to swiftly increase agricultural productivity and quality. In floriculture, larger gower/exporters like Essar and Century are propagating top-grafts of roses of excellent quality, and in the same city, Pune, one of the world's most important tissue culture laboratories is multiplying large quantities of plant material. India is also an exporter of flower seed. Dr. A. P. S. Gill has been the guiding light behind Punjab production of around 90 tons of flower seed grown on 300-400 hectares of land, and exported to Holland, France, England, Germany, the US and Japan.

Product of Choice

But it is India's roses which characterize floriculture exports. Refined and delicate Indian roses have become the product of choice of some Japanese and European florists. Rose exports from India to Japan fall short of demand. Indian flower exporting is profitable enough to make it a viable, alternative diversification for big Indian companies currently in steel, shipping and construction, oil refining, cement and power. At present 90% of the rose exports come from farms that are subsidiaries of large firms. These companies now see that scaling up (plans of doubling the area in the next five years) will bring economic and marketing efficiencies to floriculture exports. Some of the local state governments are supporting this growth with incentives. Industrial parks for flower farms (provided with electricity, water, roads, packing facilities) have been created. One is located near Pune in the Maharashtra Industrial Development Corporation (MIDC). Other Agri Export Zones have been set up in Sikkim, Tamil Nadu, Uttaranchal and Karanataka. International funding for expansion is being offered by the Dutch government, including up to 50% financing for joint ventures is available from the Dutch Ministry of Economic affairs, the AVD. New entrants will probably continue to be large, well-financed Indian corporations. According to Navneesh Sharma, DGM-APEDA, leading corporations like Reliance, ITC, Tata Tea, Bharti Group/Field Fresh, Thapar Group, etc., have already responded to invitations and are planning investment into the sector.

7

Floriculture Industry in India

Floriculture, or flower farming, is a discipline of horticulture concerned with the cultivation of flowering and ornamental plants for gardens and for floristry, comprising the floral industry. The development plant breeding of new varieties is a major occupation of floriculturists.

Floriculture crops include bedding plants, flowering plants, foliage plants or houseplants, cut cultivated greens, and cut flowers. As distinguished from nursery crops, floriculture crops are generally herbaceous. Bedding and garden plants consist of young flowering plants (annuals and perennials) and vegetable plants.

Flowers are mainly for export. This business is growing in the world at around 6-10 per cent per annum. In 2007 the size of the industry was $80 billion.

In spite of a long tradition of Agriculture and Floriculture, India's share in the international market for these flowers is negligible. During the last ten years, taking advantage of the incentives offered by the Government, a number of Floriculture units were established in India for producing and exporting flowers to the developed countries. Most of them were located near Mumbai, Bangalore and Delhi and obtained the technical know-how from Dutch and Israeli Consultants.

Karnataka is the leader in floriculture, accounting for 75% of India's total flower production. The state has the highest area under modern cut flowers, and 40 flower growing and exporting units. The country's first and only flower auction centre is located in Karnataka.

In Karnataka, there are 18,000 hectares under floriculture cultivation. Karnataka is into floriculture for over 300 years. The Tigala community near Devanahalli and Chickaballapur are extremely good at growing flowers.

In 2003 The International Flower Auction Bangalore (IFAB), the operating company controlled by growers, has taken over the operations of the flower auction centre run by the State-owned Karnataka Agro Industries Corporation (KAIC).

An agreement to this effect was signed between the representatives of the South India Floriculture Association (SIFA) and Managing Director of KAIC, in presence of the Development Commissioner and Additional Chief Secretary of Karnataka, Mr Vijay Gore, who is also the chairman of IFAB. SIFA has a 51 per cent stake in IFAB, while the small growers hold 16 per cent.

Karuturi Networks, a little known Bangalore company, is close to acquiring the Netherlands-based Sher, the world's largest producer and supplier of roses, for about $50 million (Rs 220 crore) to emerge as the global leader in roses. Sher's greenhouses in the Netherlands, Kenya and Ethiopia produce 600 million roses annually.

Universities and institutes which cater to Floriculture are;

- University of Agricultural Sciences, Bangalore
- Indian Institute of Horticultural Research, Bangalore
- AP Agricultural University
- Haryana Agricultural University
- Punjab Agricultural University
- College of Agriculture, Maharashtra.

The high-value crops grown in green houses are cultivated under controlled conditions, protected from pests, diseases, wind and humidity. Popular flowers used in the modern floriculture that are in demand are roses, anthurium, gerbera, gladioli, orchids, carnations and bird of paradise grown in poly houses.

Flowers are also used for extracts for perfumes and natural dyes. Dry flowers are used for decoration. One can earn Rs 1 lakh per hectare per year from dry flowers. Nursery business growing plants in pots and in production of seeds are also popular.

Floriculture offers careers in production, marketing, export and research. Jobs range from cultivation or growing flowers, to seed production, dry seed production, seed production, marketing, decoration. National Horticulture Board helps one to establish a flori business. Agricultural and Processed Food Products Export Development Authority (APEDA) helps entrepreneurs with cold storage facilities and freight subsidies.

Area and Production of Horticulture Crops in Karnataka - 2003-04

Sl. No.	*Name of the Crop*	*Area in hectares*	*Production in tons.*	*Yield in Tons/hectare*	*Value in lakh Rupees*
	Commercial Flower Crops				
1	Aster	710	6857	10	526
2	Crossandra	1624	7669	5	1882
3	Marigold	4583	49777	11	2222
4	Jasmine	3451	20244	6	8265
5	Chrysanthamum	2964	36294	12	3931
6	Tube Rose				
a)	Single	925	8122	9	1557

b)	Double (lakh spikes)	121	996	8	265
7	Gladiolus (lakh spikes)	135	307	2	360
8	Rose (lakh flowers)	1530	3628	2	3041
9	Bird of Paradise (lakh flowers)	27	21	1	92
10	Golden Rod (lakh flowers)	27	85	3	39
11	Calla Lilly (lakh flowers)	9	14	2	3
	(For Sl. No. 12 to 16 Area in units)				
12	Orchids (lakh spikes)	2002	12	0.01	12
13	Carnation (lakh cut flowers)	38	186	5	1376
14	Anthurium (lakh cut flowers)	40	96	106	15
15	Gerbera (lakh cut flowers)	33	77	2	56
16	Statice				
a)	Free Flowering Variety (lakh cut flowers)	81	122	2	12
b)	Seasonal Variety (lakh cut flowers)	45	58	1	5
17	Others	2075	14343	7	928
Total - Commercial Flower Crops		18182	143306	8	24681

Floriculture is the art and knowledge of growing flowers to perfection. Being a branch of horticulture, it deals with the cultivation of flowers and ornamental crops from the time of planting to the time of harvesting. It also includes production of planting materials through seeds, cuttings, budding, grafting and marketing of flowers and flower produce.

In Meghalaya, flower lovers originally practiced floriculture as a hobby. Today, the growing demand has lead to flowers also being grown for commercial reasons. Meghalaya is known for a variety of rich flora and species of orchids that grow wild. The government has come out with a scheme to provide the growers with disease free planting material, organic/inorganic fertilizers, plant protection chemicals, garden tools and implements for a minimum area of 2000 square metres along with a package of practices for commercial production. A few of the recommended ornamental crops are orchids, chrysanthemums, gerberras, carnations, liliums, strelitzia reginae, gladiolus, asters, marigolds, statice, gomphrenas, helichyrsums, zinnias, roses and different kind of house plants.

Seasonal Flowers under Shillong Condition

In Tamil Nadu, flowers such as rose, malligai, chrysanthemum, marigold, jathi malli, marikolundu and carnations are grown. Edward Rose and Andhra Red Rose are two varieties of roses grown in the plains of the State. They need well-drained sandy loam soil.

Consumption of flowers in the southern States is much higher than in the northern States. Flowers are used by florists, flower rentals, perfume industries and as garlands. Centres where modern flowers are in demand are big cities like Mumbai, Pune, Bangalore, Chennai, Hyderabad, Delhi, Chandigarh, Lucknow and Calcutta. In the Andaman and Nicobar Islands there are around 2000 species of flowering plants, of which nearly 215 are endemic to these islands. These flowering plants include around 110 species of orchids. Here plantation crops like coconut are used as an intercrop for growing orchids. Coconut plants also provide shade to the orchids. Other flowers, ferns, cycads, succulents, bamboo and ornamental medicinal plants also grow naturally in forests.

There are about 2000 species of flowering plants, of which nearly 215 are endemic to these Islands (Rao and Srivastava, 1996).

Despite the rich natural base of floriculture, organized cultivation of flowers and ornamentals towards commercialization is yet to begin in Andaman and Nicobar Islands. Since agro-climatic conditions of these Islands are similar to those in South-East Asia like Thailand, Singapore, Malaysia and Indonesia, there is great scope for floriculture especially orchid cultivation out here in a commercial scale. Thailand alone exports orchid spikes worth $ 39 million, followed by Singapore with $ 16 million. Besides, the indigenous wild orchids are hardy, easy to propagate and grow, resistant to many diseases and insects, and therefore good germ-plasms for breeding purpose.

A few orchids of Andaman and Nicobar Islands—

- Arachne spp.
- Bulbophyllum lepidium
- Cymbidium aloifolium
- Dendrobium secundum
- Eria andamanica
- Geodorum densiflorum
- Malleola andamanica (endangered sp.)
- Rhynchostylis retusa
- Vanilla andamanica (endangered sp.).

Exotic Dendrobium hybrids, Vanda Miss Joaquim, Cattleya, Epidendrum etc. have also been successfully introduced here and performing well. Plantation crop like coconut can be used for growing orchids as intercrop, providing shade as well as standards (Sharma etal, 1998). Other than orchids, a number of flowers, ferns, cycads, succulents, bamboo and ornamental medicinal plants grow naturally in forests of the pristine Islands of Andaman and Nicobar, which need proper attention and perpetuation.

In Orissa, floriculture involves the growing of flowers like rose, crossandra, jasmine and marigold.

In Punjab, floriculture mainly includes the cultivation of different species of gladiolus. Village Majra, near Chandigarh and village Dappar near Patiala are main centres of gladiolus production in Punjab.

Flowers that are cultivated in for floricultural purposes are the rose, jasmine, chrysanthemum and crossandra. The main districts that produce these flowers are

Hyderabad, Rangareddy, Guntur, Prakasam, Kurnool, Cuddapah, Ananthapur and Chittoor.

Rose

Flowers are among the loveliest objects on this earth. Among them rose is queen. Rose is estimated to be more than thirty million years old and has figures in the myths and legends, is the poetry and music. The rose has been used in rituals, as a motif in architectural decoration, in beautiful designs in jewellery, pottery and embroidery. There are about 120 species of roses distributed in the North temperate and sub tropical parts of the both the hemispheres. As far as the garden roses of the world are concerned only right species namely Rosa chinensis, R.Damascena, R.fortida, R.gallica, R.igantea, R.moschata and R.wichuraiana have played an important role. Cultivated roses can be grouped under two categories those in cultivation before 1800 A.D. The period after 1800 A.D. was notable because of the influence of perpetually flowering types derived from R.Chinensis and R.gigantea from the Far East, crossing with new introductions gave rise to the important groups of the Noisettes and the Bourbons. Further, complex crossing ultimately gave rise to the great groups of the Tea roses and the Hybrid perpetuals. The Tea roses were prized for their continuous flowering habit while Hybrid perpetuals included varieties of magnificient size and rich perfume.

Hybrid Teas

The outstanding event in the rose world the origin of Hybrid Tea roses. The Hybrid Tea roses are the crosses between Hybrid perpetuals and Tea roses. Later they were crossed among themselves and also with some other types such as the 'Pernetianas', with the result that the present day Hybrid Teas are quite a complex group. 'Crimpson Glory'Mc Gredys Sunset', 'Peace 'and 'Super Star'are some well known examples. Some of the Hybrid Teas have 'sported'to produce climbing forms, e.g. 'Cripson Glory'climbing 'Peace'and climbing 'Shot Silk'.

Hybrid Perpetuals

Obtained by crossing the existing roses in Europe such as damasks and the gallicas with perpetually flowering introductions from China. These are characterised by tall growth, large full flowers and by rich fragrance e.g 'General Jacqueminot', 'Mrs John Laing'and 'Paul Neyron'.

Floribundas

The floribundas started with a group called Hybrid Polyanthus, Created by crossing the Hybrid Teas and dwarf Polyanthus.

Drarf Polyanthus

They produce enormous clusters of small flowers, and bloom for several months, e.g. 'Baby Faurex', 'Cameo', 'Chatillon Rose' and 'Echo'.

China Roses

Also called Bengal or monthly roses, 'common blush china', 'comtesse du cayla'. They are suitable for hedges and borders.

Miniatures

These are popularly known as Baby roses and suitable for edges, e.g. 'Baby Gold Star', 'Baby Masquerade', ' Coralin' and 'pixie'.

Planting

The roses are planted in circular pits about 60-90 cm across and 60-75 cm deep. The pits are to be prepared atleast a month before the date of planting. The best time for planting roses in the plains is during September or October. Roses can be propagated by seeds, cuttings, layers and by budding. Seed propagation is done in the production of new varieties. Budding is the best method for commercial propagation.

Budding

When a rose plant comes into flower small swellings are visible between the stem and stalk of the leaf. Select the eyes which are plump and not started to elongate and grow. Cut off the portions of the branch with the eyes selected is called as 'bud wood' with a budding knife remove a shield shaped piece of the bark and the eye. The Root stock should be kept ready by cutting the branches and side shoots which are not required. Make a sharp horizontal cut at a suitable height in the stock and then make a vertical ' T ' shaped cut in the stem of the stock. The 'eye 'should point towards the top and is tied with fibre.

After Care

Stoch sprouts should be frequently removed. Irrigations should not result in water logging. Mulching with leaf should be washed by spraying water on the foliage. A basketful of FYM is sufficient for one year. A small quantity of iron sulphate may be applied @ 50-75 g/shrub by dissolving in water.

Root Stocks

R.multiflora, Braiar and Edward are best suited. Multiflora is suited for hill stations. Braiar for medium and Edward for both.

Pruning

Pruning is does to produce best possible flowers and to maintain good shape and keep plant healthy by removing dead parts. Three to four months after planting the rose plant is first pruned. Four branches placed in four directions are chosen and these are pruned back to an outer bud leaving two to four buds on the stump. All other branches are thinned out. If the plant has only one or two weak shoots they are to be cut back to two or three buds. The first fortnight of October is usually considered to be the best time for pruning. Some hybrid Teas requires severe pruning.

In Tea roses strong shoots are pruned to 2/3 of their length. Climbing roses needs no pruning. Pruning is to produce the required number of new shoots are required number of new shoots because rose bears flowers terminally on current season wood. Too many shoots on plants will reduce the size and quantity of flowering of flowering. Limiting the number of flowers promotes flower size. Depending on the variety and severity pruning roses takes about 35 to 60 days from pruning to flowering.

Manures and Fertilizers

The best time to add organic manure is at the time of pruning. FYM, leaf-mould and oil cakes are good sources of nitrogen. It is better to apply fertilizers in a mixture such as 'roxe mix'. Rose mix can be prepared as:

Neem cake	——	5 kg
Bonemeal	——	5kg
Ammophos	——	2kg
Sulphate of Ammonia	——	1kg
Superphosphate	——	2kg
Potassium sulphate	——	1kg

100 g of this mixture can be used per plant. The chelated compounds of iron, magnesium and manganese are available in the market under names like 'Sequestrene Plus'which are helpful in bringing out the full colour of flowers. In general for each rose plant urea 20-30 g. Superphosphate 30-50 g, Potassium Sulphate 20-30 g and Cakes 1/2 kg to 1 kg, in two to three split doses is recommended.

Rose Products

Gulkand

It is prepared by pounding together a mixture of rose petals and white sugar in equal proportions. It acts as a tonic and laxative. Edward rose is mostly used for this purpose.

Pankhuri

Dried rose petals called Pankhuri are used during the hot weather for preparing cool drinks. Roses are used for many Purposes such as rose water, rose vinegar. Rose develop fruits called 'Hips' and they are good source of Vitamin C.

Plant Protection

Pests

Aphids

They occur in clusters on tender shoots, buds and flowers and sucks the plant sap. The affected flowers and buds should lightly dusted in the mornings or evenings with 9.2% Pyro dust.

Jassids

They suck the sap from the plant the parts. The leaves lose their natural colour and become dull whitish yellow and subsequently dries up and the damaged plants have a sickly appearance and these are controlled by spraying with 0.03% parathion.

Leaf Cutting Bees

They cause damage to the rose leaves by cutting neat, circular or oval patches from the leaf margin and are controlled by applying 5% Carbaryl.

Red Scale

It appears as reddish brown waxy scale like pox marks on tender shoots and stems and the affected portions dry up, for controlling it dip a cotton swap in 0.1 malathion solution and rub the affected parts. If it is on large scale, spraying of 0.1% methyl parathion or malathion should be done.

Chafer Beetle

Rose chafer beetle cut away the leaves. In severe cases the plants reduced to a more skeleton The damage of the beetle may be controlled by spraying 0.5% malalthion @ 2ml/l or 0.05% methyl parathion @ 1ml/l or endosulfan 0.07% @ 2ml/l.

Diseases

Powdery Mildew

This disease effects the tender leaves and curl up exposing the lower surface exhibiting the white powdery growth. Infected buds do not open. This can be controlled by spraying sulphur compound i.e. thiovit 2g/l of water at weekly intervals during winter season.

Black Spot

This is also very common in India and it can be distinguished by the dark brown spots which occur on either side of leaves. In severe cases leaves drop and growth ceases. This is controlled by spraying copper fungicide like Blitox, Dithane M-45 or Captan at the rate of 2 g in 1 lit of water at weekly intervals.

Rust

Yellowish to black swellings called pustules are produced on the leaves, petioles and stem imparting a rusty appearance on shoots. This can be controlled by foliar slpray with 1 kg of 'Ferbam' plus 1 kg of wettable sulphur in 500 lt of water with Sandovit as a spreader.

Die-back

Affected plants show blackening as well as death of the twig from the cut an downwadrs. Often the entire plant is killed. It is controlled by spraying Mercury fungicide known as Santar which prevents fungal infection. The cool temperate climate of Jammu and Kashmir makes it the ideal place for growing flowers. Some of the flowers cultivated here are wisterias, tulips, carnations, dahlias and lilies.

Flower Power

Flower power was a slogan used by the American counterculture movement during the late 1960s and early 1970s as a symbol of passive resistance and non-violence ideology. It is rooted in the opposition movement to the Vietnam War. The expression was coined by the American Beat poet Allen Ginsberg in 1965 as a means to transform war protests into peaceful affirmative spectacles. Hippies embraced the symbolism by dressing in clothing with embroidered flowers and vibrant colours, wearing flowers in their hair, and distributing flowers to the public, becoming known as flower children. The term later became generalized as a modern reference to the hippie movement and a culture of drugs, psychedelic music, psychedelic art and social permissiveness.

Origin

Flower Power originated in Berkeley, California as a symbolic action of protest against the Vietnam War. In his November 1965 essay titled *How to Make a March/ Spectacle*, Ginsberg advocated that protesters should be provided with "masses of flowers" to hand out to policemen, press, politicians and spectators. The use of props like flowers, toys, flags, candy and music were meant to turn anti-war rallies into a form of street theater thereby reducing the fear, anger and threat that is inherent within protests. In particular, Ginsberg wanted to counter the "specter" of the Hells Angels motorcycle gang who supported the war and had threatened to violently disrupt planned anti-war demonstrations at the University of California, Berkeley. Using Ginsberg's methods, the protest received positive attention and the use of "flower power" became an integral symbol in the counter-culture movement.

Movement

> *"The cry of 'Flower Power' echoes through the land. We shall not wilt. Let a thousand flowers bloom."*
>
> — Abbie Hoffman, *Workshop in Nonviolence*, May 1967

By late 1966, the Flower Power method of guerilla theater had spread from California to other parts of the United States. The Bread and Puppet Theater in New York City staged numerous protests which included handing out balloons and flowers with their anti-war literature. *Workshop in Nonviolence* (WIN), a magazine published by New York activists, encouraged the use of Flower Power. In May 1967, Abbie Hoffman organized the Flower Brigade as an official contingent of a New York City parade honoring the soldiers in Vietnam. News coverage captured Flower Brigade participants, who carried flowers, flags and pink posters imprinted with LOVE, being attacked and beaten by bystanders. In response to the violence, Hoffman wrote in WIN magazine, "Plans are being made to mine the East River with daffodils. Dandelion chains are being wrapped around induction centres.... The cry of 'Flower Power' echoes through the land. We shall not wilt."

On the following Sunday in May 1967, WIN activists declared the Armed Forces Day as "Flower Power Day" and held a rally in Central Park to counter the traditional parade. Turnout was low and, according to Hoffman, the rally was ineffective because guerilla theater needed to be more confrontational.

In October 1967, Hoffman and Jerry Rubin helped organize the March on the Pentagon using Flower Power concepts to create a theatrical spectacle. The idea included a call for marchers to attempt to levitate the Pentagon. When the marchers faced off against more than 2500 Army national guard troops forming a human barricade in front of the Pentagon, demonstrators placed flowers in the soldier's rifle barrels.

Flower Power, a Pulitzer-nominated photograph by Washington Star photographer Bernie Boston, became a classic image of the Vietnam War–era protests. The photo, taken on October 21, 1967 at the "March on the Pentagon", showed a young, long-haired man in a turtleneck sweater, placing carnations into the rifle barrels of military policemen. (The young man in the photo was not named by Boston and his identity is disputed. He was most commonly identified as George Edgerly Harris III, an 18-year-old actor from New York who later performed in San Francisco under the stage name of Hibiscus. The young man also has been identified by Paul Krassner as Youth International Party organizer, "Super-Joel" Tornabene.)

Legacy

The iconic center of the Flower Power movement was the Haight Ashbury district in San Francisco, California. By the mid-1960s, the area, marked by the intersection of Haight and Ashbury Streets, had become a focal point for psychedelic rock music. Musicians and bands like Jefferson Airplane, the Grateful Dead and Janis Joplin all lived a short distance from the famous intersection. During the 1967 Summer of Love, thousands of hippies gathered there, popularized by hit songs such as *San Francisco (Be Sure to Wear Flowers in Your Hair)*. A July 7, 1967, *Time* magazine cover story on "The Hippies: Philosophy of a Subculture," and an August CBS News television report on "The Hippie Temptation" as well as other major media interest exposed the hippie subculture to national attention and popularized the Flower Power movement across the country and around the world.

Flower Power Makes Tropics Cooler, Wetter

The world is a cooler, wetter place because of flowering plants, according to new climate simulation results published in the journal *Proceedings of the Royal Society B*. The effect is especially pronounced in the Amazon basin, where replacing flowering plants with non-flowering varieties would result in an 80 percent decrease in the area covered by ever-wet rainforest.

The simulations demonstrate the importance of flowering-plant physiology to climate regulation in ever-wet rainforest, regions where the dry season is short or non-existent, and where biodiversity is greatest. "The vein density of leaves within the flowering plants is much, much higher than all other plants," said the study's lead author, C. Kevin Boyce, Associate Professor in Geophysical Sciences at the University of Chicago. "That actually matters physiologically for both taking in carbon dioxide from the atmosphere for photosynthesis and also the loss of water, which is transpiration. The two necessarily go together. You can't take in CO_2 without losing water."

This higher vein density in the leaves means that flowering plants are highly efficient at transpiring water from the soil back into the sky, where it can return to Earth as rain.

"That whole recycling process is dependent upon transpiration, and transpiration would have been much, much lower in the absence of flowering plants," Boyce said. "We can know that because no leaves throughout the fossil record approach the vein densities seen in flowering plant leaves."

For most of biological history there were no flowering plants — known scientifically as angiosperms. They evolved about 120 million years ago, during the Cretaceous Period, and took another 20 million years to become prevalent. Flowering species were latecomers to the world of vascular plants, a group that includes ferns, club mosses and confers. But angiosperms now enjoy a position of world domination among plants.

"They're basically everywhere and everything, unless you're talking about high altitudes and very high latitudes," Boyce said.

Dinosaurs walked the Earth when flowering plants evolved, and various studies have attempted to link the dinosaurs' extinction or at least their evolutionary paths to flowering plant evolution. "Those efforts are always very fuzzy, and none have gained much traction," Boyce said.

Boyce and Lee are, nevertheless, working toward simulating the climatic impact of flowering plant evolution in the prehistoric world. But simulating the Cretaceous Earth would be a complex undertaking because the planet was warmer, the continents sat in different alignments and carbon- dioxide concentrations were different.

"The world now is really very different from the world 120 million years ago," Boyce said.

Building the Supercomputer Simulation

So as a first step, Boyce and co-author with Jung-Eun Lee, Postdoctoral Scholar in Geophysical Sciences at UChicago, examined the role of flowering plants in the modern world. Lee, an atmospheric scientist, adapted the National Center for Atmospheric Research Community Climate Model for the study.

Driven by more than one million lines of code, the simulations computed air motion over the entire globe at a resolution of 300 square kilometers (approximately 116 square miles). Lee ran the simulations on a supercomputer at the National Energy Research Scientific Computing Center in Berkeley, Calif.

"The motion of air is dependent on temperature distribution, and the temperature distribution is dependent on how heat is distributed," Lee said. "Evapo-transpiration is very important to solve this equation. That's why we have plants in the model."

The simulations showed the importance of flowering plants to water recycling. Rain falls, plants drink it up and pass most of it out of their leaves and back into the sky.

In the simulations, replacing flowering plants with non-flowering plants in eastern North America reduced rainfall by up to 40 percent. The same replacement in the Amazon basin delayed onset of the monsoon from Oct. 26 to Jan. 10.

"Rainforest deforestation has long been shown to have a somewhat similar effect," Boyce said. Transpiration drops along with loss of rainforest, "and you actually lose rainfall because of it."

Studies in recent decades have suggested a link between the diversity of organisms of all types, flowering plants included, to the abundance or rainfall and the vastness of tropical forests. Flowering plants, it seems, foster and perpetuate their own diversity, and simultaneously bolster the diversity of animals and other plants generally. Indeed, multiple lineages of plants and animals flourished shortly after flowering plants began dominating tropical ecosystems.

Tulip

The tulip is a perennial, bulbous plant with showy flowers in the genus *Tulipa*, which comprises 109 species and belongs to the family Liliaceae. The genus's native range extends from as far west as Southern Europe, North Africa, Anatolia, and Iran to the Northwest of China. The tulip's centre of diversity is in the Pamir, Hindu Kush, and Tien Shan mountains. A number of species and many hybrid cultivars are grown in gardens, as potted plants, or to display as fresh-cut flowers. Most cultivars of tulip are derived from *Tulipa gesneriana.*

Description

Tulips are spring-blooming perennials that grow from bulbs. Depending on the species, tulip plants can grow as short as 4 inches (10 cm) or as high as 28 inches (71 cm). The tulip's large flowers usually bloom on scapes or subscapose stems that lack bracts. Most tulips produce only one flower per stem, but a few species have up to four flowers. The colourful and attractive, cup-shaped flower has three petals and three sepals, which are often termed tepals because they are nearly identical. These six tepals are often marked near the bases with darker colourings. Tulip flowers come in a wide variety of colours, except pure blue (several tulips with "blue" in the name have a faint violet hue).

The flowers have six distinct, basifixed stamens with filaments shorter than the tepals. Each stigma of the flower has three distinct lobes, and the ovaries are superior with three chambers. The tulip's fruit is a capsule with a leathery covering and an ellipsoid to subglobose shape. Each capsule contains numerous flat, disc-shaped seeds in two rows per chamber. These light to dark brown seeds have very thin seed coats and endosperm that does not normally fill the entire seed.

Tulip stems can have up to a few leaves, with larger species tending to have multiple leaves and smaller species having none. Plants typically have 2 to 6 leaves, with some species having up to 12 leaves. The tulip's foliage is strap-shaped with a waxy coating and alternately arranged on the stem. These fleshy blades have a light to medium green colour and are linear to oblong in shape. Tulip bulbs usually grow on the ends of stolons, and the bulbs' tunicate (dry and papery) coverings may or may not have hairs.

Origin of the Name

Although tulips are often associated with The Netherlands, commercial cultivation of the flower began in the Ottoman Empire. The tulip, or *lale* as it is also called in Iran and Turkey, is a flower indigenous to a vast area encompassing parts of Africa, Asia, and Europe. The word *tulip*, which earlier appeared in English in forms such as *tulipa* or *tulipant*, entered the language by way of French *tulipe* and its obsolete form *tulipan* or by way of Modern Latin *tulîpa*, from Ottoman Turkish *tülbend* ("muslin" or "gauze"), and is ultimately derived from Persian *dulband* ("turban").

Cultivation

Tulips are indigenous to mountainous areas with temperate climates and need a period of cool dormancy. They do best in climates with long, cool springs and early summers. Although perennials, they are often replanted annually in warmer areas of the world.

The bulbs are typically planted around late summer and fall in well-drained soils, normally from 4 inches (10 cm) to 8 inches (20 cm) deep, depending on the type planted. In parts of the world that do not have long cool springs and early summers though, the bulbs are often planted up to 12 inches (300 mm) deep. This provides some protection from the heat of summer and tends to force the plants to regenerate one large bulb each year instead of many smaller non-blooming ones. This can extend the life of the plant in warmer areas by a few years, but it does not stave off degradation in bulb size and the eventual death of the plant.

Propagation

Tulips can be propagated through offsets, seeds or micropropagation. Offsets and tissue culture methods are means of asexual propagation for producing genetic clones of the parent plant, which maintains cultivar integrity. Seed-raised plants show greater variation, and seeds are most often used to propagate species and subspecies or to create new hybrids. Many tulip species can cross-pollinate with each other, and when wild tulip populations overlap with other species or subspecies, they often hybridize and create mixed populations. Most tulip cultivars are complex hybrids and actually sterile. Those plants that do produce seeds most often have offspring dissimilar to the parents.

Growing salable tulips from offsets requires a year or more of growth before plants are large enough to flower. Tulips grown from seeds often need five to eight years of growth before plants are flowering size. Commercial growers usually harvest the tulip bulbs in late summer and grade them into sizes; bulbs large enough to flower are sorted and sold, while smaller bulbs are sorted into sizes and replanted. Holland is the main producer of commercially sold plants, producing as many as 3 billion bulbs annually.

Introduction to Western Europe

Although it is unknown who first brought the tulip to Northwestern Europe, the most widely accepted story is that it was Oghier Ghislain de Busbecq, an ambassador for Ferdinand I of Germany to Suleyman the Magnificent of the Ottoman Empire. He remarked in a letter that he saw "an abundance of flowers everywhere; Narcissus, hyacinths and those in Turkish called Lale, much to our astonishment because it was almost midwinter, a season unfriendly to flowers." However, in 1559, an account by Conrad Gessner described tulips flowering in Augsburg, Bavaria in the garden of Councillor Herwart. Due to the nature of the tulip's growing cycle, tulip bulbs are generally removed from the ground in June and must be replanted by September to endure the winter. While possible, it is doubtful that Busbecq could successfully have had the tulip bulbs harvested, shipped to Germany, and replanted between his first sighting of them in March 1558 and Gessner's description the following year. As a result, Busbecq's account of the supposed first sighting of tulips by a European is possibly spurious.

Carolus Clusius planted tulips at the Imperial Botanical Gardens of Vienna in 1573 and later at the Leiden University's newly established Hortus Botanicus, where he was appointed director. There he planted some of his tulip bulbs in late 1593. As a result, 1594 is considered the official date of the tulip's first flowering in The Netherlands, despite reports of the flowers being cultivated in private gardens in Antwerp and Amsterdam two or three decades earlier. These tulips at Leiden would eventually lead to both Tulip mania and the commercial tulip industry in Holland.Another account of the origin of the tulip in Western Europe is of Lopo Vaz de Sampaio, governor of the Portuguese possessions in India. After attempting to usurp power from the rightful governor, Sampaio was forced to return to Portugal in disgrace. Supposedly, he took tulip bulbs back to Portugal with him from Sri Lanka. This story does not hold up to scrutiny though because tulips do not occur in Sri Lanka and the island itself is far from the route Sampaio's ships would have likely taken.

Regardless of how the flower originally arrived in Europe, its popularity soared quickly. Carolus Clusius is largely responsible for the spread of tulip bulbs in the final years of the sixteenth century. He finished writing the first major work on tulips in 1592, and he made note of the variations in colour that help make the tulip so admired. While occupying a chair as a faculty member in the school of medicine at the University of Leiden, Clusius planted both a teaching garden and private plot of his ownwith tulip bulbs. In 1596 and 1598, Clusius suffered thefts from his garden, with over a hundred bulbs stolen in a single raid.

Between 1634 and 1637, the early enthusiasm for the new flowers triggered a speculative frenzy now known as the tulip mania. Tulips would become so expensive that they were treated as a form of currency. Around this time, the ceramic tulipiere was devised for the display of cut flowers stem by stem (bouquets displayed in vases were rare until the 19th century, although such vases and bouquets, usually including tulips, often appeared in Dutch still-life painting). To this day, tulips are associated with The Netherlands, and the cultivated forms of the tulip are often called "Dutch tulips." In addition to the tulip industry and tulip festivals, The Netherlands has the world's largest permanent display of tulips at Keukenhof, although the display is only open to the public seasonally.

Tulip Mania

Tulip mania or tulipomania (Dutch names include: *tulpenmanie, tulpomanie, tulpenwoede, tulpengekte* and *bollengekte*) was a period in the Dutch Golden Age during which contract prices for bulbs of the recently introduced tulip reached extraordinarily high levels and then suddenly collapsed. At the peak of tulip mania, in February 1637, some single tulip bulbs sold for more than 10 times the annual income of a skilled craftsman. It is generally considered the first recorded speculative bubble (or economic bubble), although some researchers have noted that the *Kipper- und Wipperzeit* episode in 1619–22, a Europe-wide chain of debasement of the metal content of coins to fund warfare, featured mania-like similarities to a bubble. The term "tulip mania" is now often used metaphorically to refer to any large economic bubble (when asset prices deviate from intrinsic values).

The event was popularized in 1841 by the book *Extraordinary Popular Delusions and the Madness of Crowds*, written by British journalist Charles Mackay. According to Mackay, at one point 12 acres (5 ha) of land were offered for a *Semper augustus* bulb. Mackay claims that many such investors were ruined by the fall in prices, and Dutch commerce suffered a severe shock. Although Mackay's book is a classic that is widely reprinted today, his account is contested. Many modern scholars believe that the mania was not as extraordinary as Mackay described, with some arguing that the price changes may not have constituted a bubble.

Research on the tulip mania is difficult because of the limited data from the 1630s—much of which comes from biased and anti-speculative sources. Although these explanations are not generally accepted, some modern economists have proposed rational explanations, rather than a speculative mania, for the rise and fall in prices. For example, other flowers, such as the hyacinth, also had high prices on the flower's introduction, which then fell dramatically. The high prices may also have been driven by expectations of a parliamentary decree that contracts could be voided for a small cost—thus lowering the risk to buyers.

History

The tulip was introduced to Europe in the mid-16th century from the Ottoman Empire, and became very popular in the United Provinces (now the Netherlands). Tulip cultivation in the United Provinces is generally thought to have started in earnest around 1593 after the Flemish botanist Charles de l'Écluse had taken up a post at the University of Leiden and established the *hortus academicus*. There, he planted his collection of tulip bulbs—sent to him from Turkey by the Emperor's

(Ferdinand I, Holy Roman Emperor) ambassador to the Sultan, Ogier de Busbecq—which were able to tolerate the harsher conditions of the Low Countries, and it was shortly thereafter they began to grow in popularity.

The flower rapidly became a coveted luxury item and a status symbol, and a profusion of varieties followed. They were classified in groups; one-coloured tulips of red, yellow, or white were known as *Couleren,* but it was the multicoloured *Rosen* (red or pink on white background), *Violetten* (purple or lilac on white background), and, to a lesser extent, the *Bizarden* (red, brown or purple on yellow background) that were the most popular. These spectacular and highly sought-after tulip bulbs would grow flowers with vivid colours, lines, and flames on the petals, as a result, it is now understood, of being infected with a tulip-specific virus known as the "Tulip breaking virus", a type of mosaic virus.

Growers named their new varieties with exalted titles. Many early forms were prefixed *Admirael* ("admiral"), often combined with the growers' names—*Admirael van der Eijck* was perhaps the most highly regarded of about fifty so named. *Generael* ("general") was another prefix that found its way into the names of around thirty varieties. Later came varieties with even more superb names, derived from Alexander the Great or Scipio, or even "Admiral of Admirals" and "General of Generals". However, naming could be haphazard and varieties highly variable in quality. Most of these varieties have now died out, though similar "broken" tulips continue in the trade.

Tulips grow from bulbs, and can be propagated through both seeds and buds. Seeds from a tulip will form a flowering bulb after 7–12 years. When a bulb grows into the flower, the original bulb will disappear, but a clone bulb forms in its place, as do several buds. Properly cultivated, these buds will become bulbs of their own. The mosaic virus spreads only through buds, not seeds, and so cultivating the most appealing varieties takes years. Propagation is greatly slowed down by the virus. Tulips bloom in April and May for only about a week, and the secondary buds appear shortly thereafter. Bulbs can be uprooted and moved about from June to September, and thus actual purchases (in the spot market) occurred during these months.

During the rest of the year, traders signed contracts before a notary to purchase tulips at the end of the season (effectively futures contracts). Thus the Dutch, who developed many of the techniques of modern finance, created a market for durable tulip bulbs. Short selling was banned by an edict of 1610, which was reiterated or strengthened in 1621 and 1630, and again in 1636. Short sellers were not prosecuted under these edicts, but their contracts were deemed unenforceable.

As the flowers grew in popularity, professional growers paid higher and higher prices for bulbs with the virus. By 1634, in part as a result of demand from the French, speculators began to enter the market. In 1636, the Dutch created a type of formal futures markets where contracts to buy bulbs at the end of the season were bought and sold. Traders met in "colleges" at taverns and buyers were required to pay a 2.5% "wine money" fee, up to a maximum of three florins, per trade. Neither party paid an initial margin nor a mark-to-market margin, and all contracts were with the individual counter-parties rather than with the exchange. No deliveries were ever made to fulfill these contracts because of the market collapse in February 1637. This trade was centered in Haarlem during the height of a bubonic plague epidemic, which may have contributed to a culture of fatalistic risk-taking.

The contract price of rare bulbs continued to rise throughout 1636. That November, the contract price of common bulbs without the valuable mosaic virus also began to rise in value. The Dutch derogatorily described tulip contract trading as *windhandel* (literally "wind trade"), because no bulbs were actually changing hands. However in February 1637, tulip bulb contract prices collapsed abruptly and the trade of tulips ground to a halt.

Available Price Data

The lack of consistently recorded price data from the 1630s makes the extent of the tulip mania difficult to estimate. The bulk of available data comes from anti-speculative pamphlets by "Gaergoedt and Warmondt" (GW) written just after the bubble. Economist Peter Garber collected data on the sales of 161 bulbs of 39 varieties between 1633 and 1637, with 53 being recorded by GW. Ninety-eight sales were recorded for the last date of the bubble, February 5, 1637, at wildly varying prices. The sales were made using several market mechanisms: futures trading at the colleges, spot sales by growers, notarized futures sales by growers, and estate sales. "To a great extent, the available price data are a blend of apples and oranges," according to Garber.

Mackay's Madness of Crowds

Goods allegedly exchanged for a single bulb of the **Viceroy**

Two lasts of wheat	448*f*
Four lasts of rye	558*f*
Four fat oxen	480*f*
Eight fat swine	240*f*
Twelve fat sheep	120*f*
Two hogsheads of wine	70*f*
Four tuns of beer	32*f*
Two tons of butter	192*f*
1,000 lb. of cheese	120*f*
A complete bed	100*f*
A suit of clothes	80*f*
A silver drinking cup	60*f*
Total	2500*f*

The modern discussion of tulip mania began with the book *Extraordinary Popular Delusions and the Madness of Crowds,* published in 1841 by the Scottish journalist

Charles Mackay; he proposed that crowds of people often behave irrationally, and tulip mania was, along with the South Sea Bubble and the Mississippi Company scheme, one of his primary examples. His account was largely sourced to a 1797 work by Johann Beckmann titled *A History of Inventions, Discoveries, and Origins*. In fact, Beckmann's account, and thus Mackay's by association, was primarily sourced to three anonymous pamphlets published in 1637 with an anti-speculative agenda. Mackay's vivid book was popular among generations of economists and stock market participants. His popular but flawed description of tulip mania as a speculative bubble remains prominent, even though since the 1980s economists have debunked many aspects of his account.

According to Mackay, the growing popularity of tulips in the early 17th century caught the attention of the entire nation; "the population, even to its lowest dregs, embarked in the tulip trade". By 1635, a sale of 40 bulbs for 100,000 florins (also known as Dutch guilders) was recorded. By way of comparison, a ton of butter cost around 100 florins, a skilled laborer might earn 150 florins a year, and "eight fat swine" cost 240 florins. (According to the International Institute of Social History, one florin had the purchasing power of •10.28 in 2002.)

By 1636, tulips were traded on the exchanges of numerous Dutch towns and cities. This encouraged trading in tulips by all members of society; Mackay recounted people selling or trading their other possessions in order to speculate in the tulip market, such as an offer of 12 acres (49,000 m^2) of land for one of two existing *Semper Augustus* bulbs, or a single bulb of the *Viceroy* that was purchased for a basket of goods (shown at right) worth 2,500 florins.

Many individuals grew suddenly rich. A golden bait hung temptingly out before the people, and, one after the other, they rushed to the tulip marts, like flies around a honey-pot. Every one imagined that the passion for tulips would last for ever, and that the wealthy from every part of the world would send to Holland, and pay whatever prices were asked for them. The riches of Europe would be concentrated on the shores of the Zuyder Zee, and poverty banished from the favoured clime of Holland. Nobles, citizens, farmers, mechanics, seamen, footmen, maidservants, even chimney sweeps and old clotheswomen, dabbled in tulips.

The increasing mania contributed several amusing, but unlikely, anecdotes that Mackay recounted, such as a sailor who mistook the valuable tulip bulb of a merchant for an onion and grabbed it to eat. The merchant and his family chased the sailor to find him "eating a breakfast whose cost might have regaled a whole ship's crew for a twelvemonth". The sailor was jailed for eating the bulb. This anecdote is unlikely if only because, like most Liliaceae bulbs, tulips are somewhat poisonous, and taste quite differently from onions.

People were purchasing bulbs at higher and higher prices, intending to re-sell them for a profit. However, such a scheme could not last unless someone was ultimately willing to pay such high prices and take possession of the bulbs. In February 1637, tulip traders could no longer find new buyers willing to pay increasingly inflated prices for their bulbs. As this realization set in, the demand for tulips collapsed, and prices plummeted—the speculative bubble burst. Some were left holding contracts to purchase tulips at prices now ten times greater than those on the open market, while others found themselves in possession of bulbs now worth a fraction of the price they had paid. Mackay claims the Dutch devolved into distressed accusations and recriminations against others in the trade.

In Mackay's account, the panicked tulip speculators sought help from the government of the Netherlands, which responded by declaring that anyone who had bought contracts to purchase bulbs in the future could void their contract by payment of a 10 percent fee. Attempts were made to resolve the situation to the satisfaction of all parties, but these were unsuccessful. The mania finally ended, Mackay says, with individuals stuck with the bulbs they held at the end of the crash—no court would enforce payment of a contract, since judges regarded the debts as contracted through gambling, and thus not enforceable by law. According to Mackay, lesser tulip manias also occurred in other parts of Europe, although matters never reached the state they had in the Netherlands. He also claimed that the aftermath of the tulip price deflation led to a widespread economic chill throughout the Netherlands for many years afterwards.

Modern Views

Mackay's account of inexplicable mania was unchallenged, and mostly unexamined, until the 1980s. However, research into tulip mania since then, especially by proponents of the efficient market hypothesis, who are more skeptical of speculative bubbles in general, suggests that his story was incomplete and inaccurate. In her 2007 scholarly analysis *Tulipmania*, Anne Goldgar states that the phenomenon was limited to "a fairly small group", and that most accounts from the period "are based on one or two contemporary pieces of propaganda and a prodigious amount of plagiarism". Peter Garber argues that the bubble "was no more than a meaningless winter drinking game, played by a plague-ridden population that made use of the vibrant tulip market."

While Mackay's account held that a wide array of society was involved in the tulip trade, Goldgar's study of archived contracts found that even at its peak the trade in tulips was conducted almost exclusively by merchants and skilled craftsmen who were wealthy, but not members of the nobility. Any economic fallout from the bubble was very limited. Goldgar, who identified many prominent buyers and sellers in the market, found fewer than half a dozen who experienced financial troubles in the time period, and even of these cases it is not clear that tulips were to blame. This is not altogether surprising. Although prices had risen, money had not exchanged hands between buyers and sellers. Thus profits were never realized for sellers; unless sellers had made other purchases on credit in expectation of the profits, the collapse in prices did not cause anyone to lose money.

Rational Explanations

There is no dispute that prices for tulip bulb contracts rose and then fell in 1636–37, but even a dramatic rise and fall in prices does not necessarily mean that an economic or speculative bubble developed and then burst. For tulip mania to have qualified as an economic bubble, the price of tulip bulbs would need to have become unhinged from the intrinsic value of the bulbs. Modern economists have advanced several possible reasons for why the rise and fall in prices may not have constituted a bubble.

The increases of the 1630s corresponded with a lull in the Thirty Years' War. Hence market prices (at least initially) were responding rationally to a rise in demand. However, the fall in prices was faster and more dramatic than the rise. Data on sales largely disappeared after the February 1637 collapse in prices, but a few other data points on bulb prices after tulip mania show that bulbs continued to lose value for decades thereafter.

Volatility in Flower Prices

Garber compared the available price data on tulips to hyacinth prices at the beginning of the 19th century—when the hyacinth replaced the tulip as the fashionable flower—and found a similar pattern. When hyacinths were introduced florists strove with one another to grow beautiful hyacinth flowers, as demand was strong. However, as people became more accustomed to hyacinths the prices began to fall. The most expensive bulbs fell to 1–2 percent of their peak value within 30 years. Garber also notes that, "a small quantity of prototype lily bulbs recently was sold for 1 million guilders ($480,000 at 1987 exchange rates)", demonstrating that even today flowers can command extremely high prices. Additionally, because the rise in prices occurred after bulbs were planted for the year, growers would not have had an opportunity to increase production in response to price.

Legal Changes

UCLA economics professor Earl A. Thompson argues in a 2007 paper that Garber's explanation cannot account for the extremely swift drop in tulip bulb contract prices. The annualized rate of price decline was 99.999%, instead of the average 40% for other flowers. He provides another explanation for Dutch tulip mania. The Dutch parliament was considering a decree (originally sponsored by Dutch tulip investors who had lost money because of a German setback in the Thirty Years' War) that changed the way tulip contracts functioned:

On February 24, 1637, the self-regulating guild of Dutch florists, in a decision that was later ratified by the Dutch Parliament, announced that all futures contracts written after November 30, 1636 and before the re-opening of the cash market in the early Spring, were to be interpreted as option contracts. They did this by simply relieving the futures buyers of the obligation to buy the future tulips, forcing them merely to compensate the sellers with a small fixed percentage of the contract price.

Before this parliamentary decree, the purchaser of a tulip contract—known in modern finance as a futures contract—was legally *obliged* to buy the bulbs. The decree changed the nature of these contracts, so that if the current market price fell, the purchaser could opt to pay a penalty and forgo receipt of the bulb, rather than pay the full contracted price. This change in law meant that, in modern terminology, the futures contracts had been transformed into options contracts. This proposal began to be debated in the fall of 1636, and if it became clear to investors that the decree was likely to be enacted, prices probably would have risen.

This decree allowed someone who purchased a contract to void the contract with a payment of only 3.5 percent of the contract price (or about 1/30th the contract). Thus, investors bought increasingly expensive contracts. A speculator could sign a contract to purchase a tulip for 100 guilders. If the price rose above 100 guilders, the speculator would pocket the difference as profit. If the price remained low, the speculator could void the contract for only 3½ guilders. Thus, a contract nominally for 100 guilders, would actually cost an investor no more than 3½ guilders. In early February, as contract prices reached a peak, Dutch authorities stepped in and halted the trading of these contracts.

Thompson states that actual sales of tulip bulbs remained at ordinary levels throughout the period. Thus, Thompson concludes that the "mania" was a rational response to changes in contractual obligations. Using data about the specific payoffs present in the futures and option contracts, Thompson argues that tulip bulb contract prices hewed closely to what a rational economic model would dictate, "Tulip contract prices before, during, and after the 'tulipmania' appear to provide a remarkable illustration of 'market efficiency'."

Critiques

Other economists believe that these elements cannot completely explain the dramatic rise and fall in tulip prices. Garber's theory has also been challenged for failing to explain a similar dramatic rise and fall in prices for regular tulip bulb contracts. Some economists also point to other factors associated with speculative bubbles, such as a growth in the supply of money, demonstrated by an increase in deposits at the Bank of Amsterdam during that period.

Social Mania and Legacy

The popularity of Mackay's tale has continued to this day, with new editions of *Extraordinary Popular Delusions* appearing regularly, with introductions by writers such as financier Bernard Baruch (1932), financial writers Andrew Tobias (1980), psychologist David J. Schneider (1993), and Michael Lewis (2008). At least six editions are currently in print.

Goldgar argues that although tulip mania may not have constituted an economic or speculative bubble, it was nonetheless traumatic to the Dutch for other reasons. "Even though the financial crisis affected very few, the shock of tulipmania was considerable. A whole network of values was thrown into doubt." In the 17th century, it was unimaginable to most people that something as common as a flower could be worth so much more money than most people earned in a year. The idea that the prices of flowers that grow only in the summer could fluctuate so wildly in the winter, threw into chaos the very understanding of "value".

Many of the sources telling of the woes of tulip mania, such as the anti-speculative pamphlets that were later reported by Beckmann and Mackay, have been cited as evidence of the extent of the economic damage. These pamphlets, however, were not written by victims of a bubble, but were primarily religiously motivated. The upheaval was viewed as a perversion of the moral order—proof that "concentration on the earthly, rather than the heavenly flower could have dire consequences". Thus, it is possible that a relatively minor economic event took on a life of its own as a morality tale.

Nearly a century later, during the crash of the Mississippi Company and the South Sea Company in about 1720, tulip mania appeared in satires of these manias. When Johann Beckmann first described tulip mania in the 1780s, he compared it to the failing lotteries of the time. In Goldgar's view, even many modern popular works about financial markets, such as Burton Malkiel's *A Random Walk Down Wall Street* (1973) and John Kenneth Galbraith's *A Short History of Financial Euphoria* (1990; written soon after the stock market crash of 1987), used the tulip mania as a lesson in morality. Also in Oliver Stone's drama *Wall Street: Money Never Sleeps* from 2010, a sequel to the classical 1987 film *Wall Street*, the tulip mania is referenced. Gordon Gekko, played by Michael Douglas, uses a historical chart displaying the market value of tulips and compares it to the Financial crisis of 2007–2010.

Tulip mania again became a popular reference during the dot-com bubble of 1995–2001. Most recently, journalists have compared it to the subprime mortgage crisis. Despite the mania's enduring popularity, Daniel Gross of *Slate* has said of economists offering efficient market explanations for the mania, that "If they're correct... then business writers will have to delete Tulipmania from their handy-pack of bubble analogies."

Introduction to the United States

It is believed the first tulips in the United States were grown near Spring Pond at the Fay Estate in Lynn and Salem, Massachusetts. From 1847 to 1865, a historic land owner named Richard Sullivan Fay, Esq., one of Lynn's wealthiest men, would settle on 500 acres (2.0 km^2) located partly in present-day Lynn and partly in present-day Salem. While there, Mr. Fay imported many different trees and plants from all parts of the world and planted them among the meadows of the Fay Estate.

Diseases

Botrytis tulipae is a major fungal disease affecting tulips, causing cell death and eventually the rotting of the plant. Other pathogens include anthracnose, bacterial soft rot, blight caused by *Sclerotium rolfsii*, bulb nematodes, other rots including blue molds, black molds and mushy rot.

Variegated varieties admired during the Dutch tulipomania gained their delicately feathered patterns from an infection with the tulip breaking virus, a mosaic virus that was carried by the green peach aphid, *Myzus persicae*. These aphids were common in European gardens of the seventeenth century. While the virus produces fantastically colourful flowers, it also causes weakened plants that die slowly. Today the virus is almost eradicated from tulip growers' fields. Tulips that are affected by mosaic virus are called "broken tulips"; while such tulips can occasionally revert to a plain or solid colouring, they will remain infected with the virus. While some modern varieties also display multicoloured patterns, the patterns result from intentional breeding. In these tulips, natural differences in the upper and lower layers of pigment in the flower are responsible for the patterns.

In Art and Culture

During the Ottoman Empire, the tulip became very popular in Ottoman territories and was seen as a symbol of abundance and indulgence. In fact, the era during which the Ottoman Empire was wealthiest is often called the Tulip era or *Lale Devri* in Turkish. In classic and modern Persian literature, special attention has been given to these beautiful flowers, and in recent times, tulips have featured in the poems of Simin Behbahani.

However, the tulip was a topic for Persian poets as far back as the thirteenth century. Musharrifu'd-din Saadi, in his poem *Gulistan*, described a visionary, garden paradise with 'The murmur of a cool stream/bird song, ripe fruit in plenty/bright multicoloured tulips and fragrant roses...'

The Black Tulip is the title of a historical romance by the French author Alexandre Dumas, père. The story takes place in the Dutch city of Haarlem, where a reward is offered to the first grower who can produce a truly black tulip.

Today, Tulip festivals are held around the world, including in The Netherlands, Spalding, England. Every spring, there are several tulip festivals in North America, including the Tulip Time Festival in Holland, Michigan, the Skagit Valley Tulip Festival in Skagit Valley, Washington, the Tulip Time Festival in Orange City and Pella, Iowa, and the Canadian Tulip Festival in Ottawa, Canada. Tulips are now also popular in Australia and several festivals are held in September and October, during the Southern Hemisphere's spring. In the film *The Whole Nine Yards*, Bruce Willis' character is a killer nicknamed "the Tulip", because he used to send tulips to his victims' funerals.

Classification

In horticulture, tulips are divided up into fifteen groups (Divisions) mostly based on flower morphology and plant size.

- Div. 1: Single early - with cup-shaped single flowers, no larger than 8 cm across (3 inches). They bloom early to mid season. Growing 15 to 45 cm tall.
- Div. 2: Double early - with fully double flowers, bowl shaped to 8 cm across. Plants typically grow from 30–40 cm tall.
- Div. 3: Triumph - single, cup shaped flowers up to 6 cm wide. Plants grow 35–60 cm tall and bloom mid to late season.
- Div. 4: Darwin hybrid - single flowers are ovoid in shape and up to 8 cm wide. Plants grow 50–70 cm tall and bloom mid to late season. This group should

not be confused with older Darwin tulips, which belong in the Single Late Group below.

- Div. 5: Single late - cup or goblet-shaded flowers up to 8 cm wide, some plants produce multi-flowering stems. Plants grow 45–75 cm tall and bloom late season.
- Div. 6: Lily-flowered
- Div. 7: Fringed (Crispa)
- Div. 8: Viridiflora
- Div. 9: Rembrandt
- Div. 10: Parrot
- Div. 11: Double late
- Div. 12: Kaufmanniana
- Div. 13: Fosteriana (Emperor)
- Div. 14: Griegii
- Div. 15: Species (Botanical)
- Div. 16: Multiflowering - not an official division, these tulips belong in the first 15 divisions but are often listed separately because they have multiple blooms per bulb.

They may also be classified by their flowering season:

- Early flowering: Single Early Tulips, Double Early Tulips, Greigii Tulips, Kaufmanniana Tulips, Fosteriana Tulips, Species Tulips
- Mid-season flowering: Darwin Hybrid Tulips, Triumph Tulips, Parrot Tulips
- Late season flowering: Single Late Tulips, Double Late Tulips, Viridiflora Tulips, Lily-flowering Tulips, Fringed Tulips, Rembrandt Tulips.

Selected Species

- *Tulipa acuminata* (Horned Tulip)
- *Tulipa agenensis* (Eyed Tulip)
- *Tulipa aleppensis* (Aleppo Tulip)
- *Tulipa armena*
- *Tulipa aucheriana*
- *Tulipa batalinii*
- *Tulipa bakeri*
- *Tulipa biflora*
- *Tulipa borszczowii*
- *Tulipa botschantzevae* (Botschantzeva's tulip)
- *Tulipa butkovii*
- *Tulipa carinata*

- *Tulipa celsiana*
- *Tulipa clusiana* (Lady Tulip)
- *Tulipa cretica*
- *Tulipa cypria*
- *Tulipa dasystemon*
- *Tulipa didieri.*

How to Plant Tulips

Tulips are one of the most beautiful flowers that can mesmerize any viewer with its astounding beauty. When in full bloom, tulip gardens and farms look as though they are carpeted with the velvety flowers. Planting tulips in your backyard might be a dream, but there is every chance of making that dream come true. Planting tulips in your garden is not all that difficult. All you need is some helpful tips on how to plant tulip bulbs so that they bloom into lovely plants. Check out some useful information on how to plant tulips.

How to Plant Tulip Bulbs

- You need to be careful while choosing the variety of tulips you want to plant. Hybrid tulips give beautiful blooms for the first time but the quality deteriorates as and when time goes on. Some species like Tulipa Greigii and T. Kaufmanniana produce flowers for many years and may even produce more than one flower per bulb.
- You can buy high-quality bulbs, which should be free of cuts, bruises and soft spots. The best place to look for tulip bulbs is your local nursery. You may also want to check reputed websites and mail-order catalogs.
- The best seasons to start planting tulip bulbs is in early fall for cold winter regions and late fall to early winter for regions that have mild temperature. Make sure you have the bulbs at least some five to six weeks prior to actual planting. Depending on the type of tulip bulb you have, you may need to refrigerate it in order to have high quality blooms.
- You need to select a location in your garden where there is plenty of sun light and apt drainage. Dig in 2 inches of compost before you start planting the bulbs.
- Take the size of the bulb into consideration and dig holes around 6-8 inches. You need to dig holes in a pattern that it looks clustered, not lined up. Instead of digging individual holes, you can dig one big hole or a trench and then plant the bulbs.
- Mix some rock phosphate at the bottom of the hole and blend it roughly with the soil. Try not to use fertilizers that may comprise of fish bone since it attracts squirrels or rats that may nibble the bulb and damage it.
- After planting the bulbs, cover it up with the top soil but don't cover it too

tight. The bulbs may die of lack of air if too much soil is put on top in a tight way. Water regularly and by spring, you should have your first beautiful bunch of tulips!

Valley of Tulips

This summer, besides the splendid mountains, valleys and lakes, Kashmir has an added attraction in the form of a colourful tulip garden. The Indira Gandhi Memorial Tulip Garden is among the largest in Asia. Thousands of tourists, as well as locals, are thronging the garden, which is in full bloom.

Situated ideally on the foothills of the mighty Zabarwan Mountains, the garden is just a stone's throw from the famous Dal Lake. Tourists are effusive in their praise of this visual feast.

Intended to be a terraced garden, three terraces have been laid out already while two more are expected to be ready next year, complete with fountains and other decorations.

Flowering Season

This is the time of year when Kashmir bursts into a riot of blossoming flowers. Everywhere one can see the almond, cherry and peach trees in full bloom. For visitors this is a delightful addition to the many attractions on offer in this richly-endowed State. "I did not know anything about the Tulip Garden in Kashmir. Somebody told us here that it is a beautiful garden to visit," says Asha Mathur, a tourist from central India. "Thank God I visited this beautiful garden whose memories I will cherish for years to come."

The authorities also plan to cultivate other varieties of all-season flowers to attract visitors year round, as the tulip season usually does not last beyond two months.

The brainchild of Jammu and Kashmir Chief Minister Ghulam Nabi Azad, the tulip garden's development was accorded importance from the highest offices of the State. No effort was spared by the authorities to make the garden one of the best of its kind. Nearly 900,000 tulip bulbs were imported from Holland and an army of about 400 gardeners and labourers have been working in triple shifts to bring the dream garden to life in a record period of one-and-a-half years.

Long Association

Experts say tulips have more than thousand years of history in Kashmir where it was cultivated in unique ways. "In Srinagar most of the houses in the olden days used to have thatched roofs, which was spread with soil. The tulips used to be grown on these roofs and it was a delightful sight during the spring season," says Farooq Dar, an elderly person. "Many travellers have captured the image of tulips in full bloom during summer in their travelogues."

Besides the locals and tourists, the tulip garden has found an ardent fan in Sonia Gandhi, who turned nostalgic during her recent visit to the garden. "Indira Gandhi loved Kashmir in its different manifestations like streams, meadows, snow-capped mountains and especially flowers," she reminisced.

She recalled how Kashmir used to be a regular holiday destination for Jawaharlal Nehru, Indira Gandhi, Rajiv Gandhi and Sanjay Gandhi and that the sheer magic of the beauty of Kashmir had lessened their tensions of politics and governance. "Such was the passion that in October 1984, Indira Gandhi came to Kashmir just to see the Chinars," she said.

The idea for the tulip garden was warmly welcome by the tourism industry too. In Kashmir the tourist season normally starts in May and there was a general feeling that tourists had nothing to look forward to during March and April. The tulip garden is now expected to fill that void and add two full months to the tourist season. The Chief Minister has asked tourism department officials to publicise the tulip garden countrywide to help attract the maximum number of tourists.

While generating approval all round, the project has also had its share of controversies. The renaming of the tulip garden from Siraj Bagh to Indira Gandhi Memorial Tulip Garden has irked many Kashmiris who feel the garden project should have steered clear of politics. But far-removed from such controversies the floriculture department is working single-mindedly to make the garden truly world-class. Their dream is to match the world-famous tulip gardens of Holland and China, both of which attract a quarter of a million tourists each year during the flowering season.

Indian Floriculture: Flower Power

The Indian floriculture industry, growing at a compound annual growth rate (CAGR) of 25 per cent over the past decade, has blossomed into a $230 million business.

While exports remain the prime motivator for cultivators, local demand is also growing by leaps and bounds, especially in cities. Modernisation and growing western cultural influences has resulted in consumers – especially the young – buying flowers on occasions like Valentine's Day, Friendship day, Mother's day, Father's day and so on. Of course, there is a huge spurt in demand for flowers during religious festivities.

Flower retailing is also undergoing a change. Many of the new shopping malls and large format retailers have exclusive flower shops that witness demand roundthe-year.

But many cultivators enter the business with a focus on exports and an eye on the global industry. Internationally, the floriculture business is estimated to be worth $11 billion and growing at 15 per cent per annum. The Flower Council of Holland projects this would top $16 billion by 2010. India's flower exports have been growing by a robust 25 per cent CAGR. It added up to $90 million in 2006-07 and is projected to touch $160 million by 2010.

Fresh flowers from India have already gained entry into nearly 90 countries including Australia, Japan, the UK, the Netherlands and the UAE. The world's biggest flower markets include USA ($12.5 billion), Japan ($5.46 billion) and Italy ($4.27 billion), while the highest per capita consumption of flowers is in Norway ($146) followed by Switzerland ($126) and Germany ($88).

While international markets offer a lot of opportunities for Indian exporters, they also face substantial challenges.

Indian exports of cut-flowers face import tariffs of up to 17 per cent (calculated on a free on board+freight basis) in Europe between May and October, considered non-peak months. It's only when Europe requires imported flowers during the peak season that import tariff is reduced by five to six per cent.

Despite these challenges, the floriculture sector has been doing well.

According to sources in the trade, much of the credit for growth of the industry should go the government. Post-liberalisation, the government identified it as a priority sector and accorded it 100 per cent 'export oriented unit' (EOU) status.

This meant tax exemptions to new EOUs and tax holidays on income from floriculture. Additionally, there were substantial duty exemptions for imports including cut flowers, flower seeds and tissue-cultured plants. Also, allowing 100 per cent FDI (foreign direct investment) in the industry has facilitated joint ventures, ensuring a steady inflow of capital and modern technology.

Further subsidies have been offered for air-freight for export of cut-flowers and tissue-cultured plants, and direct subsidy of up to 50 per cent has been offered for establishing cold storage, pre-cooling units, refrigerated vans and green houses.

Subsidy is also given by the Agricultural and Processed Food Products Export Development Authority (APEDA), for using improved packaging material. Refinance assistance is available from the National Bank for Agriculture and Rural Development (NABARD) to a number of hi-tech units at reasonable interest rates. The Union Ministry of Agriculture has also set up 11 model floriculture centre units, two large centres and 20 tissue culture units.

Six agri-export zones have also been set up for floriculture where large flower farms share common infrastructural facilities, thus bringing into effect the benefits of scale. Walk-in cold storages have been allowed at international airports like Delhi, Mumbai Hyderabad, Chennai and Thiruvananthpuram and similar facilities are expected at other airports soon.

Auction centres are also being set up at major cities, helping in price discovery and trading. In fact, APEDA has even set up a Market Facilitation Centre at Aalsmeer, in the Netherlands, the world's largest flower auction centre. The authority aims to set up similar centres in the Middle East and the Far East in due course, to help Indian flower exporters.

Najeeb Ahmed, president of the South Indian Floriculture Association, says India's share of the world rose market is still tiny. But the country is making its presence felt in several countries including Australia, Japan, Britain and Germany. "The quality of Indian roses is good and there is substantial demand in competitive markets around the world," he explains. "We can grow roses round the year without artificial heating, lighting and cooling systems, and there is abundant availability of manpower."

S.K. Gutgutia, managing director, Florance Flora, Bangalore, points out that "India needs to develop indigenous technology as the cultivation of flowers in greenhouses here is dependent on foreign technology." But initial investment is very high, so many growers opt for joint ventures with international partners.

The huge potential for growth for the floriculture sector has attracted top Indian corporates. They include Reliance, ITC, Tata Tea, Bharti Group and the Thapar Group. Their plans are being driven by demand from overseas retailers like TESCO, Sainsbury, Wal-Mart, Asda, Sears, Carrefour, Metro and K-Mart.

International retailers buy in large volumes, so scale of operations and quality control will be of paramount importance.

This is encouraging domestic corporates to enter the sector. "Floriculture in India is becoming an attractive, commercially viable diversification option," says industry expert S. Jafar Naqvi. "Several companies involved in agro-business are set to venture into this sector." Some of the smaller players are looking for innovative methods to take advantage of benefits that scale and size provides, by forming cooperative societies. By doing so, they hope to collectively buy farming materials, fertilisers and other supplies as a single entity helping them drive down costs through bulk purchases. With corporate players getting in on the action, and smaller players consolidating their efforts, India's floriculture sector can expect to blossom over the coming years.

In Full Bloom

The floriculture business in India is expected to get a boost with the government deciding to set up various facilities for the trade. Floriculture regional centres, agro-export zones for floriculture and flower clusters are being planned in different parts of the country.

Floriculture regional centres are coming up – or are being planned – in cities like Bangalore, Chennai, Thiruvananthapuram, Pune, Lucknow, Mohali, Srinagar, Kolkata and Gangtok.

These centres would not only conserve important varieties of flowers of the region, but would also encourage large-scale multiplication.

The National Horticulture Mission aims to promote half a dozen agro-export zones and flower clusters to encourage floriculture and the export of flowers. About 150 export oriented units are being set up in and around the major floriculture hubs including Pune, Chandigarh, Bangalore, Delhi and Hyderabad.

The Agricultural and Processed Foods Export Development Authority (APEDA) has set up half a dozen agro-export zones in Maharashtra, Sikkim, Tamil Nadu, Uttarakhand and Karnataka. It has also set up export facilitation centres in the Netherlands, Dubai and Tokyo, and plans flower auction centres in Mumbai, Bangalore,

Kolkata and Noida. At present, over half the floriculture units are located in south India. Kochi in Kerala, for instance, recently hosted the first international anthurium cutflower show in the country, attracting participants – and buyers – from around the world.

Pune in Maharashtra has also emerged as a major flower-growing region. The city, and some of the neighbouring areas, accounts for a third of flower exports from India. Pune and Bangalore are the two largest flower clusters in the country, with about 40 units operating from the two cities.

Indian floriculturists have also been making bold forays abroad. Ramakrishna Karuturi, a US-educated entrepreneur, and founder, Karuturi Global, started a 100 per cent EOU in Bangalore in 1995. He also launched Rose Bazaar. com, an internet auction portal, a few years later.

Today, the company is not only the biggest producer of roses in India, but is a global leader in cut rose production and has expanded abroad, setting up a presence in Ethiopia.

Karuturi Global today manages 220 hectares of rose farms in India, Ethiopia and Kenya and is adding 450 hectares in Ethiopia. This year, the company will produce 650 million stems and by 2010, hopes to cross the one billion-mark.

Karuturi, who raised $250 million recently to acquire additional land in Ethiopia, also acquired a Dutch firm for $70 million last year.

The huge growth opportunities in the floriculture business in India will see the blossoming of many more such entrepreneurs, cultivating both on domestic and foreign soil.

Indian Floriculture: Waiting to Blossom

A nascent floriculture industry in India is cashing in on growing demand for roses overseas. Anjana Pasricha has a report from New Delhi.

It was just over a decade ago that a handful of entrepreneurs began to cultivate roses commercially in southern India, encouraged by the warm climate and a booming market for flowers worldwide.

Those investments have paid dividends. About 50 farms along a "floral corridor" between the southern cities of Hosur and Bangalore exported as many as 15 million red roses ahead of Valentine's Day earlier this month.

The head of the South Indian Floriculture Association, Najeeb Ahmed, says India's share of the world rose market is still tiny. But he says the country is making its presence felt in countries stretching from Australia and Japan to Britain and Germany.

"Indian quality of roses is as competitive in world market as roses from any other country like Africa, Ecuador, Colombia," he said. "We can grow roses round the year without artificial heating, lighting and cooling systems, and with abundant availability of manpower - this industry requires a lot of manpower, and we are able to meet these requirements." It is not just overseas markets that are driving the demand for roses. The domestic market is also booming, as younger people begin to celebrate events such as Valentine's Day with enthusiasm, encouraged by expanding middle class incomes and growing Western influence. Buying flowers for other occasions is also gaining in popularity in big cities.

The blossoming industry is encouraging floriculture companies to expand overseas. Last October, India's largest grower of roses, Karuturi Networks, purchased the world's largest rose producer a Kenyan business, for $72 million. Kenya leads the world in rose cultivation.

Prasanna Pai, Karuturi's chief financial officer says business is soaring.

"This Valentine's Day has been very good for us. Compared to last year we have shipped four times the quantity we did last year," said Prasanna Pai.

Indian rose exports added up to approximately $80 million last year. Industry experts estimate that floriculture is expanding by more than 25 percent a year.

With the declaration of "floriculture as an extreme focus area by the Ministry of Commerce and Industry, Government of India: The sector has acquired a special status in the basket of India and her export commodities.

India produces a wide variety of floriculture products, which inter alia include flowers and foliage, both fresh flowers and dried, like roses carnations chrysanthemums and orchids. A major growth has been in the export of cut flowers for export.

The International Flora Expo 2006 & Landscape Expo-2006 reviewed the growth of the industry and showcased India's floriculture wealth and exposed the stakeholders to new technologies and scientific advancements, a platform for Indian exporters to interact with international buyers of cut flowers.

"With World's Fastest Growing Retail Market, Second Largest Consumer base and unlimited opportunities for growth, Indian Floriculture is today a force to recon with and Flora Expo 2007 and Landscape 2007 are the only platform to interact with flower growers globally to make recognize India as a 'Flower Power'. This is part of a new National Vision for Floriculture, the brainchild of President APJ Abdul Kalam. The floriculture industry is targeting an annual $1-billion in export of floriculture products by 2010!"

Currently, flower trade has attracted the largest demand from an estimated 300 million middle-class flower-loving people with consumption in the cities and major towns at 40 per cent per annum. Flower retail shops have mushroomed all over the place from major metros to market shops and flower boutiques. Further the supermarket/hypermarket retail chains have fueled the growth in the consumption. Cashing in on this trend, the Minister of State for Commerce also feels that floriculture is all about creating new employment opportunities in far flung areas rather than talking about Dollars, the focus should be on a million jobs!

Thanks to the variety of agro climatic zones within, India today is in a unique position to grow a large number of flowers, including temperate flowers in high altitude states. In India, we can now cultivate a wider variety on a commercial basis– an impressive range — of carnation, gerbera, lilium, orchids, anthurium and many others. Interestingly, we have genetically designed flowers – ones with unique shape, petal size, pleasing colours with longer shelf life and of course pleasant fragrance – a rose, for instance, that will be "equally attractive to the humans and the bees!"

Hence the growth in floriculture cultivation has been phenomenal in the last decade or so and the area under flower cultivation has doubled from 53,000 hectares (1993-94) to 103,000 hectares (2001-02).

Six Agri Export Zones have been set up in the states of Sikkim, Tamil Nadu, Uttaranchal, Karanataka and Maharashtra. APEDA has also taken number of measures to facilitate floriculture exports. Besides setting up cold storage and cargo handling facilities at the key airports of New Delhi, Mumbai, Hyderabad, Bangalore, Chennai, Trivandrum and Cochin.:Proposals are under consideration for setting up of such facilities at Goa, Calicut and Coimbatore Airport.

Some key Indian airports like New Delhi, Mumbai, Hyderabad, Bangalore, Chennai, Thiruvananthapuram and Cochin now have cold storage and cargo handling facilities. More airports will have these facilities in the future. Among other things, flower Auction Centres are also coming up in Bangalore, Mumbai, Noida, near Delhi, and Kolkata. These are ready made market facilities for trading and price discovery for a variety of flowers, both for export and domestic markets. India has to achieve the ambitious export target of Rs.1,000 crores per annum over the next 5 years, a paradigm shift is required. The key issues that need to be addressed in the Indian context are Economies of scale, Product range/Latest varieties, Year round exports, Quality control and Certification, Cold chain management. APEDA has been addressing these issues through various forums on a concerted basis given its mandate to promote floriculture exports from India.

Co relation with DFC (Dubai Flower Center) the trans-shipment facility for perishable goods in the region is gearing up to tap the Indian flower export market, which is expected to exceed $1 billion by 2010. DFC can act as a hub for Indian growers and traders so that they can reach out to regional, European and American markets. A DFC delegation had recently visited India to create awareness about the centre and its unique facilities. During the visit, the delegation met officials in major cities and held discussions with flower growers and exporters. Currently India produces 2,00,000 tonnes of loose flowers and 500 million tonnes of cut flowers according to India's Agricultural and Processed Food Products Export Development Authority (APEDA).

8

Effective Supply Chain Management for Indian Floriculture Industry

Integrated Cold Chain Management in Floriculture

Will India be able to meet the ambitious target of Rs. 1000 Crore of floriculture exports by 2010? This a big question looming large, despite the growing optimism about this sector. The value of floriculture exports has steadily grown over the years. India's export of floriculture has increased from Rs. 299.41 crore (USD Million 67.63) in 2005-06 to Rs. 649.83 crore (USD Million 144.15) in 2006-07, which is nice progressive development and reflects efforts of agencies like APEDA in making floriculture a thrust sector. But overall this growth is far from being satisfactory, considering the potential our country has in terms of floriculture business. Several reasons are cited for this shortfall: poor infrastructure, average quality of planting material, inferior production technology, unavailability of basic inputs at the right time, lack of cold-storage facilities, lack of proper marketing efforts and so on. Out of all these, one problem which is not only affecting the future of Indian floriculture trade, but also having a negative impact on the present produce and on its marketability is Cold-Chain Management.

Effect of Temperature on Flowers Flowers like other horticultural products are perishable in nature. The vase-life of the flowers decides the marketability both in the domestic as well as in the export market. In a research done, it was found that even 'seemingly small temperature differences can bring about very noticeable flower quality changes.' The experiments revealed that there was a huge negative impact on the vase-life of the flowers with the increase in temperature during transportation and this problem exaggerated with increase in the duration for which the flowers were exposed to such high temperatures. In India, owing to inadequate cold-chain management, the postharvest losses are very high. 'No systematic study/survey has been conducted to assess these losses. However, the estimated losses range even up to 40%, depending upon the commodity at various stages.' Often the problem starts from the grower and the rest of the damage is done during transportation and storage.

Cut-flowers must be cooled to temperature ranging between 330 F to 410 F without wasting too much time. Higher temperatures not only reduce the vase-life but also increase the respiration-rate, increase the heat production and also increase the damages caused by ethylene. In research done by Fisum G. Celikel and Michael S. Reid (2002) on gerbera (Gerbera jamesonii H. Bolus ex Hook.f. 'Vesuvio' and sunflower (Helianthus annus L.), they found that the 'respiration of cutflowers increased exponentially with increasing storage temperature. Post -Harvest vase life and negative graviotropic bending of the neck of flowers were both strongly affected by simulated transport at higher temperatures. Vase life and stem bending after dry storage showed highly significant linear relationships (negative and positive respectively) with the rate of respiration during storage. The data indicate the importance of maintaining temperatures close to the freezing-point during commercial handling and transport of these important commercial cut-flower crops for maximum vase life.' Proper cold-chain management of flowers is useful in improving their marketability.Cool storage enables quality stems to be held for longer periods before sale and ensures that the flowers still have a good vase life when they reach the market-place.

Cold Chain Management

What is a Cold Chain exactly? Cold Chain is a logistic system that provides a series of facilities for maintaining ideal storage conditions for perishables, from the point of production to the point of consumption in the entire chain.

The various components of a cold-chain are :

- Procurement and delivery systems
- Pre-cooling facilities
- Refrigerated vehicles
- Chilled Store (Normally for temporary storage)
- CA Store (Normally for long term storage)
- Grading, sorting and packaging
- Warehouses
- Information systems.

Status of Cold Chain Management

The cold-chain management is perhaps still the biggest glaring-gap in India's agro-logistics. Cold storage facility is there only for 10% of the produce. As per existing reports, over 3,500 existing cold storage warehouses have only around 13 mn ton storage capacity with old storage technology.

Some imminent problems with the cold chain in our country are:

- The modern Post harvest management is almost non-existent in our country, except for parts.
- Non-existence of world class Integrated Cold Supply Chain across the country-isolated stores without logistics support

- The cold storages mainly work on old technology and hence are not that effective.
- Cold stores are often used for storing low value products.
- Many a time cold storages opt for short duration storages dictated by market imbalances.
- Many a time cold storages opt for short duration storages dictated by market imbalances.
- There is absence of cold storage facilities at most of the airports. At present only six international airports i.e. New Delhi, Mumbai, Chennai, Hyderabad, Bangalore and Thiruvananthapuram have these facilities for perishable cargo. Under such circumstances several hours (sometimes even days) would lapse before the cargo could be loaded to the connecting flight. During this time the cargo would remain under high temperatures and low humidity. The result is very obvious.
- Constricted cargo space on flights: Airlines do not prefer to carry cargo which is seasonal, perishable, voluminous yet not in such large quantity. This is the exact nature of floral cargo. Because of the small volume of flowers they mostly have to be transported on regular flights which have limited cargo-space.

Developing an efficient cold-chain: Strategic Advantage for Indian Floriculture The real spirit of the cold chain process is to get the finest products on time, every time, in perfect or precise condition. Appropriate or new technology that can enhance the performance of the cold chain process is no doubt important. However, implementing the right technology is not enough. Cold chain managers must take into consideration that even a minor variation in the temperature can cause damage to the product and maintaining the quality level of the product will be the key. In the times to come, the cold chain will definitely be a strategic competitive advantage, and India has a lot to gain in terms of growth in profits as well as reputation in efficiency and quality in the global floriculture market. The need of the hour is to develop an model which provides a unbroken cold chain to flowers right from the point-of-origin (growers) to the point-of-consumption (customers).

Grower : The grower is a key player in the entire cold chain. Research studies indicate that in many cases the problem may actually start at this point of origin. Specially in a country like India, where the awareness level of the farmers is not very high and considering that commercial flower cultivation has not really caught-up in a big way like cereals and pulses, there seems to be considerable room for extensive extension activities. The manner and the stage at which a flower crop is harvested may be critical in the chain. Similarly awareness and facilities for modern post-harvest management are the need of the hour.

Collection Centres : The primary task at the collection centre is pre-cooling to takeoff the field heat. The flowers cannot be and must not be transported in the field condition and are required to be pre-cooled before shifting them onto refrigerated vans/trucks/containers.

Refrigerated Transport : In case of road transport we have to use refrigerated vans, and in case of rail transport we may use refrigerated containers. In our country, we need both refrigerated vans and containers in adequate quantity, especially at smaller centres and not just in metros.

Terminal/Holding Cold Storages : Normally terminal or holding cold storages must have a pre-cooling facility to reduce the quantity of harmful gases and also a humidity optimizing unit to maintain the right level of moisture level. They also have grading, sorting and packaging facilities. The terminal cold storages at Airports also need to be revamped both in terms of capacity and number. Again the focus has to be on relatively smaller airports rather then focusing only on major airports. Only then a truly intensive network can be developed.

Air-Cargo : There is a critical need to create adequate cargospace, both on domestic as well as international circuits. Without that both on domestic and primarily on export front we cannot emerge as a floriculture force to reckon-with. I remember listening to a extension-worker from Srinagar airing his concern an International Floriculture Conference way back in 2005. He said that a freight of flowers was lying at Srinagar airport for about a week, since no airline was ready to carry the cargo. His frustration was palpable, since he was actively involved in motivating the farmers there to take-up flower cultivation. Another problem at airports is the enormous handling and loading time that is taken. This has also to be reduced considerably. APEDA's praiseworthy efforts: APEDA is building cold storage facilities at more than a dozen airports in the country. Besides the six international airports which already have cold-storage facilities, more airports will be covered. Bigger airports like Kochi, Kolkata, Guwahati will have elaborate cold storage facilities, including a receiving area for perishable produce, weighing, X-ray, palletization and storage. Smaller airports such as Tiruchirapalli, Srinagar, Jammu and later on Imphal, Dimapur, Agartala will have refrigerated containers that can store 8- 10 tonnes of perishable products. The facilities may cost the government in excess of Rs. 55 crore but the more is expected to increase the exports of perishable food products by 25% in a year's time.

Receiving Airports : Cold storage facilities must also be ensured at receiving airports. In case of shipping cargo, refrigerated cargo container terminals may serve the purpose. Adequate and swift transportation to the markets must also be ensured.

Market/Floral Retail Outlets : The cold chain must not be broken even here. Although in India it may not sound so common, internationally even at florist outlets, cold display boxes are there which have regulated temperature and humidity to maintain the freshness of the flowers. The boxes are transparent to ensure display at the same time. One such product is refrigerated display case for flowers and the like. They are open-top, refrigerated case which may be mounted on a cart, booth or the like. The display case includes a plurality of elongated bins placed side by side in a stepped array from an upper bin at the rear of the case to a blower bin at the front of the case. The bins are thus advantageously arranged to provide a best possible display of merchandise within them. At the same time, the stepped arrangement permits a slow flow of refrigerated air as from back of the case, from the upper bin, and forwardly into the lower bin with the cool air cascading over merchandise especially within the lower bin.

The integrated cold chain seeks not only a change in infrastructure but also a change in mind-set. The cold chain management in India has to undergo a transformation, both in structure and approach. Let us look through the prism some of the major players and their roles (future) for:

Cost may come as a limitation but research on successful cold chain models in other countries suggests that the cost factor starts wearing off with economies of scale. It is just like getting out of the present vicious circle into a virtuous cycle that we are dreaming off. So with improvement in cold chain logistics the overall trade (both domestic as well as export) will get a boost, leading to large volumes and improved bottom-lines. The key is to integrate various elements and stakeholders of cold chain management, forge workable partnerships (even public-private partnerships wherever possible) to realize the vision for an efficient and integrated cold chain management for flowers throughout the country. Of course this has to be followed by aggressive marketing, to see that the demand for Indian flowers goes up both in domestic as well as export markets. This would definitely be a very strong and concrete step forward towards our ex-president Dr. Kalam's dream of seeing a blooming floriculture industry but also gradually make India a key player on the global floriculture scene.

The U.S. Floriculture Industry: Structural Changes, Marketing Practices, and Economic Impacts

The environmental horticulture industry, also known as the *Green Industry*, is comprised of a variety of businesses involved in production, distribution and services associated with ornamental plants, landscape and garden supplies and equipment. Segments of the industry include wholesale nursery, greenhouse and sod growers, landscape architects, contractors and maintenance firms, retail garden centres, home centres and mass merchandisers with lawn and garden departments, and marketing intermediaries such as brokers, horticultural distribution centres, and re-wholesalers. Environmental horticulture is one of the fastest growing segments of the nation's agricultural economy, often experiencing growth and expansion even during recessionary periods. The nursery and greenhouse sector has experienced considerable growth in the last two decades, albeit slowing somewhat in recent years. The landscape design, construction, and maintenance sector has also expanded due to strong economic conditions and robust building activity. Retail sales of horticultural goods have increased for both independent and chain-store type retailers, with considerable consolidation occurring due to the increased presence of home centres and mass merchants in the lawn and garden marketplace. The outlook for the Green Industry is promising, yet there are several challenges that will increase competitive pressures.

Economic impacts for the U.S. Green Industry are estimated at $147.8 billion (Bn) in output, 1,964,339 jobs, $95.1 Bn in value added, $64.3 Bn in labour income, and $6.9 Bn in indirect business taxes, with these values expressed in 2004 dollars.

For the production and manufacturing sectors, including nurseries/greenhouses, lawn and garden equipment manufacturers, and greenhouse manufacturers, total output impacts were $34.6 Bn, employment impacts were 300,677 jobs, and value added impacts were $20.8 Bn. For the horticultural services sectors of landscape services and landscape architects, total output impacts were $57.8 Bn, employment impacts were 753,557 jobs, and value added impacts were $39.0 Bn. For the wholesale/retail trade sectors, total output impacts were $55.5 Bn, employment impacts were 910,104 jobs, and value added impacts were $35.3 Bn. The largest individual sectors in terms of employment and value added impacts were landscaping services (704,875 jobs, $35.6 Bn), lawn and garden stores (347,916 jobs, $14.8 Bn), nursery and greenhouses (261,408 jobs, $18.1 Bn), florists (200,451 jobs, $4.0 Bn), and building material supply stores (123,591 jobs, $6.5 Bn). Other sectors with large value added impacts were general merchandise stores ($4.0 Bn), landscape architects ($3.5 Bn), lawn and garden equipment manufacturers ($2.6 Bn), lawn and garden equipment wholesalers ($2.7 Bn), wholesale flower, nursery stock and florist supplies ($1.9 Bn), and food & beverage stores ($1.4 Bn).

The Nature of Floriculture Production

Participants engaged in producing Green Industry products include growers of floriculture crops, nursery crops, and turfgrass sod. *Floriculture crops* include bedding plants, potted flowering plants, foliage plants, cut cultivated greens, and cut flowers. As distinguished from nursery crops, floriculture crops are generally herbaceous. Bedding and garden plants consist of young flowering plants (annuals and perennials) and vegetable plants. They are grown in flats, trays, pots, or hanging baskets, usually inside a controlled greenhouse environment, and sold largely for gardens and landscaping.

Potted flowering plants are largely sold in pots for indoor use. The major potted flowering plants are poinsettias, orchids, florist chrysanthemums, and finished florist azaleas. Foliage plants are also sold in pots and hanging baskets for indoor and patio use, including larger specimens for office, hotel, and restaurant interiors. Cut flowers are usually sold in bunches or as bouquets with cut foliage. The most popular cut flowers are roses, carnations, gladioli, and chrysanthemums. Leather leaf ferns are the leading cut foliage.

Combining cut flowers and cut greens in bouquets or other flower arrangements is a value-added retail option.

The market outlets for floriculture crops are florists, garden centres, mass merchandisers, supermarkets, chain stores, discount stores, home improvement centres, hardware stores, landscape contractors, and re-wholesalers.

Other retail outlets include farmers markets, flea markets, and street vendors. Since cut flowers are perishable and live floral crops are sensitive to variations in temperature, they usually require cool transportation and storage conditions that preserve and prolong their quality before final sale. The demand for floral crops, especially cut flowers, is highly seasonal. Sales are normally highest from February through May and in the fall. Sales of cut flowers peak during holidays such as Valentine's Day and Mother's Day. Poinsettia plants are sold mostly from Thanksgiving to Christmas. Cut flowers and foliage plants, however, are increasingly popular throughout the year as indoor home and workplace decorations.

Wholesale sales of floriculture products are usually handled by salespersons who have established relationships with large buyers. Marketing programs often include numerous trade shows, advertising in trade publications, catalogs, and direct mail. Close planning with large buyers (referred to as partnering) is required to secure long-term markets and to ensure that the right product mix is produced; however, demand for different products can still vary substantially from year to year. Sales and many variable expenses (costs-of-goods-sold) are highly seasonal, with up to 50 percent of sales in the second quarter of a typical year. Cash flow is uneven throughout the year so cash management is important. Technical knowledge of plants and pests is important for greenhouse management, although many of the everyday tasks (cultural practices) are routine and do not require highly specialized labour.

However, automation has proven to be effective but expensive. Greenhouse operations can often be very sophisticated, with automatic irrigation and fertilization (sometimes referred to as fertigation), and heating/cooling air and lighting systems driven by a variety of sensors. Innovations demanded by big-box retailers (such as custom packaging and labelling, bar codes, scanners, and electronic data interchange between suppliers and buyers) are now used by many producers.

In recent years, there has been considerable consolidation among large growers, largely in response to consolidation occurring at the retail level. The rise of large, nationwide plant retailers like home centres and mass merchandisers has created a marketing opportunity for growers who can supply the volume these customers require. Some firms have "expanded" their operations through acquisitions during the past decade, mainly to service these big customers. Large growers are geared to service big customers (big-box retailers and large landscape installation/distribution companies) by handling large volumes, which has the effect of discouraging smaller-volume buyers. Independent garden centres, retail nurseries, and smaller landscape firms tend to be supplied by small-to-medium sized growers. Proximity and premium product quality are more important to these smaller-sized buyers than low price because their targeted end consumer is more interested in quality and the breadth of retail selection. Keeping plants alive and healthy is a challenge for many consumers, and smaller retail operations typically have more technically knowledgeable staff than mass retailers to assist customers with expert plant care advice.

The Current Floriculture Situation

Despite relatively healthy growth of the U.S. economy, sales for ornamental crops in 2005 were modest. A 2-percent gain for all greenhouse and nursery receipts was evidenced following 1.3 percent growth in 2004. Sharply higher energy and gasoline prices in the last half of 2005 dampened consumer spending on discretionary goods such as ornamental crops. Sales declines are projected for cut flowers and cut cultivated greens and no growth for indoor foliage plants. Continued strong sales for bedding and garden plants and propagative material were largely responsible for the 2.3-percent gain in floriculture crop receipts in 2005.

Although sales of greenhouse and nursery crops in 2004 were stronger for growers in the Midwest and Northeast than in the South or West—the opposite of what happened in 2003—a more even growth distribution between regions occurred in 2005 since sales in all four U.S. regions are driven largely by bedding and garden annuals and herbaceous perennials. Half of total floriculture receipts in 2005 were from bedding and garden plants compared with 44 percent in 2000. The share of these crops in total grower receipts is larger in the Midwest and Northeast than in the South and West, but the gap is narrowing.

While growers in the Western and Southern States have larger average sales, their sales per acre of total production area is smaller than for growers in the Midwest and Northeast. The difference is accounted for by more intensive propagation of bedding and garden plants in the Midwest and Northeast, as well as potted flowering plants in the Northeast, compared with other floriculture crops and relative to Western and Southern States. Operations in the colder States of the Midwest and Northeast also utilize greenhouses for crop production to a greater extent given that their greenhouse-covered area represents a larger share in total production area than in the South or West. Floriculture sales per acre of total production area averaged nearly $135,000 for producers in the Midwest in 2004, followed by $112,000 in the Northeast.

In the South and West, where growers utilize significantly more shade cover and open field for crop production, average sales per acre were lower at $82,500 and $88,500, respectively. Production costs, however, are higher in the Midwest and Northeast because of more extensive use of greenhouses for growing ornamental crops. Higher energy costs expended in heating greenhouses in the colder States in 2005 will push up their production costs and crop prices.

Average wholesale prices of domestic floriculture crops were up less than 1 percent in 2005 as cheaper imports of cut flowers and flowering plants raise their share in total supply. Indeed, import prices are anticipated to be lower than in 2004 as import volumes climb faster than import values in 2005. As a result, projected composite prices for cut flowers and flowering plants were somewhat down in 2005. Composite prices for floriculture crops were up 3.4 percent in 2004 as the dollar's lower exchange rate raised U.S. import prices.

The effect of higher fuel prices in 2005 was to weaken consumer demand for domestic ornamental crops as spending is constrained by higher energy and gasoline expenses.

Prices of U.S. grown floriculture crops in 2005 were forecast to reach 5 percent above prices in 2000, which is half the pace of import prices. Import prices of cut flowers and flowering plants in 2004 were 14 and 12 percent higher than in 2000, boosted by the weaker dollar. Prices of domestic potted flowering plants, fresh cut flowers, and especially herbaceous perennial plants are all expected to exceed 5 percent above their levels in 2000. Led by potted hardy/garden chrysanthemums, perennial bedding plant prices in 2004 were already 13 percent higher than in 2000. Only bedding and garden annual plant prices have remained flat since 1999 as production supply jumped 30 percent based on the number of containers sold through 2004. Prices of foliage plants for indoor or patio use are likely to rise due to widespread crop damage in Florida's Miami-Dade County caused by hurricane Katrina in early September 2005. Estimated losses amount to more than $300 million in Miami-Dade, including destroyed shade houses and possible root rot caused by flooding. Although prices of foliage plants cannot be directly determined due to the absence of quantity sold data, the effect of any net loss is to reduce total foliage plant receipts in 2005.

Nevertheless, producers of potted foliage plants averaged $424,600 in sales at wholesale in 2004, up 4 percent from 2003. The average size of their operations, which are largely located in Florida and California, more than doubled from only $204,500 in 1993.

Bedding and Garden Plants – Prices of annuals among bedding and garden plants are expected to remain relatively unchanged, rising only 0.5 percent in 2005. Prices of potted annuals, projected to be about 12 percent higher in 2005 than in 2000, have climbed twice as fast as prices of flats or hanging baskets since 2000. Nevertheless, average prices of herbaceous perennials lead all bedding and garden plants with 15 percent anticipated growth since 2000, largely due to higher prices for potted hardy/garden chrysanthemums.

Grower sales of herbaceous perennials were expected to be up by 4.5 percent in 2005, following 8 percent growth in 2004. Since 2000, perennials' sales will be up 65 percent in 2005, not unlike sales of all propagative materials. Sales of perennials by growers in Western States, particularly California, have shot up by 121 percent since 2000, about double the sales growth in the South and Midwest, and 3.5 times that in the Northeast. Sales of annuals, on the other hand, will experience only 15 percent expansion since 2000. Total grower sales of bedding and garden plants have almost doubled in the past decade, the fastest among the floriculture crop groups. While herbaceous perennials account for a large part of this growth (partly due to more plant varieties surveyed by NASS in 2000), sales of potted annuals and hanging baskets have significantly increased over the past decade, eclipsing annuals in flats. In fact, the number of annuals in flats sold in 2004 was back down to its level in 1995. These trends coincide roughly with average sales per US household in which purchases of annuals in pots and hanging baskets have grown sharply in contrast to lower spending for annuals in flats in recent years.

Potted Flowering Plants – Sales of potted flowering plants are projected at $831.4 million in 2005, up 2 percent from 2004. These sales are driven by orchids, poinsettias, florist chrysanthemums, and spring flowering bulbs.

These plants also lead in the number of pots sold. In average price per pot sold, the plants that exceed $4 are florist azaleas, Easter lilies, and poinsettias. Orchids averaged $7.70 per pot sold at wholesale in 2004, the highest among all major flowering plants. Average prices for all major plants in 2004 were highest in the Midwest at $4.25 per pot. Producers in the West and South received $4.07 per pot on average in 2004, and $3.48 in the Northeast. Price inflation per pot was most favourable to growers in the Northeast and Midwest—advancing by 17 and 15 percent respectively, from 2000 to 2004. For growers in the South and West, prices received per pot in 2004 were 4 percent lower than in 2000.

Prices of potted flowering plants have been relatively flat since 2001 as the number of pots sold annually has remained around 238 million on average from 1992 to 2005, within a narrow range of plus or minus 4 million pots. The stagnant demand for these plants is in part due to competition from less expensive bedding and garden annuals as well as perennials, and in part due to increasing imports. Imported orchid plants are rising rapidly in value and volume in 2005 despite a decline in 2004, an indicator of strong consumer demand. However, import prices are down sharply as cheaper plants from Thailand, South Korea, and Canada are penetrating the expanding U.S. market.

Growers of potted flowering plants averaged $374,300 sales at wholesale in 2004, up 6 percent from 2003. The average size of their operation has expanded by nearly 50 percent from $252,500 in 1996. However, for producers of orchids, average sales were $564,600 in 2004, up 19 percent from 2003. This value is almost twice as much as in 1996 when orchids were first surveyed by NASS. In contrast, growers of Easter lilies averaged only $63,100 in sales in 2004, the lowest among major potted flowering plants.

Thus, as would be expected, the number of growers producing Easter lilies have fallen by a third since 1996 while those growing orchids have increased 29 percent.

Cut Flowers — Growers of fresh cut flowers sold 784 million stems, bunches, and blooms worth $422 million in 2004. Both value and quantity sold were down from 2003 as growers continue to reduce production volume of roses, carnations, gladioli, and snapdragons despite higher or unchanged unit value prices. Although the average U.S. household spent $10 on cut flowers in 2004, up from $9.18 in 2003, about 62 cents out of each dollar spent were paid to importers and foreign growers. Imported cut flowers reached $700 million in 2004, up more than 16 percent but import value is expected to remain unchanged in 2005 despite a projected 3 percent gain in import quantity. Import prices are projected down as import quantity has grown faster than corresponding import value in January-June 2005.

The import share in total supply of cut flowers based on sales value was 62 percent on average in 2004, up from 59 percent in 2003. The import share for roses was 85 percent; pompon chrysanthemums was 75 percent; standard carnations 95 percent; and orchids 42 percent. Based on quantity sold, the import share in the supply of major cut flowers was 78 percent in 2004, which is much higher than the 62 percent based on value because U.S. growers are producing more higher-value cut flowers that face less competition from imports.

The number of fresh cut flower growers in the United States has dwindled to 536 in 2004 from 829 in 1997 and 667 in 2000. While growers' total sales receipts and quantity sold continue to decline, average sales per grower is still growing—$786,600 in 2004, indicating that growers have larger operations on average. Most of these large growers are located in California, whose cut flower sales in 2004 were $303.6 million. Growers in other States sold $118 million worth of cut flowers in 2004, less than half of California's sales receipts. Average cut flower sales per grower are highest for roses, lilies, gerbera daisies, gladioli, and tulips—all above or close to $400,000 in 2004. Except for roses and gladioli, total grower sales and quantity sold of these cut flowers have also grown since 2000.

Floriculture Industry Outlook

Competitive rivalry in the floriculture industry is intensifying, especially at the retail level. Lowe's, Home Depot, Wal-Mart and Kmart, the mass merchandising stores [often referred to as box store or chain stores] that sell truckloads of plants at "every-day-low-prices," are continuing to ramp up their presence in the lawn and gardening industry. Their largest competitors include thousands of independent garden centres across the nation that charge higher prices but typically provide far more service and variety. Box stores recognize that knowledge and service are key success factors and are attempting to reconcile deficiencies in that arena. For example, Home Depot Inc., North America's largest seller of lawn and garden products, has embraced a computerized gardening and landscaping training course organized by the University of Georgia's horticulture department to help increase the level of horticultural acumen their employees can provide to customers. These trained salespeople, in turn, offer workshops on such topics as weed control and garden planting much like the chain offers clinics on laying tile or hanging wall paper. Home Depot is also installing larger garden center formats in new stores and expanding centres at some of their older, established stores. And in a move to help it overcome its traditionally poor care of its plants, Home Depot is changing its relationship with its plant vendors though "pay-by-scan" compensation systems where vendors now will get paid only for plants that actually sell (much like a consignment system), putting the onus on them to make sure the flowers and shrubs sitting on the chain's shelves are attractive and in good shape. They will be expected to visit its 2,000+ stores daily with fresh deliveries if need be.

The other titan in garden goods, Lowe's Companies, Inc., also has developed aggressive strategies for capturing for market share. Lowe's is working with selected breeders to devise exclusive plant varieties for its own premium plant brand, Garden Club Select. It also has joined with rose-industry pioneer Jackson & Perkins to sell J&P-branded roses, a relationship it hopes to broaden by obtaining rights to "cause" roses, such as the Lady Diana rose, that would be unique to the 1,225-store chain.

While big box stores and mass merchants have captured over half of the amount Americans spend a year on lawn and garden plants, the independent garden centres aren't surrendering. They do not have the volume to compete effectively on price, so they attempt to compete with better selection and more value-added services, which are especially attractive to new gardeners.

They are introducing their own branded plants sold under house names or nationally available labels such as Proven Winners, Simply Beautiful, Flower Fields, Miracle Grow, and Plants that Work. Specific products have also reached national brand status like the Wave Rave petunia campaign by Pan American Seed. Notable gardening personalities such as Martha Stewart and P.Allen Smith are marketing their own collection of gardening products and plants. Some garden centres are going further, adding cafes, coffee bars, meeting spaces, or other amenities. Others offer extensive how-to workshops or provide free landscaping advice and/or designs or plant shrubs for customers (though usually at an extra charge) and should a plant have problems, in-house "experts" will assist in diagnosing what went wrong.

Structural Changes in the Industry — There is little doubt that the impacts of mass marketers on the nursery and floricultural industry are farreaching. To their credit, many would argue that the box store chains have exposed many more consumers to nursery and floral products, thereby increasing the "size of the pie". Undoubtedly this is true, as the presence of mass marketers has opened not only the consumers' eyes to the industry's products but provided additional marketing opportunities for growers as well. Obviously, earlier discussion leads to conclusion that one of the impacts of the mass marketing of floricultural crops has been to promulgate an increase in overall size of growing operations. The capital requirements needed to afford the greenhouse infrastructures required to produce mass quantities of product in a confined marketing window exceed those that this industry has historically managed. Most firms have been able to generate the capital on their own, but the industry also has seen examples of investment brokers entering the industry to help finance some of these production operations. The financial returns of many of these venture capital acquisitions have seldom met Wall Street expectations however.

In some instances, chain store buyers have limited the number of approved vendors with whom they deal in any market area, as chains have come to realize certain procurement and merchandising efficiencies if fewer vendors are utilized. Chains have begun asking vendors to provide plant care for instore displays, especially during the peak of the bedding/garden plant season, something that is easier to request if there are only a few firms handling most of the merchandise. Whether or not producers are adequately rewarded for the additional expense of providing fully managed retail displays is debatable, but some growers report that the improved product care leads to additional product turns, which in turn provide the needed financial (gross margin) results.

There are also several instances of large producers partnering (usually on a contract basis) with several smaller firms in order to handle the volumes required to supply large retail chains. In some instances, there may be several dozen growers involved in cross-docking activities to satisfy one chain's product supply needs in one market area. Depending upon the arrangements, this helps to spread the risk among several producers. Still, there are numerous examples of producers who supply 50 to 75 percent of their output to one chain. When asked about the market risk, these growers often respond with discussions about production efficiencies, economies of scale, and questions about what they could do even if they wanted to change, noting that their competitors would be more than happy to take over the account.

In contrast, the focus on mass marketers by large growers has created opportunities for smaller growers to develop niches serving independent retailers/landscapers or to go into retailing themselves, selling directly to the consumer. In a recent national survey of growers, it was found that 59 percent of several thousand producers surveyed did retail some portion of their product mix on their own, ranging from 1 to 80 percent of their production. Smaller growers appeared to sell higher percentages, on average, of their production on a retail basis either from drive-in customer traffic at the nursery/greenhouse or at wholly-owned retail sales yards. Yet, some larger producers have also used their own retail outlets as a tactic for risk diversification.

The other impact of mass marketers has been consolidation within the production sector. In recent years, grower numbers have appeared to decline from year to year, or at best, remain stable. One could debate as to the reasons producer numbers are diminishing, but many would argue that the stresses of either supplying mass marketers or competing with them as an independent grower-retailer are taking their toll. The capitalization requirements, increased input costs (e.g. fuel), reduced margins, increased demands from buyers, and the market power associated with fewer numbers of buyers have all created intense market pressures and heightened competitive rivalry among larger producers. The struggle to remain competitive in a viable niche for smaller producers can be equally trying in markets being inundated by competing chains.

The long-term consequences are uncertain, but the need to recognize and closely monitor consumer expectations, tastes, and preferences is imperative. If consumers were, for some reason, to develop a negative impression of the industry's products or for how those products are presented by a particular retail giant, this could have dire consequences for future growth in the entire industry. Keeping the consumer intrigued and sufficiently motivated is important, and will likely require collaborative efforts among participants in each sector of the industry-wide value chain.

Is the Industry Showing Signs of Maturing?

A maturing industry is one that is moving from rapid growth to significantly slower growth. An industry is said to be mature when nearly all potential buyers are already users of the industry's products. Market demand consists mainly of replacement sales to existing users, with growth hinging on the industry's ability to attract new buyers and convince existing buyers to increase their usage. In the case of consumer goods industries, maturity means that they typically have a growth rate under 5 percent – roughly equal to the growth of the customer base or economy as a whole.

An industry's transition to maturity does not begin on an easily predicted schedule. Industry maturity can be forestalled by the emergence of new technological advances, frequent product innovations, or other driving forces that keep rejuvenating market demand. Nonetheless, when growth rates do slacken, the onset of market maturity usually produces fundamental changes in the industry's competitive environment including some of which we are seeing in the floriculture industry:

- Slowing growth in demand is generating more head-to-head competition for market share. Firms that want to continue on a rapid-growth track are starting to look for ways to entice customers away from competitors. In such situations, price cutting, increased advertising, and other aggressive tactics to gain market share are common.
- Buyers are becoming more sophisticated, often driving a harder bargain or requiring additional services in order to repeat purchases. Since buyers have experience with the product and are familiar with competing brands, they are better able to evaluate different brands and can use their market power (knowledge) to negotiate a better deal with vendors.
- Competition is producing a greater emphasis and cost and service. As growers all begin to offer the product attributes buyers prefer, buyer choices increasingly depend on which grower offers the best combination of price and service.
- Growers are experiencing a topping-out problem in adding new facilities. Reduced rates on industry growth mean slowdowns in capacity expansion for input manufacturers and slowdowns in new store growth for retails chains. With slower industry growth, adding too much capacity too soon can create oversupply conditions that adversely affect company profits well into the future.
- Product innovation and new end-user applications are becoming harder to come by. Breeders/growers find it increasingly difficult to create new product features and sustain buyer excitement.
- International competition continues to increase. Growth-minded domestic firms are starting to seek out sales opportunities in foreign markets. Some companies, looking for ways to cut costs, relocate growing operations to countries with lower wage rates. Industry leadership passes to companies that succeed in building strong competitive positions in most of the world's major geographic markets and in winning the biggest global market shares.
- Industry profitability is being influenced by tighter margins. Slower growth, increased competition, more sophisticated [and maybe fewer] buyers that place greater demands on their vendors, stagnant price levels, and occasional periods of overcapacity have put pressure on industry profit margins. Weaker, less-efficient firms are usually the hardest hit.
- Stiffening competition has induced a number of mergers and acquisitions among former competitors, driving the weakest firms out of the industry, and producing industry consolidation in general. Inefficient firms and firms with weak competitive strategies can achieve respectable results in a fast-growing industry with booming sales. But the intensifying competition of a maturing market throws second-and-third-tier competitors into a survival-of-the-fittest contest.

Strategic Moves in a Maturing Industry

As the new competitive character of floriculture industry maturity begins to hit full force, any of several strategic moves can strengthen a growers' competitive position including pruning the product line, improving value chain efficiency, trimming costs, accelerating marketing and sales promotion efforts, and acquiring struggling competitors.

Pruning Marginal Products and Sizes – A wide selection of plant materials and sizes sometimes has competitive value during the growth stage, when buyers needs are still evolving. But such variety can become too costly as price competition stiffens and profit margins are squeezed.

Maintaining many product "versions" works against achieving production economies at the growing level and can increase inventory stocking costs for re-wholesale distributors and retailers. In addition, the prices of slow-selling plants may not cover their true costs. Pruning marginal products from the product mix opens the door for cost savings and permits more concentration on items whose margins are highest and/or where the firm has a competitive advantage.

More Emphasis on Value Chain Innovation – Efforts to reinvent the industry value chain can have a fourfold payoff: Lower costs, better product or service quality, greater capability to turn out multiple or customized product variations (upgrades), and shorter design-to-market cycles. Growers can mechanize high-cost activities, re-design production practices to improve labour efficiency, build flexibility into the "assembly" process so that customized product versions can be easily produced, and increase use of advanced technology (robotics, computerized controls, and automated/guided vehicles.) Suppliers of parts and components, input manufacturers, distributors, and buyers can collaborate on the use of internet technology and e-commerce techniques to streamline various value chain activities and implement cost-saving innovations.

A Stronger Focus on Cost Reduction – Stiffening price competition gives growers extra incentive to drive down unit costs. Company costreduction initiatives can cover a broad front. Some of the most frequently pursued options are pushing suppliers for better prices, implementing tighter supply chain management practices, cutting low-value activities out of the value chain, developing more economical product designs, streamline order processing and pulling, reengineering internal processes using e-commerce technology, and shifting to more economical distribution arrangements and systems (e.g. racking systems).

Increasing Sales to Present Customers – In a mature market, growing by taking customers away from rivals may not be as appealing as expanding sales to existing customers. Strategies to increase purchases by existing customers can involve providing complementary items and ancillary services, and finding more ways for customers to use the product. Developing deeper relationships with key buyers (through personal visits, customized mailings, etc.) will more often than not pay big dividends.

Purchasing Rival Firms at Bargain Prices – In a maturing market, sometimes a firm can acquire the facilities and assets of struggling rivals quite cheaply. Bargain-priced acquisitions can help create a low-cost position if they also present opportunities for greater operating efficiency. In addition, an acquired firm's customer base can provide expanded market coverage and opportunities for greater scale economies. The most desirable acquisitions are those that will significantly enhance the acquiring firm's competitive strength.

Building New Or More Flexible Capabilities – The stiffening pressures of competition in a maturing or already mature market can often be combated by strengthening the company's resource base and competitive capabilities. This can mean adding new competencies or capabilities (by either making or buying them), deepening existing competencies to make them harder to imitate, or striving to make core competencies more adaptable to changing customer requirements and expectations.

Strategic Pitfalls in Maturing Industries

Perhaps the biggest strategic mistake a company can make as the floricultural industry matures is steering a middle course between low cost, differentiation, and focusing; blending efforts to achieve low cost with efforts to incorporate differentiating features and efforts to focus on a limited (niche or cache) target market. Such strategic compromises typically result in a firm ending up stuck in the middle, with a fuzzy strategy, too little commitment to winning a competitive advantage, an average image with buyers, and little chance of springing into the leading ranks of the industry.

Other strategic pitfalls include being slow to adapt existing competencies and capabilities to defend against stiffening competitive pressures, concentrating more on protecting short-term profitability than on building or maintaining long-term competitive position, waiting too long to respond to price cutting by rivals, over-expanding in the face of slowing growth, overspending on advertising and sales promotion efforts in a losing effort to combat the growth slowdown, and failing to pursue cost reductions and/or production efficiencies soon enough or aggressively enough.

Export Opportunities for Indian Floriculture

But for the Indian floriculture exporters, the party was over—at least for this year. This is because roses account for 99 per cent of total Indian flower exports, which, according to Agriculture and Processed Food Products Export Development Agency (APEDA), aggregated to Rs 340.14 crore ($84.49 million) in 2007-08. "Traders held back the orders due to uncertainty. One of the leading traders, who normally buys six million roses for Valentine's Day, had firmed up orders for just one million by February 6," says Ramakrishna Karuturi, Managing Director, Karuturi Global, one of the largest players in the industry, with a presence in both India and Africa. Adds Anne Ramesh, Chairman, Suvarna Florex, and a veteran in the industry: "Every year, we used to get firm orders for 70 per cent of the export quantity by February 1. This year, even as late as February 9, we had orders for less than 20 per cent of the volume. This meant that a huge quantity went into auction and prices, naturally, came under pressure." If at all the exporters managed to liquidate the stocks, it was at the cost of realisation, which fell by 15-20 per cent, the sharpest decline ever since the industry began to export flowers some 15 years ago.

This development has a much larger impact on the sector as the ideal climatic condition for producing best quality roses in India is between December and March. It is only during this period that the flowers effectively compete, in quality terms, with other foreign firms in the global market and the exporters can earn maximum value. In fact, the money they make during this period helps them compensate for lower income during rest of the year. Inability to make money during the Valentine's Day celebrations will impact the industry's efforts to expand, develop infrastructure and experiment with newer varieties. Economies of scale are crucial to defray fixed costs. Indian farms are small and fragmented compared to those in Africa. Infrastructure facilities such as green houses, cold room and refrigerated vans need constant investments. More importantly, offering newer varieties is crucial for better prices. But it is a costly proposition as it involves paying royalty fees as high as Rs 50 a stem to the breeders. In fact, lack of new varieties is one of the reasons (apart from recession in Japan and competition from Kenya) that India lost the demanding Japanese market. In 2006-07, floriculture exports from India to Japan was Rs 325.54 crore. It crashed to Rs 32.77 crore in 2007-08.

Floriculture or flower farming, is a discipline of horticulture concerned with the cultivation of flowering and ornamental plants for gardens and floristry, comprising the floral industry. The development plant breeding of new varieties is a major occupation of floriculturists.

Floriculture crops include bedding plants, flowering plants, foliage plants or houseplants, cut cultivated greens, and cut flowers. As distinguished from nursery crops, floriculture crops are generally herbaceous. Bedding and garden plants consist of young flowering plants (annuals and perennials) and vegetable plants. They are grown in cell packs (in flats or trays), in pots, or in hanging baskets, usually inside a controlled environment, and sold largely for gardens and landscaping. Geraniums, impatiens, and petunias are the best-selling bedding plants. Chrysanthemums are the major perennial garden plant in the United States.

Flowering plants are largely sold in pots for indoor use. The major flowering plants are poinsettias, orchids, florist chrysanthemums, and finished florist azaleas. Foliage plants are also sold in pots and hanging baskets for indoor and patio use, including larger specimens for office, hotel, and restaurant interiors.

Cut flowers are usually sold in bunches or as bouquets with cut foliage. The production of cut flowers is specifically known as the cut flower industry. Farming flowers and foliage employs special aspects of floriculture, such as spacing, training and pruning plants for optimal flower harvest; and post-harvest treatment such as chemical treatments, storage, preservation and packaging. In Australia some species are harvested from the wild for the cut flower market.

Area of Cultivation

Of the four zones identified as potential centres for flower production namely Bangalore, Hyderabad, Pune and New Delhi, the area around Bangalore and Pune have got the advantage of ideal climatic conditions where the temperature ranges between 15 to 30°C. In view of this, the units established in these locations do not require either cooling or heating system.

The units located around New Delhi, though are near to the to the international airport but have the disadvantage of high as well as low temperatures resulting in additional capital investment for installation of cooling and heating system, besides incurring additional running cost for the power and energy Hyderabad has got the disadvantage of experiencing high summer temperature, necessitating the airport forcing the units to send flowers through Bangalore of Mumbai airport, which resulted in additional cost compared to the sectors like Pune and Bangalore.

Among the four major sectors from the point of view of agro-climatic conditions and cost effectiveness Bangalore is considered the most ideal location for setting up of floriculture units for export followed by Pune. This is evident from maximum number of units being established in these two sectors.

Rose

Plenty of light, humid and moderate temperature ranging from 150°C to 200°C is considered as an ideal condition for production of roses in the tropical and sub-tropical climate of India. When roses are grown at a temperature below 150°C, the interval between flushes becomes long. At higher temperature above 300°C roses can be grown provided high humidity is maintained and evaporation rate is brought down.

Rose crop requires to be grown in a soil with good drainage properties rich in organic matter and oxygen. Organic matter as high as 30% in the top 30cm of the growing beds is provided. The pH of the soil should be around 6 to 6.5, i.e., slightly acidic soil is considered more suitable for rose cultivation.

In case the plant is in green house, the area is normally divided into 2 parts A&B, consisting roughly ½ ha, each. When the green house area is 1 ha. Each area is having about 91 meters length is which 56 beds of 1.2 meter each and 0.4 metes for furrow path area are provided. Therefore, the size of each bed is of 26.25-meter length, 1.2-meter width.

The plants are planted in the bed by providing spacing of 0.6 meter between rows and 0.17 meter between the plants. With this spacing, the plant density comes to 700000 plants per ha. As against the 2 rows system of planting followed in many units, some of the companies like Unit-11, Unit-20, Unit-19 etc. established with the collaboration of M/s Flodac have gone for 3 rows system which has resulted in high rate of fungal disease and also difficulty faced in carrying out inter-cultural operations. These units have removed the 3rd row in the second year of planting hence; it is advisable to go only for 2 rows system of planting for better inter-cultural operation and also to avoid attack of fungal diseases.

Rose plants under protected green house condition require approximately 1 litre of water per plant per day. Except for the very old units, most of the other units are adopting drip irrigation system from the reputed companies to deliver the above quantity of water. A drainage line is also provided below the beds for disposal of excess water.

Organic manure is required to be added so that top 30 cms of the soil has 30% of organic matter content. A dose of 25 kg of manure per square meter will take care of this requirement. Applications of inorganic fertilizers including micronutrients are supplied during irrigation time alongwith the Drip lines. Majority of the units are having fertigation tanks and supplying the required quantity of both major and minor nutrients as per the requirement of the plant based on the soil test reports.

Nutritional requirements, pruning methodology, planting distance and use of growth-regulating chemicals in field-grown roses cv. Super Star and Raktagandha for production of first grade flowers with maximum yield have been standardized at the IARI, New Delhi. A fertilizer dose comprising 520 kg N, 868 kg P2O5 and 694 kg K2O has been recommended/ha. Among the secondary nutrients, application of 50 kg MgSO4/ha was found to be beneficial. Out of the several micronutrients, spray of FeSO4 1.0% and ZnSO4 0.5% during November and January was most effective. Plants pruned at 45 cm height from ground level retaining only 4 healthy shoots during mid-October to mid-November, an spaced at a distance of 30 cm x 40 cm gave maximum flower yield per unit area.

Gladiolus

Gladiolus is one of the most important bulbous flowering crops grown commercially for cut-flower trade in India. Fertilizer trials have revealed that best results an be obtained with 400 g N, 100g P and 200 g K/ha in cv Snow Princess at the PAU, Ludhiana; 500 g N, 200 g P and 300 g K/ha in cv Vinks Glory at the BCKV, West Bengal; 400 g N, 300 g P and 200 K/ha in cv Friendship at the IIHR, Bangalore, and 500 g N, 200 P and 200 g K/ha in cv Tropic Sea at the BCKV, Kalimpong.

Chrysanthemum

Chrysanthemums, often called mums or chrysanths, are of the genus (*Chrysanthemum*) constituting approximately 30 species of perennial flowering plants in the family Asteraceae which is native to Asia and northeastern Europe.

Etymology

The name Chrysanthemum is derived from the Greek, *chrysos* (gold) and *anthos* (flower).

Taxonomy

The genus once included a larger number of species, but was split several decades ago into several genera. The naming of the genera has been contentious, but a ruling of the International Code of Botanical Nomenclature in 1999 resulted in the defining species of the genus being changed to *Chrysanthemum indicum*, thereby restoring the economically important florist's chrysanthemum to the genus *Chrysanthemum*. During the period between the splitting of the genus and the ICBN ruling, these species have customarily been included under the genus name *Dendranthema*.

The other species previously included in the narrow view of the genus *Chrysanthemum* are now transferred to the genus *Glebionis*. The other genera separate from *Chrysanthemum* include *Argyranthemum*, *Leucanthemopsis*, *Leucanthemum*, *Rhodanthemum*, and *Tanacetum*.

Chrysanthemum are herbaceous perennial plants growing to 50–150 cm tall, with deeply lobed leaves with large flower heads that are generally white, yellow or pink in the wild and are the preferred diet of larvae of certain lepidoptera species.

History

Chrysanthemums were first cultivated in China as a flowering herb as far back as the 15th century BC. An ancient Chinese city (Xiaolan Town of Zhongshan City) was named Ju-Xian, meaning "chrysanthemum city". The plant is particularly significant during the Double Ninth Festival. It is believed that the flower may have been brought to Japan in the 8th century CE, and the Emperor adopted the flower as his official seal. There is a "Festival of Happiness" in Japan that celebrates the flower.

The flower was brought to Europe in the 17th century. Linnaeus named it from the Greek word *chrysous*, "golden" (the colour of the original flowers), and *anthemon*, meaning flower.

Economic Uses

Ornamental Uses

Modern chrysanthemums are much more showy than their wild relatives. The flowers occur in various forms, and can be daisy-like, decorative, pompons or buttons. This genus contains many hybrids and thousands of cultivars developed for horticultural purposes. In addition to the traditional yellow, other colours are available, such as white, purple, and red. The most important hybrid is *Chrysanthemum* × *morifolium* (syn. *C.* × *grandiflorum*), derived primarily from *C. indicum* but also involving other species. Chrysanthemums are broken into two basic groups, Garden Hardy and Exhibition. Garden hardy mums are new perennials capable of being wintered over in the ground in most northern latitudes. Exhibition varieties are not usually as sturdy. Garden hardies are defined by their ability to produce an abundance of small blooms with little if any mechanical assistance (i.e., staking) and withstanding wind and rain. Exhibition varieties on the other hand require staking, over-wintering in a relatively dry cool environment, sometimes with the addition of night lights.

The Exhibition varieties can be used to create many amazing plant forms; Large disbudded blooms, spray forms, as well as many artistically trained forms, such as: Thousand Bloom, Standard (trees), Fans, Hanging Baskets, Topiary, Bonsai, and Cascades.

Chrysanthemum blooms are divided into 13 different bloom forms by the US National Chrysanthemum Society, Inc., which is in keeping with the international classification system. The bloom forms are defined by the way in which the ray and disk florets are arranged.

Chrysanthemum blooms are composed of many individual flowers (florets), each one capable of producing a seed. The disk florets are in the center of the bloom head, and the ray florets are on the perimeter. The ray florets are considered imperfect flowers, as they only possess the female productive organs, while the disk florets are considered perfect flowers as they possess both male and female reproductive organs.

Irregular Incurve: These are the giants of the chrysanthemum world. Quite often disbudded to create a single giant bloom (ogiku), the disk florets are completely concealed, while the ray florets curve inwardly to conceal the disk and also hang down to create a 'skirt'.

Reflex: The disk florets are concealed and the ray florets reflex outwards to create a mop like appearance.

Regular Incurve: Similar to the irregular incurves, only usually smaller blooms, with nearly perfect globular form. Disk florets are completely concealed. They used to be called 'Chinese'.

Decorative: Similar to reflex blooms without the mop like appearance. Disk florets are completely concealed, ray florets usually don't radiate at more than a 90 degree angle to the stem.

Intermediate Incurve: These blooms are in-between the Irregular and Regular incurves in both size and form. They usually have broader florets and a more loosely composed bloom. Again, the disk florets are completely concealed.

Pompon: *Note the spelling, it is not pompom. The blooms are fully double, of small size, and almost completely globular in form.

Single/Semi-Double: These blooms have completely exposed disk florets, with between 1 and 7 rows of ray florets, usually radiating at not more than a 90 degree angle to the stem.

Anemone: The disk florets are prominently featured, quite often raised and overshadowing the ray florets.

Spoon: The disk florets are visible and the long tubular ray florets are spatulate.

Quill: The disk florets are completely concealed, and the ray florets are tube like.

Spider: The disk florets are completely concealed, and the ray florets are tube like with hooked or barbed ends, hanging loosely around the stem.

Brush & Thistle: The disk florets may be visible. The ray florets are often tube like, and project all around the flower head, or project parallel to the stem.

Exotic: These blooms defy classification as they possess the attributes of more than one of the other twelve bloom types.

Chrysanthemum leaves resemble its close cousin, the mugwort weed — so much so that mugwort is sometimes called wild chrysanthemum — making them not always the first choice for professional gardeners.

Culinary Uses

Yellow or white chrysanthemum flowers of the species *C. morifolium* are boiled to make a sweet drink in some parts of Asia. The resulting beverage is known simply as "chrysanthemum tea". Chrysanthemum tea has many medicinal uses, including an aid in recovery from influenza. In Korea, a rice wine flavoured with chrysanthemum flowers is called *gukhwaju*.

Chrysanthemum leaves are steamed or boiled and used as greens, especially in Chinese cuisine. Other uses include using the petals of chrysanthemum to mix with a thick snake meat soup in order to enhance the aroma.

Insecticidal Uses

Pyrethrum (*Chrysanthemum* [or *Tanacetum*] *cinerariaefolium*) is economically important as a natural source of insecticide. The flowers are pulverized, and the active components called pyrethrins, contained in the seed cases, are extracted and sold in the form of an oleoresin. This is applied as a suspension in water or oil, or as a powder. Pyrethrins attack the nervous systems of all insects, and inhibit female mosquitoes from biting. When not present in amounts fatal to insects, they still appear to have an insect repellent effect. They are harmful to fish, but are far less toxic to mammals and birds than many synthetic insecticides, except in consumer airborne backyard applications. They are non-persistent, being biodegradable and also breaking down easily on exposure to light. They are considered to be amongst the safest insecticides for use around food. (Pyrethroids are synthetic insecticides based on natural pyrethrum, e.g., permethrin.

Environmental Uses

Chrysanthemum plants have been shown to reduce indoor air pollution by the NASA Clean Air Study.

Medicinal Uses

Extracts of Chrysanthemum plants (stem and flower) have been shown to have a wide variety of potential medicinal properties, including anti-HIV-1, antibacterial and antimycotic.

Chrysanthemum is recognized as a potent flower crop in India. It is used as a cut-flower for interior decoration and as a loose flowers for making garlands, wreaths and for religious offerings. Technology for year-round blooming in chryasanthemum has been standardized at the NBRI, Lucknow. About 150 cultivators have been screened at the institute for their photo-induction requirement and these have been classified into 7 response groups ranging from 7 to 13 weeks. The miniculture technique standardized at the NBRI, shows how pot culture of chrysanthemum can be made cheaper, easier, aesthetically rewarding and commercially more profitable.

Bougainvillea

Bougainvillea is a genus of flowering plants native to South America from Brazil west to Peru and south to southern Argentina (Chubut Province). Different authors accept between four and 18 species in the genus. The plant was classified by Europeans in Brazil in 1768, by Philibert Commerçon, a French botanist accompanying French Navy admiral and explorer Louis Antoine de Bougainville during his voyage of circumnavigation.

They are thorny, woody vines growing anywhere from 1-12 meters tall, scrambling over other plants with their hooked thorns. The thorns are tipped with a black, waxy substance. They are evergreen where rainfall occurs all year, or deciduous if there is a dry season. The leaves are alternate, simple ovate-acuminate, 4-13 cm long and 2-6 cm broad. The actual flower of the plant is small and generally white, but each cluster of three flowers is surrounded by three or six bracts with the bright colours associated with the plant, including pink, magenta, purple, red, orange, white, or yellow. *Bougainvillea glabra* is sometimes referred to as "paper flower" because the bracts are thin and papery. The fruit is a narrow five-lobed achene.

Bougainvillea are relatively pest-free plants, but may suffer from worms, snails and aphids. The larvae of some Lepidoptera species also use them as food plants, for example the Giant Leopard Moth.

Cultivation and Uses

Bougainvilleas are popular ornamental plants in most areas with warm climates, including Ethiopia, Indonesia, Aruba, the Philippines, Thailand, Pakistan, India, Sri Lanka, Taiwan, Vietnam, Malaysia, Australia, Greece, Turkey, Cyprus, Singapore, the Mediterranean region, the Caribbean, Central America, Mexico, South Africa, Kuwait, the United Arab Emirates and the United States in Arizona, California, Florida, Hawaii, Louisiana, South Carolina, and southern Texas. Locarno in Switzerland, with its mild Mediterranean climate, is famous for its bougainvilleas.

Although it is frost-sensitive and hardy in U.S. Hardiness Zones 9b and 10, bougainvillea can be used as a houseplant or hanging basket in cooler climates. In the landscape, it makes an excellent hot season plant, and its drought tolerance makes bougainvillea ideal for warm climates year-round. Bougainvillea has a high salt tolerance, which makes it a natural choice for colour on coastal regions. As a woody clambering vine, bougainvillea will stand alone and can be pruned into a standard, but it is perfect along fence lines, on walls, in containers and hanging baskets, and as a hedge or an accent plant. Its long arching branches are thorny, and bear heart-shaped leaves and masses of papery bracts in white, pink, orange, purple, and burgundy. Many cultivars, including double flowered and variegated, are available. Also it needs 60 degrees in the winter for its growing condition

Twenty years after Commerçon's discovery, it was first published as 'Buginvillea' in Genera Plantarum by A.L. de Jusseau in 1789. The genus was subsequently split in several ways until it was finally corrected to 'Bougainvillea' in the Index Kewensis in the 1930s. Originally, B. spectabilis and B. glabra were hardly differentiated until the mid 1980s when botanists recognized them to be totally distinct species. In early 19th century, these two species were the first to be introduced into Europe, and soon, nurseries in France and England did a thriving trade providing specimens to Australia and other faraway countries. Meanwhile, Kew Gardens distributed plants it had propagated to British colonies throughout the world. Soon thereafter, an important event in the history of bougainvillea took place with the discovery of a crimson bougainvillea in Cartagena, a Spanish port in the Mediterranean, by Mrs. R.V. Butt. Originally thought to be a distinct species, it was named B. buttiana in her honour. However, it was later discovered to be a natural hybrid of a variety of B. glabra and possibly B. peruviana a "local pink bougainvillea" from Peru. Natural hybrids were soon found to be common occurrences all over the world. For instance, around the 1930s, when the three species were grown together, many hybrid crosses were created almost spontaneously in East Africa, India, the Canary Islands, Australia, North America, and the Philippines.

Many of today's bougainvillea are the result of interbreeding among only three out of the eighteen South American species recognized by botanists. Currently, there are over 300 varieties of bougainvillea around the world. Because many of the hybrids have been crossed over several generations, it's difficult to identify their respective origins. Natural mutations seem to occur spontaneously throughout the world; wherever large numbers of plants are being produced, bud-sports will occur. This had led to multiple names for the same cultivar (or variety) and has added to the confusion over the names of bougainvillea cultivars.

The growth rate of *Bougainvillea* vary from slow-growing to rapid, depending on the particular variety. Bougainvillea tend to flower all year round in equatorial regions. Elsewhere, they are seasonal bloomers. They grow best in somewhat dry, fertile soil. Bloom cycles are typically four to six weeks. Bougainvillea grow best in very bright full sun and with frequent fertilization, but the plant requires little water once established. As indoor houseplants in temperate regions, they can be kept small by bonsai techniques. If overwatered, *Bougainvillea* will not flower and may lose leaves or wilt, or even die from root decay.

Symbolism

Various species of *Bougainvillea* are the official flowers of the island of Grenada, Guam, Lienchiang and Pingtung Counties in Taiwan, Ipoh, Malaysia and the cities of Tagbilaran, Philippines; Camarillo, California; Laguna Niguel, California; San Clemente, California; and Naha, Okinawa.

Toxicity

The sap of the *Bougainvillea* can cause skin rashes similar to *Toxicodendron* species.

Species

- *Bougainvillea* × *buttiana* Holttum & Standl. (*B. glabra* × *B. peruviana*)

- *Bougainvillea glabra* Choisy
- *Bougainvillea peruviana*
- *Bougainvillea spectabilis* Willd.
- *Bougainvillea spinosa.*

Bougainvillea is an important and popular flowering plant grown widely throughout the country. To obtain dwarf, bushy and attractive pot plants, 8 varieties of bougainvillea were treated with cycocel and SADH as foliar sprays once at 4,000 an 8,000 ppm, and twice within 8,000 ppm. Soil drench with cycocel was done at 2,000 and 4,000 ppm once, and twice with 4,000 ppm. Considering the reduction in plant height and improvement in flowering, double application of SADH at 8,000 ppm as foliar spray was recommended for dwarfing different varieties of bougainvillea in pots. SADH treated plants also produced larger number of axillary branches and flowering shoots.

Jasmine

Jasmine is a genus of shrubs and vines in the olive family (Oleaceae), with about 200 species, native to tropical and warm temperate regions of the Old World. The leaves can be either evergreen (green all year round) or deciduous (falling in autumn).

Species

Species include:

- *Jasminum abyssinicum* Hochst. ex DC. – Forest jasmine
- *Jasminum adenophyllum* Wall. – Pinwheel Jasmine, Bluegrape jasmine, Princess jasmine, Che vang, Lai la co tuyen
- *Jasminum angulare* Vahl
- *Jasminum dichotomum* Vahl – Gold Coast Jasmine
- *Jasminum didymum* (indigenous to Samoa Islands)
- *Jasminum grandiflorum* L. – Spanish Jasmine, Royal Jasmine, Catalonian Jasmine
- *Jasminum humile* L. – Italian Yellow Jasmine
- *Jasminum lanceolarium* Roxb.
- *Jasminum mesnyi* Hance – Japanese Jasmine, Primrose Jasmine, Yellow Jasmine
- *Jasminum multiflorum* Hance – Indian Jasmine, Winter Jasmine
- *Jasminum multipartitum* Hochst. - Starry Wild Jasmine
- *Jasminum nervosum* Lour.
- *Jasminum odoratissimum* L. – Yellow Jasmine
- *Jasminum officinale* L. – Common Jasmine, Poet's Jasmine, jasmine, jessamine
- *Jasminum parkeri* Dunn – Dwarf Jasmine

- *Jasminum polyanthum Franch.*
- *Jasminum sambac* (L.) Aiton – Arabian Jasmine
- *Jasminum sinense* Hemsl.
- *Jasminum urophyllum* Hemsl.

Cultivation and Uses

Widely cultivated for its flowers, jasmine is enjoyed in the garden, as a house plant, and as cut flowers. The flowers are worn by women in their hair in southern and southeast Asia. The delicate jasmine flower opens only at night and may be plucked in the morning when the tiny petals are tightly closed, then stored in a cool place until night. The petals begin to open between six and eight in the evening, as the temperature lowers.

Jasmine Tea

Jasmine tea is consumed in China, where it is called jasmine-flower tea. *Jasminum sambac* flowers are also used to make so-called jasmine tea, which often has a base of green tea, but sometimes an Oolong base is used. Flowers and tea are "mated" in machines that control temperature and humidity. It takes four hours or so for the tea to absorb the fragrance and flavour of the jasmine blossoms, and for the highest grades, this process may be repeated as many as seven times. Because the tea has absorbed moisture from the flowers, it must be refired to prevent spoilage. The spent flowers may or may not be removed from the final product, as the flowers are completely dry and contain no aroma. Giant fans are used to blow away and remove the petals from the denser tea leaves. If present, they simply add visual appeal and are no indication of the quality of the tea.

Jasmine Syrup

The French are known for their jasmine syrup, most commonly made from an extract of jasmine flowers. In the United States, this French jasmine syrup is used to make jasmine scones and marshmallows.

Jasmine Essential Oil

Jasmine essential oil is in common use. Its flowers are either extracted by the labour-intensive method of enfleurage or through chemical extraction. It is expensive due to the large number of flowers needed to produce a small amount of oil. The flowers have to be gathered at night because the odour of jasmine is more powerful after dark. The flowers are laid out on cotton cloths soaked in olive oil for several days and then extracted leaving the true jasmine essence. Some of the countries producing jasmine essential oil are India, Egypt, China and Morocco.

Jasmine Absolute Used in Perfume and Incense

Its chemical constituents include methyl anthranilate, indole, benzyl alcohol, linalool, and skatole. Many species also yield an absolute, which is used in perfumes and incense.

Cultural Importance and other Information

- In Syria, jasmine is the symbolic flower of Damascus, which is called the City of Jasmine.
- In Thailand, jasmine flowers are used as a symbol of the mother.

Jasmine is the national flower of the following countries/states:

- Tunisia, it's the national flower
- Indonesia, where the variety *Jasminum sambac* is the "*puspa bangsa*" (national flower), and goes by the name "*Melati Putih*". It is the most important flower in wedding ceremonies for ethnic Indonesians, especially in the island of Java.
- Pakistan, where *Jasminum officinale* is known as the "*Chambeli*" or "*Yasmine*" is the national flower.
- Philippines, where it is known as "*Sampaguita*", and is usually strung on garlands which are then used to adorn religious images.
- Hawaii In Hawaii, Arabian Jasmine is called Pikake and is perhaps the most popular of flowers. It is often strung in leis and is the subject of many songs.

Cultural Importance of Jasmine in India

- The jasmine flower, depending on variety, has different names in many languages in India, and only one name in others. Some of those names are:
  * *Malati* mâlatî or *mallika* mallikâ in Sanskrit.
  * *Chameli* chamélî, *juhi* Juhî, or *motiya* motiyâ in Hindi. The latter is a thicker variety on smaller shrubs which may also grow as a climbers. Its name, *motiyâ*, refers to the flower being compared to a pearl in looks and beauty, since it is white, round and beautiful, *motî* meaning "pearl" in Hindi (from *muktâ*, *muktâmaGi*, or *mauktika* in Sanskrit, *muktâ* also meaning "free" in the sense of "unbounded").
  * *Jaaie*, *juie*, *saayalie*, *chamelie* or *mogaraa* in Marathi. The latter is equivalent to *motiyaa* in Hindi. The former has smaller leaves and larger petals, while the second has larger leaves and smaller petals; the third is creamy in texture (*saay* meaning "cream" in Marathi); and the fourth is yet another variety.
  * In Oriya Jasmine is known as *Mahli*.Jasminum officinale is known as Jai and Jasminum auriculatum is known as Juhi.
  * In Oriya culture Jasmine plays a big role people use in rituals like marriage, Puja and Lord Jagannatha's festivals in Chandan Yatra of lord Jagannath takes bathe with water flavoured in sandalwood paste and jasmine.
  * *Juie* in Bengali.
  * *Mallige, Malle, jaaji,* and *iruvantige* in Kannada. The double-flowered cultivar is called *mooru or eLu suttina mallige.*
  * *Malle* in Telugu. The double-flowered cultivar is called *gundu malle.*

  * *Malligai* in Tamil.
  * *Jaisara and Chameli* in [(Nepali)].

- In Tamil Nadu, jasmine is produced mainly in Madurai District, and is transported to Mumbai/Bombay for use there, as well as being exported to other countries. The city of Madurai is called *Malligai Maanagar* ("City of Jasmine").
- Jasmine is cultivated at Pangala, in Karnataka, India, and exported to Middle Eastern countries.
- Jasmine flowers are worn in girls' and womens' hair mainly for their beauty and fragrance. They are also used in floral decorations for marriages and other important functions.
- Throughout most of India, especially in the western and southern states, including Andhra Pradesh, Karnataka, Maharashtra, Tamil Nadu, etc., jasmine (along with many other flowers, including roses) is cultivated in private homes, within gardens or as potted plants. These flowers are used in regular worship at home as well as for hair ornaments (for the girls and women of the house). Jasmine is also cultivated commercially, for both the domestic purposes discussed above and other purposes (such as use in the perfume industry).
- Jasmine flower sellers (vendors) selling ready-made garlands of jasmine, or in the case of the thicker *motiyaa* (in Hindi) or *mogaraa* (in Marathi) varietal, bunches of jasmine, as well as flowers by weight, are a common sight on city streets in many parts of India. They may be found around entrances to temples, on major thoroughfares, and in major business areas (including bus stands). This is common as far north as Mumbai, and generally from Maharashtra downward through all of South India. Jasmine vendors may also be found in Kolkata, though roadside sales are fewer there, since in North India women and girls generally by tradition do not wear flowers in their hair.

As a Weed

Jasminum fluminense, which is sometimes known by the inaccurate name "Brazilian Jasmine", is an invasive species in Hawaii and Florida. *J. dichotomum*, also called Gold Coast Jasmine, is also an invasive weed in Florida.

Other Info

Jasmine gave name to the jasmonate plant hormones as methyl jasmonate isolated from the jasmine oil of Jasminum grandiflorum led to the dicovery of the molecular structure of jasmonates.

Jasmine finds a very important place both in the perfume industry and flower market. Research work on standardization of agro-technology in jasmine was carried in different parts of India; notable among these are the IIHR, Bangalore and the TNAU, Coimbatore.

A complete agro-technology has been worked out for Jasminum grandiflorum at the IIHR, Bangalore, through large number of field trials. For maximum yield of flowers and high recovery of essential oil, the plants are required to be spaced at a distance of 1.80 m^2, pruned at 90 cm height from ground level, retaining only 10 shoots in mid-December, and fertilized with 100 g N, 150 g P2O5 and 100 g K2O/plan/year over a basal dose of 10 kg of FYM. Among the defolians, pentachlorophenol at 3,000 ppm and potassium iodide at 4,000 ppm have been identified for their beneficial effect on growth and flowering of J. grandiflorum.

Tuberose

The tuberose (*Polianthes tuberosa*) is a perennial plant of the agave family Agavaceae, extracts of which are used as a middle note in perfumery. The common name derives from the Latin *tuberosa*, meaning swollen or tuberous in reference to its root system. It consists of about 12 species. Polianthes means "many flowers" in Greek language.

The tuberose is a night-blooming plant thought to be native to Mexico along with every other species of Polianthes. The Aztecs called it Omixochitl [oh me' zu che' tl] or bone flower.

It is a prominent plant in Indian culture and mythology. The flowers are used in wedding ceremonies, garlands, decoration and various traditional rituals. Its Hindi name is "Rajnigandha", though it is sometimes mistakenly referred to as "Raat ki Rani" ("Queen of the Night"), which is really Cestrum nocturnum. The name Rajnigandha means "night-fragrant" (rajni=night; gandha=fragrance). In Bengali, it is called "Rajoni-Gandha", meaning "Scent of the Night". In Marathi, it is called "Nishi-Ghanda". In parts of South India, it is known as "Sugandaraja", which translates to "king of fragrance/smell". In Indonesia it is called "bunga sedap malam", meaning night fragrant flower. In Tamil Nadu it is called as Sambangi or nilasambangi and traditionally used in all type of garlanding especially in south Indian marriages. In Cuba it is called "azucena" which is the name given to amaryllis in Mexico.

In Iran the tuberose is known as "Gole Maryam" ("Mary flower") and the oil extracted from the flower is used as a perfume.

The tuberose is also used traditionally in Hawaii to create leis and was considered a funeral flower in Victorian times. Its scent is described as a complex, exotic, sweet, floral.

The tuberose grows in elongated spikes up to 45 cm (18 in) long that produce clusters of fragrant waxy white flowers that bloom from the bottom towards the top of the spike. It has long, bright green leaves clustered at the base of the plant and smaller, clasping leaves along the stem.

Members of the closely-related genus *Manfreda* are often called "tuberoses".

Tuberose is grown commercially in India for cut-flower and loose flower trade. Planting during April-March was suitable for Single variety. Pre-planting treatment of tuberose Single bulbs with GA3 significantly increased the rachis length. Trials conducted at the Indian Botanical Garden, Howrah, West Bengal, revealed that planting of 3.0 to 3.5 cm diameter bulbs at a depth of 4 cm, spaced at a distance of 25 cm each way and fertilized with 20 g N, 40 g/m^2 each of P2O5 and K2O, resulted in maximum number of flower spike production, with more number of florets on long rachis.

The effective herbicide in controlling weeds in tuberose was found to be Alachor 2.0 kg/ha, Pendimenthalin 1.25 kg/ha and Metalachlor 2.0 kg/ha. Treatment with Pendimethalin 1.25 kg/ha resulted in higher flower yield and more number of bulblets/ plant.

Double Flowering Tuberose Planting Information

Fragrance of the Gods

If you're a gardener who loves scent, you'll flip head over heels for tuberose. This bulb produces tall flower spikes with numerous tubular blooms and a fragrance that is rich, sultry and wonderful. With Mexican origins it's not hard to understand how this plant is ideally suited to warm weather gardening, but northern gardeners with a warm sunny site can grow tuberose, too.

Give your tuberose sun, heat and time. You'll get back beautiful blooms with a fragrance fit for the gods.

Outdoor Beds

1. Find a location where the soil drains well. If there are still water puddles 5-6 hours after a hard rain, scout out another site. Or amend the soil with the addition of organic material to raise the level 2"-3" to improve the drainage. Peat moss, compost, ground bark or decomposed manure all work well and are widely available. Tuberose prefer drier soils and will not survive in those that are water logged.
2. Site your bulbs where they will receive full sun.
3. The bulbs we sell are actually clusters. Plant the whole clump with 2-3" of soil above the top and place clumps 8-10" apart.
4. After planting, water your tuberose generously, soaking the soil. Roots and top growth will form within a few weeks. Water regularly during the growing season if rain does not occur regularly. Tuberose are heavy feeders and appreciate applications of 8-8-8 fertilizer while they are in active growth.
5. Tuberose flower in mid to late summer, typically 90-120 days after planting. When in bloom, feel free to clip stems for bouquets. This will not hurt the plants and will provide a supply of extraordinarily fragrant flowers.
6. After blooming has finished for the season leave the foliage in place; don't cut it off. The leaves will gather sunlight and provide nourishment for next year's show. Water as needed. Leaves may be removed late in the season, when they yellow. (In zones 8 and cooler, to save your tuberose for next year, dig

the bulbs carefully after the first frost. Let them air dry for several days. Then store in a cool location in paper bags or boxes filled with peat moss.)

7. Your tuberose will rest for a few months before beginning the next growing cycle.

Pots, Tubs & Urns

1. Fill your containers with good quality, well-drained soil. Almost any commercially available potting medium will work fine. Make sure there are adequate drainage holes; tuberose bulbs must never sit in waterlogged soil or they will rot.
2. Site your bulbs where they will receive full sun.
3. Plant the whole clump with 2-3" of soil above the top and place clumps 8-10" apart.
4. After planting, water your tuberose generously, soaking the soil. Roots and top growth will form within a few weeks. Water regularly during the growing season if rain does not occur regularly. Tuberose are heavy feeders and appreciate applications of 8-8-8 fertilizer while they are in active growth.
5. Tuberose flower in mid to late summer, typically 90-120 days after planting. When in bloom, feel free to clip stems for bouquets. This will not hurt the plants and will provide a supply of extraordinarily fragrant flowers.
6. After blooming has finished for the season leave the foliage in place; don't cut it off. The leaves will gather sunlight and provide nourishment for next year's show. Water as needed. Leaves may be removed late in the season, when they yellow. (In zones 8 and cooler, to save your tuberose for next year, dig the bulbs carefully after the first frost. Let them air dry for several days. Then store in a cool location in paper bags or boxes filled with peat moss.)
7. Your tuberose will rest for a few months before beginning the next growing cycle.

Orchids

Orchidaceae, commonly referred to as the Orchid family, is a morphologically diverse and widespread family of monocots in the order Asparagales. It is currently believed to be the second largest family of flowering plants (only the Asteraceae is larger), with between 21,950 and 26,049 currently accepted species, found in 880 genera. The number of orchid species equals more than twice the number of bird species, and about four times the number of mammal species. It also encompasses about 6–11% of all seed plants. The largest genera are *Bulbophyllum* (2,000 species), *Epidendrum* (1,500 species), *Dendrobium* (1,400 species) and *Pleurothallis* (1,000 species).

The family also includes *Vanilla* (the genus of the vanilla plant), *Orchis* (type genus) and many commonly cultivated plants such as *Phalaenopsis* and *Cattleya*. Moreover, since the introduction of tropical species in the 19th century, horticulturists have produced more than 100,000 hybrids and cultivars. The complex mechanisms which orchids evolve to achieve cross-pollination were investigated by Charles Darwin and described in his 1862 book *Fertilisation of Orchids*.

Etymology

The name comes from the Greek "órkhis," literally meaning "testicle," because its root has a similar shape. The term was introduced in 1845 by John Lindley in "School Botany.They are also know as Winston Churchills Favourite flower."

Distribution

Orchidaceae are cosmopolitan, occurring in almost every habitat apart from glaciers. The great majority are to be found in the tropics, mostly Asia, South America and Central America, but they are also found above the Arctic Circle, in southern Patagonia, and even two species of *Nematoceras* on Macquarie Island, close to Antarctica.

The following list gives a rough overview of their distribution:

- tropical America: 250 to 270 genera
- tropical Asia: 260 to 300 genera
- tropical Africa: 230 to 270 genera
- Oceania: 50 to 70 genera
- Europe and temperate Asia: 40 to 60 genera
- North America: 20 to 26 genera.

Taxonomy

The taxonomy of this family is in constant flux, as new studies continue to identify more classificatory elements. The Orchidaceae is currently placed in the order Asparagales by the APG III system of 2009.

Five subfamilies are recognised.

Characteristics

Orchids are easily distinguished from other plants, as they share some very evident apomorphies. Among these: bilaterally symmetric (zygomorphic), many resupinate, one petal (*labellum*) is always highly modified, stamens and carpels are fused, and the seeds are extremely small.

Stem and Roots

All orchids are perennial herbs and lack any permanent woody structure. Orchids can grow according to two patterns:

- Monopodial: The stem grows from a single bud, leaves are added from the apex each year and the stem grows longer accordingly. The stem of orchids with a monopodial growth can reach several metres in length, as in *Vanda* and *Vanilla*.

- Sympodial: The plant produces a series of adjacent shoots which grow to a certain size, bloom and then stop growing, to be then replaced. Sympodial orchids grow laterally rather than vertically, following the surface of their support. The growth continues by development of new leads, with their own leaves and roots, sprouting from or next to those of the previous year, as in *Cattleya*. While a new lead is developing, the rhizome may start its growth again from a so-called 'eye', an undeveloped bud, thereby branching.

Terrestrial orchids may be rhizomatous or form corms or tubers. The root caps of terrestrials are smooth and white.

Some sympodial terrestrials, such as *Orchis* and *Ophrys*, have two subterranean tuberous roots. One is used as a food reserve for wintry periods, and provides for the development of the other one, from which visible growth develops.

In warm and humid climates, many terrestrial orchids do not need pseudobulbs.

Epiphytic orchids have modified aerial roots that can sometimes be a few meters long. In the older parts of the roots, a modified spongy epidermis called *velamen* has the function to absorb humidity. It is made of dead cells and can have a silvery-grey, white or brown appearance. In some orchids the velamen includes spongy and fibrous bodies near the passage cells. These structures are named tilosomes.

The cells of the root epidermis grow at a right angle to the axis of the root to allow them to get a firm grasp on their support. Nutrients mainly come from animal droppings and other organic detritus on their supporting surface.

The base of the stem of sympodial epiphytes, or in some species essentially the entire stem, may be thickened to form what is called a *pseudobulb* that contains nutrients and water for drier periods.

The pseudobulb has a smooth surface with lengthwise grooves and can have different shapes, often conical or oblong. Its size is very variable; in some small species of *Bulbophyllum* it is no longer than two millimeters, while in the largest orchid in the world, *Grammatophyllum speciosum* (giant orchid), it can reach three meters. Some *Dendrobium* have long, canelike pseudobulbs with short, rounded leaves over the whole length, some other orchids have hidden or extremely small pseudobulbs, completely included inside the leaves.

With ageing the pseudobulb sheds its leaves and becomes dormant. At this stage it is often called a *backbulb*. A pseudobulb then takes over, exploiting the last reserves accumulated in the backbulb, which eventually dies off too. A pseudobulb typically lives for about five years.

Leaves

Like most monocots, orchids generally have simple leaves with parallel veins, although some Vanilloideae have a reticulate venation. Leaves may be ovate, lanceolate, or orbiculate, and very variable in size. Their characteristics are often diagnostic. They are normally alternate on the stem, often plicate, and have no stipules. Orchid leaves often have siliceous bodies called *stegmata* in the vascular bundle sheaths (not present in the Orchidoideae) and are fibrous. The structure of the leaves corresponds to the specific habitat of the plant. Species that typically bask in sunlight, or grow on sites which can be occasionally very dry, have thick, leathery leaves and the laminas are covered by a waxy cuticle to retain their necessary water supply. Shade species, on the other hand, have long, thin leaves.

The leaves of most orchids are perennial, that is they live for several years, while others, especially those with plicate leaves, shed them annually and develop new leaves together with new pseudobulbs, as in *Catasetum*.

The leaves of some orchids are considered ornamental. The leaves of the *Macodes sanderiana*, a semiterrestrial or lithophyte, show a sparkling silver and gold veining on a light green background. The cordate leaves of *Psychopsiella limminghei* are light brownish green with maroon-puce markings, created by flower pigments. The attractive mottle of the leaves of Lady's Slippers from tropical and subtropical Asia, (*Paphiopedilum*) is caused by uneven distribution of chlorophyll. Also *Phalaenopsis schilleriana* is a pastel pink orchid with leaves spotted dark green and light green. The Jewel Orchid (*Ludisia discolor*) is grown more for its colourful leaves than its white flowers.

Some orchids, as *Dendrophylax lindenii* (Ghost Orchid), *Aphyllorchis* and *Taeniophyllum* depend on their green roots for photosynthesis and lack normally developed leaves, as do all of the heterotrophic species.

Flowers

Orchidaceae are well known for the many structural variations in their flowers.

Some orchids have single flowers but most have a racemose inflorescence, sometimes with a large number of flowers. The flowering stem can be *basal*, that is produced from the base of the tuber, like in *Cymbidium*, *apical*, meaning it grows from the apex of the main stem, like in *Cattleya*, or *axillary*, from the leaf axil, as in *Vanda*.

As an apomorphy of the clade, orchid flowers are primitively zygomorphic (bilaterally symmetrical), although in some genera like *Mormodes*, *Ludisia*, *Macodes* this kind of symmetry may be difficult to notice.

The orchid flower, like most flowers of monocots, has two whorls of sterile elements. The outer whorl has three sepals and the inner whorl has three petals. The sepals are usually very similar to the petals (and thus called *tepals*), but may be completely distinct.

The upper medial petal, called the *labellum* or lip, is always modified and enlarged. The inferior ovary or the pedicel usually rotates 180 degrees, so that the labellum, goes on the lower part of the flower, thus becoming suitable to form a platform for pollinators. This characteristic, called resupination occurs primitively in the family and is considered apomorphic (the torsion of the ovary is very evident from the picture). Some orchids have secondarily lost this resupination, e.g. *Zygopetalum* and *Epidendrum secundum*.

The normal form of the sepals can be found in *Cattleya*, where they form a triangle. In *Paphiopedilum* (Venus slippers) the lower two sepals are fused together into a synsepal, while the lip has taken the form of a slipper. In *Masdevallia* all the sepals are fused.

Orchid flowers with abnormal numbers of petals or lips are called *peloric*. Peloria is a genetic trait, but its expression is environmentally influenced and may appear random.

Orchid flowers primitively had three stamens, but this situation is now limited to the genus *Neuwiedia*. *Apostasia* and the Cypripedioideae have two stamens, the central one being sterile and reduced to a staminode. All of the other orchids, the clade called *Monandria*, retain only the central stamen, the others being reduced to staminodes. The filaments of the stamens are always adnate (fused) to the style to form cylindrical structure called the *gynostemium* or column. In the primitive Apostasioideae this fusion is only partial, in the Vanilloideae it is more deep, while in Orchidoideae and Epidendroideae it is total. The stigma is very asymmetrical as all of its lobes are bent towards the centre of the flower and lay on the bottom of the column.

Pollen is released as single grains, like in most other plants, in the Apostasioideae, Cypripedioideae and Vanilloideae. In the other subfamilies, that comprise the great majority of orchids, the anther, carries and two pollinia.

A pollinium is a waxy mass of pollen grains held together by the glue-like alkaloid viscin, containing both cellulosic stands and mucopolysaccharides. Each pollinium is connected to a filament which can take the form of a *caudicle*, like in *Dactylorhiza* or *Habenaria* or a *stipe*, like in *Vanda*. Caudicles or stipes hold the pollinia to the *viscidium*, a sticky pad which sticks the pollinia to the body of pollinators.

At the upper edge of the stigma of single-anthered orchids, in front of the anther cap, there is the *rostellum*, a slender extension involved in the complex pollination mechanism.

As aforementioned, the ovary is always inferior (located behind the flower). It is three-carpelate and one or, more rarely, three-partitioned, with parietal placentation (axile in the Apostasioideae).

Fruits and Seeds

The ovary typically develops into a capsule that is dehiscent by 3 or 6 longitudinal slits, while remaining closed at both ends. The ripening of a capsule can take 2 to 18 months.

The seeds are generally almost microscopic and very numerous, in some species over a million per capsule. After ripening they blow off like dust particles or spores. They lack endosperm and must enter symbiotic relationship with various mycorrhizal basidiomyceteous fungi that provide them the necessary nutrients to germinate, so that all orchid species are mycoheterotrophic during germination and reliant upon fungi to complete their lifecycle.

As the chance for a seed to meet a fitting fungus is very small, only a minute fraction of all the seeds released grow into an adult plant. In cultivation, germination typically takes weeks, while there is a report of one paphiopedilum that took fifteen years.

Horticultural techniques have been devised for germinating seeds on a nutrient-containing gel, eliminating the requirement of the fungus for germination, greatly aiding the propagation of ornamental orchids.

The main component for the sowing of orchids in artificial conditions is the agar agar. The substance is put together with some type of carbohydrate (actually, some kind of glucose) which provides qualitative organic feed. Such substance may be banana, pineapple, peach or even tomato puree or coconut milk. After the *cooking* of the agar agar (it has to be *cooked* in sterile conditions) the mix is poured into test tubes or jars where the substance begins to jelly. The seeds have to be put in the dish above boiling water, in the steam because that secures sterile conditions. The test tubes are put diagonally after that.

Reproduction

Pollination

Orchids have developed highly specialized pollination systems and thus the chances of being pollinated are often scarce. This is why orchid flowers usually remain receptive for very long periods and why most orchids deliver pollen in a single mass; each time pollination succeeds thousands of ovules can be fertilized.

Pollinators are often visually attracted by the shape and colours of the labellum. The flowers may produce attractive odours. Although absent in most species, nectar may be produced in a spur of the labellum, on the point of the sepals or in the septa of the ovary, the most typical position amongst the Asparagales.

In orchids that produce pollinia, pollination happens as some variant of the following. When the pollinator enters into the flower, it touches a viscidium, which promptly sticks to its body, generally on the head or abdomen. While leaving the flower, it pulls the pollinium out of the anther, as it is connected to the viscidium by the caudicle or stipe. The caudicle then bends and the pollinium is moved forwards and downwards. When the pollinator enters another flower of the same species, the pollinium has taken such position that it will stick to the stigma of the second flower, just below the rostellum, pollinating it. The possessors of orchids may be able to reproduce the process with a pencil, small paintbrush, or other similar device.

Some orchids mainly or totally rely on self-pollination, especially in colder regions where pollinators are particularly rare. The caudicles may dry up if the flower hasn't been visited by any pollinator and the pollina then fall directly on the stigma. Otherwise the anther may rotate and then enter the stigma cavity of the flower (as in *Holcoglossum amesianum*).

The labellum of the Cypripedioideae is poke-shaped and has the function to trap visiting insects. The only exit leads to the anthers that deposit pollen on the visitor.

In some extremely specialized orchids, like the Eurasian genus *Ophrys*, the labellum is adapted to have a colour, shape and odour which attracts male insects via mimicry of a receptive female. Pollination happens as the insect attempts to mate with flowers.

Many neotropical orchids are pollinated by male orchid bees, which visit the flowers to gather volatile chemicals they require to synthesize pheromonal attractants. Each type of orchid places the pollinia on a different body part of a different species of bee, so as to enforce proper cross-pollination.

An underground orchid in Australia, *Rhizanthella slateri*, never sees the light of day and depends on ants and other terrestrial insects to pollinate it.

Catasetum, a genus discussed briefly by Darwin actually launches its viscid pollinia with explosive force when an insect touches a seta, knocking the pollinator off the flower.

After pollination the sepals and petals fade and wilt, but they usually remain attached to the ovary.

Asexual Reproduction

Some species, as some *Phalaenopsis*, *Dendrobium* and *Vanda*, produce offshoots or plantlets formed from one of the nodes along the stem, through the accumulation of growth hormones at that point. These shoots are known as *keiki*.

Evolution

A study in the scientific journal *Nature* has hypothesized that the origin of orchids goes back much longer than originally expected. An extinct species of stingless bee, *Proplebeia dominicana*, was found trapped in Miocene amber from about 15-20 million years ago. The bee was carrying pollen of a previously unknown orchid taxon, *Meliorchis caribea*, on its wings. This find is the first evidence of fossilised orchids to date.

The extinct orchid *M. caribea* has been placed within the extant tribe Cranichideae, subtribe Goodyerinae (subfamily Orchidoideae).

This indicates that orchids may have an ancient origin and have arisen 76 to 84 million years ago during the Late Cretaceous. In other words, they may have co-existed with dinosaurs. It shows also that at that time insects were active pollinators of orchids. According to M.W. Chase *et al.* (2001) the overall biogeography and phylogenetic patterns of Orchidaceae show that they are even older and may go back roughly 100 million years.

Using the molecular clock method, it was possible to determine the age of the major branches of the orchid family. This also confirmed that the subfamily Vanilloideae is a branch at the basal dichotomy of the monandrous orchids, and must have evolved very early in the evolution of the family. Since this genus occurs worldwide in tropical and subtropical regions, from tropical America to tropical Asia, New Guinea and West Africa, and the continents began to split about 100 million years ago, significant biotic exchange must have occurred after this split (since the age of Vanilla is estimated at 60 to 70 million years).

Ecology

A majority of orchids are perennial epiphytes, which grow anchored to trees or shrubs in the tropics and subtropics. Other species are lithophytes, growing on rocks or very rocky soil, or are terrestrial. Nearly all temperate orchids are terrestrial.

Some orchids, like *Neottia* and *Corallorhiza*, lack chlorophyll and are unable to photosynthesize. Instead, these species obtain energy and nutrients by parasitising soil fungi through the formation of orchid mycorrhizas. The fungi involved include those that form ectomycorrhizas with trees and other woody plants, parasites such as *Armillaria*, and saprotrophs.

These orchids are known as myco-heterotrophs, but were formerly (incorrectly) described as saprophytes due to the belief that they gained their nutrition by breaking down organic matter. While only a few species are achlorophyllous holoparasites, all orchids are myco-heterotrophic during germination and seedling growth and even photosynthetic adult plants may continue to obtain carbon from their mycorrhizal fungi.

Uses

One orchid genus, *Vanilla*, is commercially important, used as a flavouring.

The underground tubers of terrestrial orchids (mainly *Orchis mascula* (Early Purple Orchid)) are ground to a powder and used for cooking, such as in the hot beverage *salep* or in the Turkish ice-cream (Dondurma - salepli dondurma). It has been claimed that the name salep comes from the Arabic expression %asyu al-tha'lab, "fox testicles", however, it appears more likely that the Turkish name salep comes directly from the Arabic name sa%labý. The similarity in appearance to testes naturally accounts for salep being considered an aphrodisiac.

The dried leaves of *Jumellea fragrans* are used to flavour rum on Reunion Island.

The scent of orchids is frequently analysed by perfumers (using headspace technology and Gas-liquid chromatography) to identify potential fragrance chemicals.

The other important use of orchids is their cultivation for the enjoyment of the flowers. Most cultivated orchids are tropical or subtropical, but quite a few which grow in colder climates can be found on the market. Temperate species available at nurseries include *Ophrys apifera* (Bee Orchid), *Gymnadenia conopsea* (Fragrant Orchid), *Anacamptis pyramidalis* (Pyramidal Orchid) and *Dactylorhiza fuchsii* (Common Spotted Orchid).

Orchids of all types have also often been sought by collectors of both species and hybrids. As such many hundreds of societies and clubs worldwide have been established. These can be small local clubs like Sutherland Shire Orchid Society or larger national organisations like American Orchid Society. Both serve to encourage cultivation and collection of orchids, but some go further by concentrating on conservation or research.

The term *botanical orchid* loosely denotes those small flowered tropical orchids belonging to several genera (not necessarily related to each other) that don't fit into the "Florist" orchid category.

A few of these genera contain enormous numbers of species. Some, such as *Pleurothallis* and *Bulbophyllum*, contain approximately 1700 and 2000 species, respectively, and are often extremely vegetatively diverse.

The primary use of the term is among orchid hobbyists wishing to describe unusual species they grow, though it is also used to distinguish naturally occurring orchid species from horticulturally created hybrids.

A few of the most common orchids found in "casual" culture are:

- *Anguloa*
- *Cattleya*
- *Cymbidium*
- *Laelia*
- *Dendrobium*
- *Phalaenopsis*
- *Paphiopedilum*
- *Oncidium*
- *Vanda*
- *Epidendrum*
- *Brassia*
- *Bulbophyllum*
- *Catasetum*
- *Sophronitis*
- *Miltonia*
- *Phaius.*

The Taiwan Orchid Nursery in Pingtung, Taiwan, and The National Orchid Garden in the Singapore Botanic Gardens are considered by some to be among the finest collections of orchids in cultivation open to the public.

Orchids, like tulips, have become a major market throughout the world. Buyers now bid hundreds of dollars on new hybrids or improved ones. Because of their apparent ease in hybridization, they are now becoming one of the most popular cut-flowers on the market.

Trade shows like the Taiwan International Orchid Show attracted more than 200,000 visitors and 3,000 foreign floral exports and buyers from abroad. As such, some governments, including the Taiwanese, considers orchid plantation as a high-tech industry and invests in science parks like the Taiwan Orchid Plantation in Tainan County.

Orchids produce remarkably attractive flowers with long lasting quality. Aerides multiflorum, a monopodial epiphyte, performed best in a substrate of hard-wood charcoal. Dendroblum moschatum, a sympodial epiphytic orchid, showed marked improvement in growth and flowering of the plants with the treatment of tree fern fibre.Cymbidium aloifolium and Phaius tankervilliae, the terrestrial orchids, responded well with organic-rich porous compost.

Best results in vegetative growth and flowering of these 2 species were obtained when plants were grown in a compost containing mixture of 1 part each of loamy soil, river sand and leaf mould, and half part each of charcoal dust and old mortar.

Marigold

The Wonderful Rich World of the Marigold Flower

The marigold flower is one of the least appreciated flowers of the world and is one of the most common flowers in the Southwest of the United States and into Mexico. Its deep oranges and yellows seem to burst into being in an explosion of colour the equal of any Fourth of July fireworks display. The powerful scent of the marigold flower is a distinctive odour that has its own secret charms to the gardener's nose.

The marigold flower has won the hearts of many around the world for its many uses and unique charms, and so it is that you can travel the populated continents and find marigolds in use everywhere you go. Here are just of few of the uses to which the marigold is put.

A Star of the Day of the Dead Celebration: In the Mexican celebration of the Day of the Dead—Dia de los Muertos—the marigold plays a central role because of its association with death. Marigolds are used in the Day of the Dead alters in Mexico because the ancient Aztecs used to commemorate their dead with the orange marigold (called the cempasúchil by the Aztecs). So to this day, you will see Mexicans creating large orange ornamented alters to those loved ones that have passed.

Latin American perfume: Although many people find the scent of marigolds too pungent for general use, in Latin America the perfume industry extracts the marigold's scent to use as the basic palette for many perfumes. The scent is not kept by itself, but has other layers of scents added to it, in the way that a chef might add spices to a specific type of meat to both temper it and to bring out certain qualities of its flavours.

Brazil, the Birthplace of the Marigold: In fact, it was in Brazil that early European explorers initially stumbled upon the beautiful orange flower. They took it back to Europe with them where it quickly became mixed up in the Spice trade, as you shall see when you get to India.

South African regenerator: In the Republic of South Africa, farmers plant marigolds in dead fields to help ward off various viral and bacterial threats to cash crops. The marigold flower produces a substance that supplants several negative disease-carrying bacteria and what are called nematodes (a type of roundworm that is lethal to cash crops). This forward-looking government sponsored plan is one of the many signs of South Africa's rebirth since the end of Apartheid.

Throughout the World as an Agricultural Aid: This disease preventing aspect of the Marigold in agriculture is the main reason why our orange beauty is also grown along side food plants like potatoes and tomatoes throughout the world. In effect, marigold is a natural insecticide, fungicide and nematode killer all in one.

The French Marigold: The French also took on the marigold flower, adapting it to their gardens where even to this day you will find it is one of the main staples of any francophone flowerbed. Don't try to convince the French that it's not a native of their republic.

The dominance of the Marigold in India: In Indian, where Portuguese traders brought the Marigold flower, the flower has found a new home. Indian farms mass produce the flower and the Hindu culture has completely adopted it as one of its own. You will find our world traveling orange friend in everything from store displays to temples to clothing accessories. It is so popular that there was a Bollywood movie a couple of years ago entitled Marigold. Its plot was about an American actress who moves to India where she falls in love.

The Marigold in my backyard: And then of course there is that wonderful, little appreciated marigold in my backyard. Only I and a handful of friends and family have appreciated its beauty, but like many gardeners who have marigolds, it is certainly one of the most beautiful sights I have ever seen.

Marigold is an important floriculture crop. Investigations on the effect of nutrients and growth retardants on African marigold at the UAS, Bangalore, showed that application of N, P K at 250:200:100 kg/ha and spraying of cycocel twice at 2,000 ppm in the second and fourth week after transplanting, resulted in higher flower yield and quality flowers. The maximum flower yiled (22.19 tonees/ha) of African marigold was recorded in the closes spacing (30 cm x 30 cm). Higher level of N (100 kg/ha) recorded a significantly higher flower yield (20.47 tonnes/ha) as compared to control at the GAU, Anand.

China Aster

Aster (syn. *Diplopappus* Cass.) is a genus of flowering plants in the family Asteraceae. The genus once contained nearly 600 species in Eurasia and North America, but after morphologic and molecular research on the genus during the 1990s, it was decided that the North American species are better treated in a series of other related genera.

After this split there are roughly 180 species within the genus, all but one being confined to Eurasia. The name *Aster* comes from the Ancient Greek word *astér*, meaning "star", referring to the shape of the flower head. Many species and a variety of hybrids and varieties are popular as garden plants because of their attractive and colourful flowers.

The genus *Aster* is now generally restricted to the Old World species, with *Aster amellus* being the type species of the genus, as well as of the family Asteraceae. The New World species have now been reclassified in the genera *Almutaster, Canadanthus, Doellingeria, Eucephalus, Eurybia, Ionactis, Oligoneuron, Oreostemma, Sericocarpus* and *Symphyotrichum*, though all are treated within the tribe Astereae. Regardless of the taxonomic change, all are still widely referred to as "asters" in the horticultural trades.

In the UK there are only two native members of the genus: Goldilocks, which is very rare, and *Aster tripolium*, the Sea aster. *Aster alpinus* spp. *vierhapperi* is the only species native to North America.

Some common species are:

- *Aster alpinus*, Alpine Aster
- *Aster amellus*, European Michaelmas Daisy or Italian Aster
- *Aster linosyris*, Goldilocks Aster
- *Aster pringlei*
- *Aster scaber*
- *Aster sibiricus*
- *Aster subulatus* Hairless Fleabane
- *Aster tataricus*, Tatarian Aster
- *Aster tongolensis*
- *Aster tripolium*, Sea Aster.

Some common North American species that have now been moved are:

- *Aster breweri* (now *Eucephalus breweri*), Brewer's Aster
- *Aster cordifolius* (now *Symphyotrichum cordifolium*), Blue Wood Aster
- *Aster divaricatus* (now *Eurybia divaricata*), White Wood Aster
- *Aster ericoides* (now *Symphyotrichum ericoides*), Heath Aster
- *Aster laevis* (now *Symphyotrichum laeve*), Smooth Aster
- *Aster lateriflorus* (now *Symphyotrichum lateriflorum*), Lady in Black, Calico Aster
- *Aster novae-angliae* (now *Symphyotrichum novae-angliae*), New England Aster
- *Aster novi-belgii* (now *Symphyotrichum novi-belgii*), New York Aster
- *Aster peirsonii* (now *Oreostemma peirsonii*), Peirson's Aster
- *Aster pilosus* (now *Symphyotrichum pilosum*), Frost Aster
- *Aster scopulorum* (now *Ionactis alpina*), Lava Aster.

The family Asteraceae have an estimated 1,150 genera and a huge number of species seen almost all over the world originating from North America, Europe, China etc. The large China Asters are the most common form grown by florists and gardeners. Native to Asia, this lovely flowering annual is worth the little extra effort it takes for growing.

It has a wide array of varieties and the height ranges from 6 inches to 3 feet. Most species of Aster are perennial and generally bloom in August. They have daisy-like or star-like flower heads (4-6" in diameter) with a yellow center on leafy, often tall, stems. Their colours vary from white to creamy yellow, pink, blue, red and purple. They do well in beds, borders or pots and are a favourite as cut flowers because of their longevity.

Propagation

Asters need fertile, sandy, well-drained soil and full sun. The propagation is through seeds. Before sowing, sterilize the seeds with organic mercuric compounds to prevent stem rot and damping-off. For sowing of seeds, it is best to use a shallow flat. Add one-forth well rotted manure and enough sand to the soil to prevent the soil from becoming hard. Fill the flat half or two-thirds full, water thoroughly, and finish filling with soil just moist enough to handle nicely. Sow the seed in rows 2 inches apart, cover with fine sand and press firmly. The wet soil in the bottom will furnish enough moisture so that the boxes will not have to be sprinkled until after the seedlings are up.

The seeds will germinate in a little over a week. As soon as several true leaves have grown, the seedlings should-be transplanted. Use rich soil in the bottom of the flats and ordinary soil above as further protection against disease. Set the seedlings 3 to 9 inches apart depending on the variety. The plants will grow to a height of 6-10 inches to 3 feet and a width of 6-18 inches.

Care

Since they are prone to wilt disease, virus diseases, foot rot, root rot, and aphids, avoid planting in the same location year after year. Pinch off faded flowers and yellowing foliage regularly to encourage new blooms. Shading plants prevents injury from the tarnished plant bug and yellows. If plants are badly diseased, burn them and stop growing China-asters for a few years. Water well in dry weather.

The experiments at the NBRI were directed to achieve loner stems and larger flower heads in China aster to improve its commercial value. Application of 100 ppm GA3 produced the desired results.

Institutional Support for Floriculture

Government of India: The Government plays a crucial role in physical planning, infrastructure creation, tax legislation and policies in respect of export oriented floriculture units through its various ministries and also provides subsidy through its specialized bodies like NHB and APEDA.The tax legislative measures that are favourable to the Indian Floriculture industry are as follows:

- Zero import duties on import of plant materials and without the need of an import license.
- Duty free imports on certain components of green house structure.

Growth Prospects

The Indian floriculture industry is poised for high growth and that is the opinion of most experts in the field. Enormous genetic diversity, varied agro climatic conditions, versatile human resources etc., are certain factors that contribute to it.

The exports have registered an impressive growth from Rs 18.83 crore in 1993-94 to Rs 298.57 crore in 2005-06. R K Mondal, Regional In-Charge, APEDA, Mumbai says,"Dry flowers are the future of this industry. It will be the greatest foreign exchange earner for the industry."

According to Dr Prasad,"Other factors like the retail revolution, contract farming of flowers for pharmaceutical and nutraceutical compounds, essential oils and other value added products would further fuel the industry."

Employment Opportunities

One can find employment in the floriculture industry as a farm/estate manager, plantation expert, supervisor, or project coordinator. Research and teaching are some other avenues of employment in the field. Marketing of floriculture products for different ventures is emerging as a potential segment in the field. Besides, one can work as consultant or landscape architect with proper training.

There are also opportunities for entrepreneurs who offer employment to others. In addition to these careers, which involve research and actual growing of crops, floriculture also provides service career opportunities, which include such jobs as floral designers, groundskeepers, landscape designers, architects and horticultural therapists. Such jobs will require direct client interaction.

Future in the Field

Literally speaking, qualified professionals have a bouquet of options to choose from. According to Megha Borse, successful new genre floriculturist and the President of Flower Growers' Association, Maharashtra, you need to really work hard on the field to make a living in this industry.

She says," Floriculture, after all, is an industry and has to be managed in a planned, methodical manner." Talking about her humble beginnings, she says," I wanted to establish an export oriented industry. It was when I came across the Green House concept, I thought of making a career in floriculture. The best part of floriculture is that it is not dependent on weather. That gives you assured produce and guaranteed income. The investment was approximately Rs 13 lakh to begin with (excluding land cost) for a 1000 square metre green house. Today, it is easy to get loans from banks.

There are also many supportive schemes from National Horticulture Board, National Horticulture Mission and Agriculture departments."

Export Prospects of Floricultural Products

With the declaration of "Floriculture" as an extreme focus area by the Ministry of Commerce & Industry, Government of India, the sector has acquired a special status in the basket of India's export commodities. The sector is still in a nascent stage of development and accounts for a negligible share in the global exports. However, the sector has attracted attention of the major giants from both within the country and overseas market. As a result, over the years, there has been a mushroom growth of floricultural units in the country. The industry thus offers tremendous potential for boosting exports of floriculture and floriculture products and the same are set to make a breakthrough in the near future.

India produces a wide variety of floricultural products which *inter alia* include flowers both fresh and dried: roses, carnations, orchids, chrysanthemums, etc., foliage of plants, potted and nursery plants, bulbs and seeds, tubers, rooted cuttings, leaves, micro propagation material and extraction of oils from flowers. A major growth in the country has been in the production of cut flowers for exports.

More than 150 export-oriented units have been approved by the Government of India and are coming up in different parts of the country, particularly around Pune, Chandigarh, Bangalore, Delhi and Hyderabad. A majority of these are being set up with collaboration of foreign technology and infrastructure support.

Cultivation of floriculture is concentrated mainly in the states of Karnataka, Tamil Nadu, Andhra Pradesh, West Bengal, Maharashtra, Rajasthan, Uttar Pradesh, Delhi and Haryana.

The Agricultural and Processed Food Products Development Authority (APEDA) over the years has been playing an important role in the export promotion and development of floriculture and floriculture products. Growers/exporters are provided assistance under its various schemes for making improvement in packaging, quality upgradation, creation of infrastructures, etc.

World Trade Scenario

World trade in floriculture is estimated at $100 billion. It has reportedly been growing at the rate of 15 per cent annum. Developed countries account for more than 90 per cent of the total world trade in floriculture products. European countries, Japan and United States are the major importers of floricultural products. On the export side, Netherlands continues to be the world leader in the export of floricultural products. Italy, USA and Israel are the other major exporting countries.

Recent Export Promotion Measures

Ever since the setting up of an Expert Group on Floricultural Development by the Government of India in 1989, several steps have been initiated for the speedy development of the sector. Some of the major steps taken in this direction are as under:

Recent initiatives taken by APEDA to boost export of floriculture: The Agriculture and Processed Food Export Development Authority (APEDA) has been designated as the nodal agency for export promotion and development activities relating to floriculture.

It has recently taken several steps to boost export of floriculture products so as to achieve an export target of Rs. 1,000 crore within five years. Some of the major ones in this direction inter *alia* include:

(i) Commissioning of Indian Institute of Packaging for standardizing export packing used for floriculture export items;

(ii) Visits of Indian Delegations to Netherland,

(iii) Finalization of a host of joint ventures with prominent foreign companies;

(iv) Visit of Dutch Delegations to India for finalizing joint ventures on protected cultivation techniques i.e. the practice of growing plants in green houses as is done in European countries and in the area of refrigeration marketing and propagation of planting;

(v) Setting up of integrated facilities for handling and storage of exportable perishable products like floricultural products cargo at International Airports in Mumbai, Delhi, Chennai, Bangalore, Hyderabad and Thiruvananthapuram;

(vi) Setting up of market cum auction centres for exports at Bangalore, Mumbai and NOIDA.

(vii) Setting up of a Market Facilitation Centre in Aalsameer, Netherlands.

(viii) Setting up of Export Processing Zones for floriculture products in Tamil Nadu, Uttaranchal, Maharashtra, Karnataka and Sikkim.

Liberalized Govt. policies to promote floriculture products: The liberalized policies of the government both at the center and state level have resulted in a mushroom growth of floricultural units in the country over the years. Some of the recent policy initiatives taken include: (i) Setting up of an exclusive agriexport zone for floricultural units in Bangalore, and (ii) According approval for setting up of 100% export oriented units by the Govt. of India to about 191 units. Of this, it has been reported that 70 100% EoUs units are operational. As a result, a large number of reputed companies have set up 100% export oriented units in the area of floriculture. Besides, there is a full-fledged Department of Floriculture in the Ministry of Agriculture which has been making all out efforts to increase area and production of floricultural produce from the country.

About 34,000 hectares are reportedly under the cultivation of floriculture in the country. The bulk of the area lies in the Southern states like Tamil Nadu, Karnataka and Andhra Pradesh.

Karuturi sets up Floriculture unit in Ethiopia: A news story under the caption *"Africa Promises Bed of Roses for Floriculture Sector"* said that Karuturi Networks, a Bangalore-based company having its Floriculture Division has set up a 50 hectare floriculture unit at Holeta near Addis Ababa and is in the process of adding 50 hectares more. A few more Indian companies including Pushpam Florabase Pvt. Ltd are in the process of setting up their floriculture operations in Ethiopia.

The Ethiopian Government, it says, offers large tracts of land for floriculture companies on a perpetual long-lease at very attractive rentals. Ethiopia offers ideal climatic conditions that help in producing premium grade roses and command a significant premium compared to the Indian roses.

Though Ethiopia offers no fiscal incentives, growers can avail themselves of funds at a low rate of interest from the World Bank-assisted Fund which has earmarked around US$350 million for the development of floriculture in Ethiopia.

New Foreign Trade Policy (FTP): Under the new FTP (2004-09) announced by the Ministry of Commerce & Industry, Government of India, a host of incentives have been given to boost agri-exports. These *inter alia* include (i) duty-free import of capital goods under the Export Promotion Capital Goods (EPCG) scheme, (ii) Duty credit scrip equivalent to 5 per cent of the f.o.b. value of exports, and (iii) Launching of *Vishesh Krishi Upaj Yojana* which is aimed at promoting agri-exports, viz., flowers, vegetables, fruits, minor forest produce, etc.

Exemption from Import Duty: The Govt. has fully exempted import duty on ornamental plants, tubers and bulbs of flowers, cutting of saplings of flower plants used for the purpose of sowing and planting.

Entry of Large Units: To meet the burgeoning demand for floricultural products both in the domestic and overseas markets, a host of major Indian companies have entered this sector. Prominent among them comprise RPG Group, Harrison Malayalam, Oriental Floriculture, a Tata company. These companies have started commercial production of roses with technical and marketing tie-ups mainly with most of the Dutch companies.

Air freight subsidy: In a move to make export of floricultural products more competitive in the international market, the Government has decided to grant air freight subsidy on cut flowers (HS Code: 06.03.1000). The scheme is being implemented by APEDA.

India's Exports

India's Exports—Segment wise: India's exports of floriculture and floriculture products in 2004-05 registered a decline of 15.07 per cent over the previous year when the same nosedived to Rs. 212.70 crore as against Rs. 250.45 crore in the previous year. "Cut flowers & flower buds suitable for bouquets/ornamental purposes, fresh, dried, dyed, bleached, impregnated, otherwise prepared" emerged as the largest segment of exports in 2004-05, accounting for a share of 44.65 per cent of India's total exports of floriculture products. However, this segment also registered a negative growth of 14.50 per cent during the period. A close analysis of segment wise exports reveals that "Foliage branches and other plant parts, grasses, moses and lichens for bouquets/ornamental purposes, fresh, dried, dyed, bleached, impregnated, otherwise prepared" which happened to be the largest segment of exports in 2003-04 slipped to second place and showed a negative growth of 19.11 per cent when its exports stooped to a level of Rs. 92.33 crore from Rs. 114.14 crore during the period. On the other the segments "Other live plants (incl. roots) cuttings and slips, mushroom spawn" and "Bulbs, tubers, tuberous roots, corms, crowns & rhizomes, dormant, in growth/in flower, chicory plants & roots excluding roots of HS Code 1212" showed a marginal growth of 0.64 per cent and 0.91 per cent respectively during the period.

India's Exports—Country wise: USA continues to be the largest market for Indian floriculture products : However, exports to this market registered a steep decline of 33.57 per cent in 2004-05 over the previous year when the same reached a level of Rs. 48.75 crore from Rs. 73.39 crore. The other markets showing a steep negative growth during the period include: Japan (51.69%), Poland (49.41%), France (34.85%), Singapore (30.61%), Italy (25.76%), New Zealand (25.13%), Switzerland (22.87%), and Greece (18.02%). On the other, the markets showing a phenomenal growth during the period comprised: Maldives (524%), China (479.41%), South Africa (360.53%), Hong Kong 123.53%), Saudi Arabia (50%), and UAE Belgium each (38.42% each).

Segment wise exports of floricultural products to major markets during the period 2001-02 and 2004-05 appear at Annexure II. Segment wise exports of major floricultural products in the select markets during the period 2003-04 and 2004-05 appear at Annexure III.

Export Prospects

Against the backdrop of India enjoying several advantages like favourable climatic conditions, availability of Arable land and skilled manpower, the country offers a tremendous potential for boosting floricultural products from the country. For tapping the potential, there is a need to adopt a coherent export strategy by initiating suitable measures, especially providing better and efficient infrastructural facilities.

Some of the measures suggested in this direction include: (i) Cultivation of floriculture at present is concentrated in the states of Karnataka, Tamil Nadu, Andhra Pradesh, West Bengal, Maharashtra, Rajasthan, Uttar Pradesh, Delhi and Haryana.

The Agricultural and Processed Food Products Development Authority (APEDA) over the years has been playing an important role in the export promotion and development of floriculture and floriculture products. Growers/exporters are provided assistance under its various schemes for making improvement in packaging, quality upgradation, creation of infrastructures, etc. Setting up of more cold storage facilities at the airports and making availability of a fleet of refrigerated vans for handling flowers from production units to airports; (ii) Greater participation in the specialized international trade fairs in floriculture; (iii) Diversifying product range of flowers as presently roses constitute main item of exports, (iv) Making improvements in the quality of the flowers; (v) Better supply chain; and (vi) Setting up of more flower auction centres.

Indian Floriculture Industry Present Status & Scope

After liberalization, the Govt. of India identified floriculture as a sunrise industry and accorded it 100 percent export oriented status. Owing to steady increase in demand of flower floriculture has become one of the important Commercial trades in Agriculture. Hence commercial floriculture has emerged as hi-tech activity-taking place under controlled climatic conditions inside greenhouse. Floriculture products mainly consist of cut flowers, pot plants, cut foilage, seeds bulbs, tubers, rooted cuttings and dried flowers or leaves. The important floricultural crops in the international cut flower trade are rose, carnation, chrysanthemum, gargera, gladious, gypsophila, liastris, nerine, orchids, archilea, anthuriu, tulip, lilies.

Present status and growing trade is still in infancy. Floriculture in India, is being viewed as a high growth Industry. Commercial floriculture is becoming important from the export angle. The liberalization of industrial and trade policies paved the way for development of export oriented production of cut flowers. The new seed policy had already made it feasible to import planting material of international varieties.

The government of India offers tax benefits to new export oriented floriculture companies in the form of income-tax holidays and exemption from certain import duties. Agricultural and Processed Food Products Export Development Authority (APEDA), responsible for export promotion and development of floriculture in India, grants subsidies for establishing cold storage, precooling units, refrigerated vans and green houses, and air freight subsidy to exports. It has been found that commercial floriculture has higher potential per unit area than most of the field crops and is therefore a lucrative business.

According to a report of the APEDA, the total area under flower crops was estimated around 34,000 hectares, which included 24,000 hectares under traditional flowers such as marigold, jasmine, aster, rose, chrysanthemum, tuberose and 10,000 hectares under modern flowers like cornation, rose, gerbera, gladiolous, anthurium.

Returns from floricultural products were estimated at Rs.205 Crores, which included Rs.105 Crores from traditional and Rs. 100 Crores from modern flowers. The data collected by National Horticultural Board (NHB) (1994-95) show that estimated area and production were 53,000 hectares and 4 lakh tonnes. In 1995-96, there were mainly 230 export-oriented units in India. More than 50% of the floriculture units are based in South zone mainly in Karnataka, Andhra Pradesh, Tamil Nadu. Also West Bengal, Maharashtra, Rajasthan have large areas under floriculture. The domestic flower production goes on increasing annually. Technical collaborations with foreign companies have been approved for India, in order to increase total share in the floriculture world trade.

Conductive Conditions

India is endowed with diverse agro-climatic conditions like good quality soils, suitable climate, abundant water supply, low labour cost, proximity to market in Japan, Russia, South-East Asia, Middle-East Countries.

Subsidy on airfreight for export of cut flowers and tissue-cultured plants is allowed by the Government. Freight rates are Rs.10 per kg for export to Europe and Rs. 6 per Kg for export to West Asia, SouthEast Asia whichever is less.

Import duties have been reduced on cut flowers, flower seeds, tissue-cultured plants, etc. Setting up of walk in type cold storage has been allowed at the International airports for storage of export produce.

Direct subsidy upto 50 percent of the precooling and cold storage units is available, as well as subsidy for using improved packaging material is given by APEDA.

Eleven-model floriculture centre units and two large centres, 20 tissue culture units have been established by Ministry of Agriculture.

Refinance assistance is available from NABARD to a number of hi-tech units at reasonable interest rate.

Export Constraints

In spite of an abundant and varied production base, India's export of floricultural product is not encouraging. The low performance is attributed to many constraints like non-availability of air space in major airlines, since most of the airline operators prefer heavy consignments.

The existing number of flights during the peak seasons is not sufficient for export purpose.

Exporters for infra-structural problems like bad interior road, inadequate refrigerated transport and storage facilities. Lack of professional backup of delivery and supporting companies, which resort into high cost of technology for Indian entrepreneurs. Tedious Phyto-sanitary certification and an unorganized domestic market.

In order to overcome these problems, attention must be focussed on –Reduction in import duty on planting material and equipment.

Airfreight should be reduced to a reasonable level. Sufficient cargo space may be provided in airlines. Establishment of model nurseries far supplying genuine planting material.

Co-operative florist organizations should be established at regional level. Training centres for diploma course on the pattern of ITI for training the personnel in floriculture should be set up.

Exporters should plan and monitor effective quality control measures right from production to post harvesting, storage, and transportation. An analysis of strengths, weaknesses, opportunities of the floricultural industry shows that India has immense potential for export of floricultural products.

Though the global floriculture industry is growing comparatively at a faster pace than in India, still a scope exists to bridge the demand and supply gap.

Prospects of Floriculture Projects in Jammu and Kashmir

Floriculture, an important branch of horticulture involves the cultivation of flowers and it includes ornamental gardening and landscaping. Gardening for aesthetic purposes has also been a part of the Indian community with special emphasis on cut flowers. This is an emerging economic field and has been in the forefront in the recent years. Many industrialists and business houses too have expanded their business in the line. The high-tech floriculturists grow flowers under controlled conditions mainly to cater to the demands of the fast expanding international market. Floricultural products constitute a small but important segment of the international trade. These are high value products that are used for their beauty and elegance and they reap very high economic returns when compared to other agricultural and horticultural products.

The floriculture industry in the world showed a dramatic increase in growth during the 1970's and 1980's. The increase in floral products is attributed to standard of living of the people. The major cut flowers that are in demand in the international market are roses, chrysanthemum, carnations, tulips, gerberas, alstroemerias etc. Many new types of flowers like anthuriums, aster, asclepias, spray carnation, gypsophila, liatris, limonium, heliconia etc. that are entering the flora trade.

Over the past decade, flower and pot plant business in the world has increased to US $ 40 billion. The annual rate of growth in the floriculture industry is about 15 percent. The floricultural products include cut flowers, which contribute about60 percent of the global trade, flowering and green potted plants and bedding plants from a small segment of the floricultural crop production worldwide. India's share in this global floriculture market is around 0.75%.

There are clear signs that India would soon play an important role in the world trade of flowers. With the liberalization of economic policies and identification of floriculture as one of the thrust areas by the Government opportunities offered by the large global market. In view of the potential for exports, a large number of export oriented units have been approved by the Government of India and many projects have been commissioned in recent years. Domestic market can absorb the non-exportable grades.

To reduce the capital cost and cost of production, it is desirable to establish export oriented units so that concessions on import of capital goods and other consumables can be availed of. J&K State, which has the agro-climate conditions suitable for various varieties of flowers, can be a good source of production and supply of the quality cut flowers to the domestic and international market.

State of Jammu & Kashmir have ideal climatic conditions for floriculture. The climate of the three regions of the State viz. Jammu, Kashmir and Leh are Sub - tropical, temperate and cold arid. In J&K State this activity is in vogue mainly in the form of beautification of existing gardens/parks in the State. Department of Garden, Parks and Floriculture develop and maintain parks and gardens at suitable locations to upgrade the aesthetic value. The department is contributing significantly by adding new varieties of flowers and beautification plants in existing gardens and parks. Some potential areas are being developed as demonstration units by the department. State Government proposes to invest an amount of Rs. Crores during 9th Five-year plan for the development of this sector. The State intends to propagate floriculture as an important economic activity and hence thrust on its commercial exploitation.

The Department of Garden, Parks and Floriculture is providing technical know-how support for plantation and proper maintenance of plants and also by way of setting up demonstrative parks therein growing greenhouse seedlings, and helps in commercializing this activity through:

- Demonstrative units
- Training Facilities
- Supply of other inputs at Subsidized rate and Market Suppor.

The demonstrative units are set up to establish the technical feasibility of the proposal. Training Facilities include providing know how about setting up of green houses and maintaining them, proper growth of cut flowers, handling equipment, post harvest treatment of cut-flowers and the packaging practices. The department also provides subsidy on purchase of bulb for gladilieus and Tulips (50% of its cost) for the maximum purchase of 2000 bulbs by individual private grower.

Department also guides and assist the entrepreneurs in marketing of the cut flowers. Department has linkage with departments and institutions engages in marketing of floriculture products at National and International level including Agriculture product Export Development Agency (APEDA), New Delhi. J&K State Industrial Development Corporation extends the promotional support as well as commercial banks/NABARD provides financial assistance for floriculture project. 25 number of floriculture growers has set up their units under the technical guideline and supervision by the experts of the Department.

The marketing of cut-flowers could be tied up in European and Gulf Countries where demand of Lily, Rose, Tata Rose, Gladilieus cut flower and other varieties is quite high, The State of Jammu and Kashmir has suitable climate conditions to grow the flowers on commercial lines for National and International markets. Under green house polyhouse conditions, the flowers can be grown through out the year.

The main varieties, which are grown in the State, include Gladilieus, Roses, Tata Roses, Lilly, Tulip, Hyceanthus, Carnation etc. Details pertaining to growing different varieties and suitable locations are as under:

Gladilieus

This variety could be grown for cut flower purposes in both Jammu and Kashmir regions. It is grown ideally within 90 days from the date of seeding of tuber. Ideal locations for cultivation are Jammu, Kud, Chenani, Poonch and Rajouri Districts in Jammu and Srinagar, Budhgam and Anantnag in Kashmir region. In Leh region such flowers can be grown under greenhouse cultivation. The total production of this variety of cut flowers in the State is estimated at about one lakh numbers.

Rose

Rose is grown in the month of November starts production from the month of April. Such plant can be grown in greenhouses under controlled temperature between 20 Deg. C. to 30 Deg. C. and also in open field. The output of Roses in greenhouses is around double compared to open field cultivation. The rose flower output is very less in comparison to its existing demand.

Lilly

The demand for this variety of flowers emanates from as well as Delhi and Mumbai markets during Christmas. The bud is ready within 75 days. The suitable locations are Panchari, Sudhmahadev (during winter), Poonch and Rajouri (during summer) in Jammu region and Srinagar, Budhgam, Pulwama in Kashmir region. The production could be carried out through out the year in Jammu region for 3 to 4 months in Kashmir. The demand is more than 1.5 lakh to 2 lakhs tubes and presently only 10 thousand bulbs is grown in the State.

Tulips

This variety of flower is grown in Kashmir Region of the State. It is sown in the month of July or August and its spike is ready in around 70 to 75 days time. The spike can be kept for 8 - 9 days and mostly sold during the Christmas. About 40 to 50 thousand tubes is grown and some no. Of spikes are exported to Delhi Market from Kashmir region. However, demand is very high compared to its supply.

Hysenthesis

Its production in the State is merged. The potential for the growth of this variety is available in the State. Kashmir, Leh and hilly regions in Jammu, Doda, Poonch district of Jammu region. More than 3 to 3.5 lakhs tubes could be grown in the three locations of State.

Marigold

There is an increasing demand of this variety of flowers by the visiting pilgrims (more than 4.2 millions) at Vaishno Devi Ji. Its total production is estimated at 600 to 700 quintals and unfulfilled demand of the State is of the order of 800 quintals. This variety of flowers are exported during the month of October to November from the State and imported to the State from Delhi etc. in July and August.

Conclusion

The practice of floriculture cultivation of commercial scale is at its initial stage by individual growers. The agro climatic conditions of the State are quite suitable for the propagation of the floriculture in the State either on open or through greenhouse technology. Floriculture is popularizing this activity and some strong initiatives by way of hi-tech floriculture project based on greenhouse.

Transgenic Plants for Floriculture

During 1990s, the interest in the production of transgenic ornamental plants also gained momentum. While in 1990, transformation procedure was available for only one ornamental i.e. Petunia, in 1995-2000, transformation procedures became available for many ornamentals including some monocots.

These cut-flower species include rose, carnation, tulip, chrysanthemum, gerbera, daffodils, gypsophilla, lily, iris, etc. In several of these cut tlowers, recent work led to successful production of transgenics with novel aesthetic properties including novel colours, fragrance, plant and tlower architecture, longer life, etc. Some of these traits having a commercial demand have been the target for genetic engineering and will be brietly discussed.

Plant Size

The ratio of concentrations of cytokinin and auxin can be altered in transgenic plants through expression of oncogenes derived from Ti/Ri plasmids. Following are some examples:

(i) Constitutive expression of roLC gene encoding cytokinin-α-glucosidase from Ri Plasmid leads to release of active cytokinins giving desired dwarf and bushy phenotype (e.g. potato becoming an ornamental).

(ii) GA concentration can be altered, so that reduced GA can give dwarf plants and increased GA can give tall plants; two genes (gal, anl) involved in GA biosynthesis can be used to alter the levels of GA; overexpression of phytoene synthase gene also leads to reduced GA level and dwarfism.

Flower Colour

The first application of genetic engineering to modify flower colour led to the production of an orange pelargonidin-producing Petunia variety, which produced flowers pale brick colour. This was achieved by the expression to the dihydroflavonal-4 reductase gene (dfr) from maize in a petunia line.

The variety was unstable due to transgene silencing and was therefore not viable in commercial market. Subsequently several transgenic plants with dfr gene were used for breeding with elite material to produce a commercially viable brown orange petunia variety in the Netherlands.

Production of blue flower in rose, tulip or carnation has also been an excitement among biotechnologists and floriculturists. It is complicated, because it needs following three factors:

(i) synthesis of 3', 5' hydroxylated anthocyanins (delphinidins);

(ii) presence of tlavonal co-pigments and

(iii) relatively high vacuolar pH. Delphinidins are not available in roses, but two genes responsible for blue colour that have been isolated from petunias include the following:

 (i) gene for enzyme flavonoid-3'-5' hydroxylase (for synthesis of delphinidins) and

 (ii) gene for tlavonal synthase (for synthesis of tlavonals).

The genes that control vacuolar pH were difficult to clone but some progress has been made. From six genes (pH l-pH6), pH6 had been cloned, and pH3 and pH4 were being cloned in 1995. Three other genes involved in regulation of anthocyanin biosynthetic pathway i.e. anl, an2 and anl1 also cause alterations in intracellular pH and can be isolated by differential screening of cDNA library.

Chalcone synthase (Chs) is another gene which has been used for production of pink, white and variegated flowers (sense and antisense genes were used) in - petunias, chrysanthemum, gerbera and roses. Transcriptional activators of anthocyanin biosynthetic pathway, namely R and CI have also been used to alter the intensity of flower colour. Mutations in regulatory sequence also gave novel pigmentation patterns in ornamentals like snapdragon. The gene Chs has also been used to produce white moneymaker chrysanthemum.

Architecture of Flower and Inflorescence

The identity of floral organs can be altered by using homeotic genes involved in determining the identity of floral whorls and the number of organs in each whorl. Mutants in Petunia and Arabidopsis have been isolated, which have altered number of floral organs. These can be used to change the architecture of the flower.

The inflorescence in ornamentals may have determinate (terminal flowers) or indeterminate growth (indefinite number of flowers). A single gene is sufficient to convert the inflorescence from indeterminate to determinate type or vice versa. Several of these genes have already been cloned.

Flowers with Longer Life

The vase life of flowers can be altered by manipulating the biosynthesis of ethylene. The enzymes ACC synthase and ACC oxidase are encoded by genes acs and aco respectively and both have been cloned from many species including carnation. Transgenic carnation having antisense aco gene or sense acs gene have been produced and exhibited longer vase-life (delay in petal senescence). They have already been approved for commercial release.

Other strategies for longer vase-life of flower including the following:

(i) overexpression of a gene for ACC deaminase, that metabolises ACC before it is converted to ethylene;

(ii) overexpression of the gene for SAM hydrolase (isolated from T3 phage);

(iii) expression of a gene for isopentenyl transferase (ipt) giving increased level of cytokinin;

(iv) isolation and use of genes involved in ethylene signal transduction pathway.

Flower Fragrance

Monoterpens are an important class of compounds that are involved in fragrance and are commercially valuable as essential oils for perfumes. These are derived from geranyl pyrophosphate by various monoterpene synthases, and a gene for one of them (S-linalool synthase) was cloned during early 1990s. Two other genes modifying monoterpenes were also cloned. These genes will be used in genetic engineering to modify floral fragrance.

9

Indian Floriculture Industry : Challenges and Strategies

POST globalization, floriculture has become one of the important commercial activities in Indian agriculture. Indian floriculture industry comprises the florist trade, nursery plants, bulb and seed production, apart from production of micropropagation material, and extraction of essential oils from flowers. The industry has been growing at a CAGR of 25 percent over the past decade, with production area growing at a CAGR of 6.89 percent since 2000-01.

A number of Export Oriented Units (EOUs) have been set up in the country; most of them are largely dependent on foreign collaborations for technological support. In the recent years, a number of large Corporate Houses such as ESSAR group, TATA group, Birla, Nagarjuna and Pariwala have also invested in the flower sector. Protected cultivation is not a common practice in India, the greenhouse designs and structures for the modern floriculture units are mostly imported from countries like Holland, Israel, France, and the USA. In India, there are three types of greenhouse production technologies, viz., low-tech units, mid-tech units and hi-tech units, with the investment costs varying significantly among the three groups. Marketing of cut flowers in India is much unorganised at present. Packaging and transportation of flowers is also very unscientific. In the metros, however, in the recent years, some modern florist show rooms have come up, where flowers are kept in controlled temperature conditions, with considerable attention to value added services. To facilitate flower trade, two auction centres have also been established one at Bangalore and the other at Mumbai.

Exports from India

Commercial floriculture in India is going through a paradigm shift, where traditional flower cultivation is giving way to modern hi-tech flower cultivation, which is evident from India's rising production and exports. Exports of floricultural products have been growing at a CAGR of 15 percent over the past decade. However, the growth of the industry has been significantly affected by the recent global recession largely due to decline in demand in all major markets. India's exports of floricultural products in the year 2007-08 decreased by 41 percent to US$ 84.5 million (Rs. 340 crores), from US$ 144 million (Rs. 653 crores) in 2006-07, which further decreased by 5.18 percent in the year 2008-09 to US$ 80.19 million. However, in 2008-09, in rupee terms, export of floriculture from India increased marginally. In the recent years, dried flowers and foliage have been forming a large part of floricultural product exports from India. During 2008-09, dried flowers constituted over 60 percent of cut flowers exports, and dried foliage constituted over 95 per cent of total foliage exports from India. Fresh cut flowers are mainly exported from Tamil Nadu, Karnataka and Maharashtra. Dried flowers are exported mainly from Tamil Nadu and West Bengal, with the later accounting for around 70 per cent of the dried flower exports from the country. Europe continues to be the largest destination for Indian floriculture exports. However, in the recent years Indian exports of floriculture products have also been to the Japanese and Australian markets.

India - Outlook

Besides the global recession affecting significantly the Indian floriculture exports, the competition in the international arena has also been increasing with the entry of new African countries, and Asian neighbours in global floriculture trade such as Ethiopia, Tanzania, China, Nepal, Sri Lanka and Pakistan. With international trade in the sector showing not so favourable scenario, the demand in the domestic markets for flowers have been growing, reportedly at 40 percent per annum.

The potential for expansion of commercial floriculture in India, including production for domestic and export sales of cut flowers and plants is unlimited provided the country expands the production of existing products as well as the product range. Further, it has been observed that except for Saudi Arabia and UAE, India has not been able to make its presence felt in the regional floriculture markets such as Japan, Singapore, Korea, and Russia, where China, Thailand and Malaysia are consistent suppliers. Therefore, more concerted efforts are required for the Indian floriculture industry to make a significant mark in the regional, and new and emerging markets, such as those in the Asian region, and in the East European region.

Some of the areas in which market opportunities exist for the Indian floriculture industry are: hi-tech cultivation and product diversification (to competitively grow Orchids, Anthurium, Gladiolus, Tuberose, and Foliage varieties, as the demand for tropical flowers and plants is increasing worldwide); tapping the landscaping opportunities (with growing real estate market worldwide, nursery industry is also opening up for supplying trees, greens and ornamental plants); floral arts-floral designing (which is increasingly getting popular in the events such as weddings and festivities); and catering to the growing demand for custom made products, dry flowers and foliage, flower seed production, flower extracts and essential oil, and natural dye.

Challenges

The Indian floriculture industry is faced with a number of challenges mainly related to trade environment, infrastructure and marketing issues such as high import tariff vis-a-vis African countries, low availability of dedicated perishable carriers, higher freight rates, inadequate support infrastructure, constraint in achieving economies of scale, and inadequate cold chain management. At the production level the industry is faced with challenges mostly related to availability of basic inputs, including quality seeds and planting materials, quality irrigation and skilled manpower, and ageing plantations. With regard to marketing, major challenges faced by the Indian flower exporters are related to low level of product diversification and differentiation, vertical integration and innovation, and challenges associated with quality and environmental issues. With increasing involvement of supermarkets in flower trade, organizing logistics is also becoming a critical factor for the Indian flower exporters.

Strategies

Several positive steps are being undertaken by India to develop the floriculture sector as a viable export sector. However, more strategies need to be adopted to consolidate the strengths of India in this sector and create efficient backward and forward linkages. The first and foremost requirement is to develop a state-of-the art Integrated Cold Chain, for flowers right from the point-of-origin (growers) to the pointof- consumption (customers). The freight cost disadvantage can partially be overcome by achieving economies of scale. To minimize the time taken and damages caused during handling, emphasis should be given on improvement of cargo handling facilities and making internal container depots suitable for floricultural products. Besides creating additional cargo space specific to floriculture, efforts may be taken to increase the frequency of international flights and chartered flights handling floriculture cargo.

Government of India may consider making additional efforts towards encouraging private sector for coinvestments in the supply-chain infrastructure. To address the import duty disadvantage faced by India in the international markets, such as Europe, the Government may enhance its efforts in negotiating preferential tariff regimes with such countries. For efficient implementation of developmental initiatives, attempts may be made in creating National Export Promotion Council for development of modern floriculture. Shifting to Integrated Supply Chain Model, integrating small and medium scale growers into large-scale producer supply chains may help in attaining economies of scale in the industry. Periodic re-plantation with appropriate re-plantation support from Government is necessary in order to meet the changing demand in global market for variety and quality.

Developing the non-traditional production areas such as South Gujarat, South Maharashtra, North Karnataka, hilly areas of Tamil Nadu and Karnataka, Uttarakhand, Himachal Pradesh, Jammu and Kashmir, and North Eastern States, which are well endowed to grow flowers and ornamental plants, may help in meeting the growing needs of both domestic and international market. In order to ensure regular supply of quality inputs such as planting material, efforts may be made to develop and propagate varieties indigenously. To help protect breeder's rights, India may consider becoming a member of the UPOV with adequate safeguard mechanism for the Indian growers. Besides, indigenous technology for modern floriculture may be developed to suit the Indian growing conditions. To address the shortage of trained manpower on modern commercial floriculture, training centres and programmes for skill development in floriculture may be established. Marketing initiatives should be suitably designed to meet the market requirements, and yet remain competitive. Attention may be required for achieving the flower labelling and certification requirements, besides, maintaining international quality standards throughout the production chain. Diversification of product mix is also essential considering the changing pattern of demand. Indian floriculture exporters should emphasize developing other diversified products and marketing them through dedicated outlets, thus, expanding the definition of floriculture from just fresh flowers to products and accessories manufactured from flowers.

Establishing a network of support systems with the involvement of Government, private sector and research institutions/universities will be another strategy for development of the floriculture sector. Indian exporters should adopt customized marketing strategies while targeting various markets. There is need to intensify marketing efforts for direct sales in emerging markets, and expand markets outside the traditional markets. Promoting strong brand recognition for Indian cut flowers and flower accessories through designated outlets at the foreign markets, and advance product positioning in these markets would help Indian flower exporters to compete in international markets effectively. A "consortium" approach may be encouraged among the units for branding, grading, packaging, transporting, quality control, supply assurance, market development, market promotion, and research and development.

Indian Flower Auctions - Few Thoughts

Karnataka Agro Industries Corporation (KAIC) has started a flower auction centre, which is currently, dealing more with domestic trade, with a volume much lesser than the capacity. However, eventually it is to be converted into an international auction centre. Maharashtra Agro Industries Corporation has also established a flower auction centre at Mumbai, which is yet to be operationalised. Under such situation, the first and foremost requirement is to ascertain what value the proposed auctions in India are likely to create. To make the auctions viable most importantly volumes of flowers to be auctioned needs to be synchronized with the auction infrastructure. Towards this, the traders and growers, currently operating in the existing markets should be facilitated to relocate to the auction centres. A great deal of attention will have to be given to the rules governing auction, transparency in the transactions, and dispute resolutions. Auction should be managed by growers and buyers, and government should act only as a facilitator. A system of reefer transportation also needs to be organized on commercial lines facilitating transfers from production sites to auctions and auctions to the market at an economical cost. Lastly, efforts should be made to develop and strengthen backward linkages on a continuous basis to benefit from export trade. Indian floriculture industry has inherent strengths such as favourable geographical situation, soil and environment, labour costs, and the positive investment sentiments of Indian corporate players. With corporate players getting in on the business, and smaller players consolidating their efforts, India's floriculture sector could expect to blossom over the coming years.

The global market for floriculture products was an estimated US$10 billion for the year 2004 with an 8% annual growth. It is expected to grow to US$ 16 billion by 2010. Fresh cut flowers account for around 2/3rd of the world trade in floriculture and has been growing at around 8-10% annually.

India.s share in international trade in floriculture is around 0.18%. India.s exports of floriculture products for 2004 was estimated at about US$ 60 million, of which US$ 11 million was contributed by fresh cut flowers.

Over 95% of Indian cut flower exports are different varieties of Rose. India is, thus at present, a negligible player in the international trade in fresh cut flowers which is dominated by the Netherlands, Columbia and Italy (accounting for about 32%, 10% and 6% respectively of the world trade), the African countries (particularly Kenya, Tanzania, Zambia, Zimbabwe, Ethiopia and Uganda), South Korea and Israel.

Europe (especially EU) is the largest market and the Netherlands is both the largest producer and exporter (including re-exports). The key non-EU competitors to Indian fresh cut flower exports in the EU market are the African Countries of Kenya, Tanzania, Zambia, Zimbabwe, Ethiopia and Uganda. The other important market for India is Japan, where India competes with local Japanese production as well as exports from South Korea and Taiwan and more recently, China.

Indian floriculture industry is still in its infancy. Rapid changes that have taken place since the early 1990s, however, suggest that India will be able to improve its position on the world market in the coming decade.

The organized floriculture industry in India is of nascent origin (slightly over a decade old) and is primarily export oriented. India mainly exports roses grown in greenhouses in Maharashtra (mainly in the Pune- Talegaon and Nasik belt) and around Bangalore in Karnataka and Hosur in Tamil Nadu.

The exports are primarily from Mumbai and Bangalore airports to various international destinations. Although Indian export of fresh cut flowers has been increasing in recent years, the volume as well as share in international trade is negligible compared to its competitors. India faces a major challenge in terms of infrastructure and awareness among small producers. The main feature of Indian floriculture is that the producers are small and fragmented.

India.s Distinctive Advantages for Development of the Floriculture sector

India has relatively better opportunities for development of the floriculture sector for the following reasons:

* Diverse agro-climatic conditions and geographical locations suited for growing various types of flowers

* Skilled manpower to absorb the technology and implement the same at a relatively low cost
* Soil and water supply at most locations
* Good radiation/sunlight leading to healthier plant growth and better quality flowers
* Light rains and salubrious climate during winter, the prime export season, leading to sustained high yields
* Good period of sunlight even during the heavy rains leading to continued plant growth and proper yield
* India is located centrally for catering to European and Far Eastern markets, as well as being close to the South East Asian and Middle East Asian markets that have high consumption requirement of flowers.

Main Markets for Indian Fresh Cut Flowers

European consumers buy flowers for different purposes. The main purpose is to give them as a gift, representing about 40 to 50 percent of European consumer spending. Another 20 to 30 percent of the flowers are bought for special occasions like weddings etc. The remaining is spent on flowers for decoration of their home. As can be seen, the two main markets for Indian Fresh cut flowers are Europe and Japan. The EU is believed to consume over 50% of the world.s flowers and includes many countries with a relatively high per capita consumption of cut flowers.

Although some of the larger EU markets like Germany, France and The Netherlands show signs of saturation, overall EU consumer sales of floriculture products, especially cut flowers, are still increasing. Germany is the biggest consumer of flowers, followed by the UK, France and Italy. Together, the two markets (Europe and Japan) account for around 75% of Indian fresh cut flower exports and are the mainstay of the Indian fresh cut flowers export industry.

Indian exports to Europe (which has been traditionally the largest market for Indian fresh cut flowers) has in recent years been falling due to the significant freight and transaction cost disadvantage that Indian exporters face vis-à-vis African countries who have now emerged as the largest (non-EU) suppliers of fresh cut flowers to Europe, with export volumes that are 8-10 times that of India.

Floriculture has tremendous potential in India. The different types of climatic conditions provide for the possibility of growing almost all the major cut flowers species of the world, whether of tropical, sub-tropical or temperate climate origin. However, flowers in India are produced in open field conditions, mostly during the mild winter months without use of any advanced technology. As a result, the quality and quantity available for marketing are heterogeneous and vary according to the prevailing weather conditions.

The consumers. demand abroad is based on high quality products, with additional requirements for the growing conditions that should be in line with environment/

preservation standards, implying reduced use of pesticides and other agro chemicals. The global demand for quality flowers is sustained throughout the year. Since fresh cut flowers are perishable, exports have to be necessarily by air, which, given the recent sharp increase in international airfreight (in turn due to the sharp increase in prices of aviation fuel), puts the Indian exporter at a significant costdisadvantage vis-à-vis it.s competitors.

Thus, though India has significant competitive advantage in terms of growing fresh cut flowers, one of the main reasons for Indian exports not being able to penetrate the large world market is the high cost of airfreight and the geographical and transaction/airfreight cost disadvantage that Indian exporters face vis-à-vis their competitors viz. Africa for the EU and Australia/New Zealand market, Colombia/ Ecuador for the North American market and South Korea for the Japanese market. Analysis indicates that Indian cut flower exporters face a 50.60% airfreight cost disadvantage vis-à-vis their competitors. The cost disadvantage arises due to geographical location, inability of Indian exporters to get airfreight discounts due to low export volumes, frequent air freight increase for Indian exports, fall in airfreight rates for competitors due to large volumes and economies of scale and import duty disadvantage to Indian exporters in the EU market). Some support from the Government and Export Promotion Agencies will go a long way in bringing down freight cost and making Indian flowers competitive in the International markets.

Important Flowers Traded in the International Market

The most important flower traded in the international market is still Rose. However, Dendranthema, Dianthus, Chrysanthemum, Carnation, Gerbera, Dahlia, Poinsettia, Orchids, Lily, Tulip are the flowers emerging as close competitors to Rose.

International Trade Channels

Growers and traders exporting cut flowers and foliage to EU send their merchandise either to a wholesaler or to an auction. In this respect, the Netherlands auctions play a pivotal role in the trade of flowers destined for both domestic and other European markets. Products handled by agents and import wholesalers are either sold directly to a wholesale buyer or submitted for auction. Export wholesalers re-export the products to other EU countries, where the flowers and foliage find their way to wholesalers and retailers.

Development at retail level show a slowly decreasing market share of traditional outlets like specialized flower shops and street and market stalls. Nevertheless, although retail chains sell more and more flowers, about 50 - 55% of flower purchases are still made at specialised flower shops (florists). It is estimated that retail chains have a market share of about 20%, while the rest is bought at other places like street vendors and gas stations.

Opportunities for India

The market for cut flowers consists of a range of product groups which offer varying opportunities for countries like India as potential suppliers. It is a highly competitive market in which importers are continually seeking new, special and different products.

They tend not to change easily from one rose supplier to another but co-operation with a company supplying a new product is considered attractive. The market is clearly searching for these kinds of novelties to take over some of the market from the species preponderant at present. A new product also offers the prospect of making higher profits than those gained from selling conventional floriculture products.

Demand for foliage varieties is still increasing in Europe, particularly for small leafed foliage for use in cheap bouquets. Furthermore, European importers do not have any reticence about using tropical foliage. Another opportunity exists for India in supplying products during periods when these products are scarce in the market.

India has a conducive environment for floriculture exports due to :

* increasing labour cost that is putting pressure on the cost of cultivation of major flower producing countries like The Netherlands, Japan, Taiwan and Israel
* environment degradation and cost of land which impede further expansion of cut-flowers production in EU countries
* increasing demand in the nearer markets of West Asia and South East Asia, where the rising standards of living are pushing-up demand for floriculture products
* dependence of Europe, which is one of the biggest markets on flowers imports during winter months
* better prices expected to be offered by the other potential markets viz. West Asia, South East Asia, Japan, Hong Kong. Singapore and Korea throughout the year than European market, provided they are supplied with good quality flowers on continuous basis.

Strategies for Developing Floriculture Exports

With a strong preference for direct marketing and private R&D for developing proprietary products, the future competitiveness of the Indian industry will depend on the unique selling proposition that the industry will develop. It can be deduced that the structure and composition of the Indian floriculture industry will undergo major changes with the emergence of medium and small size entrepreneurs in place of big corporates.

In the short run, infrastructure impediments need to be erased paving way for future growth. Developing new products to cater to niche markets is imperative.

With direct marketing as the major marketing channel, a variety of products, rather than concentration on rose production, is required. Also, in order to compete, Indian exporters must be able to supply products of consistent quality on a regular basis.

Credit Flow to the Floriculture Sector

The financial institutions and the banks have developed various financial products and necessary expertise to take care of the ever-growing demand for high tech agriculture like floriculture and horticulture.

The financial institutions have taken following initiatives.

- Identification of potential borrowers possessing good exposure to undertake the venture
  - Selection of floriculture crop with preference patterns of consumers in international market
  - Provision of assistance in preparation of project reports
  - Setting up of specialized branches for Agri- Export oriented projects
  - Simplification of loan formalities, speedy loan availability
  - Increasing accessibility to rural youth for taking up floriculture activity
  - Provision of finance by way of term loan, working capital and pre and post shipment credit for land, greenhouse, planting material, irrigation system, drainage system, grading and packaging shed, pre-cooling and cold storage, refrigerated van, office equipment, fee of Indian/foreign consultants for providing technology, cost of import of grafts of varieties which are popular abroad.

The National Horticulture Board [NHB] had undertaken a project in the late 1980.s under which loans were provided with a service charge of 4 per cent to the cooperative sector, corporate sector, registered and voluntary organisations, horticulture corporations, etc. to a maximum of Rs.1 crore. This project of NHB popularized horticulture as a viable business opportunity and laid the foundations for development of the sector. Such type of soft loans can be made available to progressive farmers in identified centres for cultivation of traditional flowers under controlled conditions. This type of soft credit inputs would also lead to adoption of the technologies by general farmers, who at a later date would be motivated to undertake cultivation of export oriented flowers. This will also induce awareness about the innovative ideas in agriculture in general and in floriculture in particular.

Even under poverty alleviation programmes like SGSY and employment generation schemes like PMRY and group activities like SHGs should have provision for floricultural activities. On the part of the credit providers, more than the credit products, the importance lies in financing the entire value chain of floriculture. Bankers should make themselves abreast with the latest innovative technologies in this sector and should provide credit input for;

i) import of most popular varieties and

ii) cultivation inside plastic greenhouse under ultra high densities, with drip irrigation, grading, packing, pre cooling, storage and transport as per European standards.

Only the production of high quality flowers at low cost can face the international competition. This should also be kept in mind while extending finance to floriculture units. In addition, since the cut flowers are generally imported for export there should be a viable insurance mechanism to hedge the risk of importing highly perishable commodity like flower. In essence, more than the credit input, lenders should pay attention to need based financing leading to successful implementation of the project.

Action Plan for Development of Export of Floriculture Products

In order to develop the Indian floriculture sector and for becoming a regular supply base for the export market, the following suggestions are offered :

* Technology for the construction of green houses/glass/poly/shade houses are to be dovetailed to Indian climatic conditions c Low cost green houses have to be developed using indigenous R&D
* Flower specific area to be identified as part of Agri Export Zone and necessary infrastructure viz., cold chains, testing & cargo handling facilities and inland container depots to be established
* Select Indian agriculture universities could establish partnership with overseas research institutes like PTC+, Holland for developing new varieties of flowers and foliages
* Import duty on inputs, especially on green house equipments currently at around 60% should be brought down to 15% to facilitate economic viability of green houses c Indian exporters to diversify into nearby markets viz., Asia-Pacific, West Asian market
* National Horticulture Board in consultation with APEDA could develop grant/ subsidy programme to fund technology up-gradation c Banks to extend long term loans (over 3 years) to growers/breeders to develop new varieties as has been the practice in EU (Rabo Bank report).
* ECGC/Insurance companies to develop appropriate risk mitigation policies for exporters/Growers
* APEDA in association with respective state agencies to orient Indian growers to the need to conform to Union for Protection of New Varieties of Plants (UPOV) and Breeders Rights
* Fresh cut flowers exporters be provided a 50% airfreight transport assistance as this would provide a strong boost for increasing fresh cut flower exports from the country and provide a level playing field for Indian exports.

To encourage the floriculture sector, attention should be given equally to production/productivity factors, quality of produce and availability of quality planting material. Shaping up the growth will require highly skilled manpower at the middle and lower levels necessitating training and development. Post-harvest management systems and infrastructure need to be developed. The shelf life of flowers needs considerable improvement. On-shelf management practices must be standardised and research should focus on lowering input costs and on improving low cost indigenous systems. Tissue culture and bio-technology hold a lot of promise and should be exploited to ensure better high yielding floral varieties with in-built resistance to plant diseases. Given the problem of pesticide residue, organic horticultural cultivation seems to offer a viable option.

Increasing quality consciousness among farmers and processors and enhancing skills in the areas of grading and standardisation will be crucial for global trade in the WTO regime. In order to survive in the competitive international markets, synchronisation between market trends and production systems is necessary. Creation of a well dispersed infrastructure and efficient storage and transport systems will be a crucial factor in harvesting the full potential of the floriculture sector.

In time to come, application of sanitary and phyto-sanitary (SPS) measures by the developed countries would be another aspect governing exports. The SPS agreement harmonises the standards internationally so as to minimise or eliminate the risk of SPS becoming a barrier for trade. Harmony with these measures calls for quality regulation which would benefit exports in the long run. Issues like plant protection organisation for surveillance, certification, and inspection, etc. for the export of flowers should also be properly addressed.

A lot has been done but we have miles to go before we can boast of achievements in the floriculture sector. As of now, all we can affirm is that slowly but steadily India is moving towards a Golden Revolution.

When it comes to consumer floral transactions, they usually take place at retail florist shops, supermarkets, and garden centres. In addition, floral sales via the Internet have also increased in the past few years. However, just as the industry continues to grow, it must also confront challenges affecting how the industry conducts business on a global scale.

According to statistics, the floral industry grows at a rate of almost $2 billion dollars per year, with the per capita spending at the retail level averaging about $55 per person. Even with the economic challenges of fluctuating values in foreign currencies, the current retail estimate of the floriculture industry is said to be over $20 billion, with Central and South American countries as the top suppliers to the United States market.

Exports of cut flowers to the United States are dominated by Columbia and Ecuador, who together account for approximately 77% of the total U.S. imports. These two countries supply 95% of the roses that are imported into the U.S., as well as 94% of chrysanthemums, and 99% of carnations. The reasons Columbia and Ecuador far surpass other countries include almost year-round growing seasons, and relatively low investment costs. The third-largest grower, the Netherlands, also has extended growing seasons due to its attention to production of cut flowers under glass.

One challenge facing the industry, as well as the rest of the country, is the green movement and conflicting definitions of "organic" and "green-friendly" products. The declining use of pesticides is seen as an asset to the industry; however, immigration issues are seen as a liability to the industry as it affects the labour force needed for harvest. Ultimately, for success in the industry, changes and challenges must be faced head-on in order for floral businesses to continue to prosper and grow.

Workers, the Environment and the Flower Industry

We often mark a special event—a birthday, an anniversary, a graduation—with a gift of flowers. Holidays like Valentine's Day and Mother's Day are times when giving flowers seems especially appropriate. Flowers brighten a hospital room, grace a banquet table, add dignified beauty to a funeral and are indispensable at weddings. Flowers mean love, respect, recognition, celebration... Rarely, however, do we think about where the flowers we buy come from. While we may perceive them as a fragile and beautiful gift of nature, most commercially sold flowers are produced through an artificial, labour-intensive process where every step, from farm to store, is carefully planned and managed.

The rose you buy from a Vancouver florist was probably harvested several days earlier in Colombia. It was then tightly rolled and semi-dehydrated by women working in a refrigerated warehouse, dipped in a preserving chemical bath, packed in a box with dozens of other roses, flown to Miami, inspected by specially trained customs agents, shipped to an auction near Toronto and later rehydrated in another chemical solution at the shop where you bought it.

The international flower trade is big business in the global economy. The variety, quantity, colour and size of the flowers for sale all reflect the changing demands of the world market. The need to produce huge quantities of highquality flowers all year round means that most of the roses, carnations and other varieties we buy are grown in countries in Latin America and Africa, where the hours of sunlight are long and labour costs are low.

International cooperation agencies, like the U.S. Agency for International Development (USAID), and institutions like the World Bank, have promoted export flower cultivation as a solution for countries in economic crisis. But while flower production, or "floriculture," as it is technically called, has been promoted to developing countries as an ideal solution for economic problems, the price tag it carries can be a high one.

The cut flower industry is a good example of the way globalization of the economy operates. Countries of the South, with low wages and weak environmental laws, produce the goods, while countries in the North hold a monopoly on the necessary technology and know how, and control the market.

The hub of the global flower industry is the Netherlands. This tiny European country leads all others in flower production and exports. It is the home of the world's largest flower auction, where 19 million cut flowers are sold during an average day. Farmers in the Netherlands began perfecting floriculture over 300 years ago. During the 17th century, Dutch tulips were already being exported at high prices to wealthy people in other parts of Europe. By 1637, the Dutch flower auction had been established. Growing and breeding techniques developed in the Netherlands have been exported around the world, along with the bedding plants used for propagating flower crops in the newer producer countries. Dutch technology and standards have set the pace for commercial floriculture wherever it is practised.

Today, the Netherlands exports 50 percent of the flowers sold on the world market. The second largest flower exporting country is Colombia, with 11 percent of the market. The number of flowers grown in the South has been increasing since the 1960's. Kenya, Ecuador, Zimbabwe, Tanzania and Zambia are all mass exporters of cut flowers. Israel also exports large quantities of flowers, especially to Europe.

In 1999, Colombia and Ecuador shipped 134,000 tons of flowers, worth about US$470 million, through Miami. About 3 percent of these flowers ended up in Canada.

The international flower trade is organized along regional lines. Asia-Pacific countries are the biggest suppliers to Japan and Hong Kong. Africa and Latin America supply the European Union, which buys 67.5 percent of the world's flowers, with the biggest share coming from Zimbabwe, Tanzania and Zambia.

The United States and Canada buy most of their flowers from Latin America, especially Colombia and Ecuador.

Ninety-two percent of carnations imported to the U.S. and Canada were produced in Colombia.

Workers in the Flower Industry

Flowers need long hours of sunlight, a stable, moderate climate and lots of water. Flower production involves many delicate operations that must be carried out by hand for every plant. Flowers do not need particularly fertile soil, but they can only be raised in quantity where there is available land on which to build large greenhouses. Ready access to an international airport is a must so that flowers can be shipped to their destination before any freshness is lost. And since flower production requires a lot of labour, it is important to have a large workforce near the plantations.

Most flower workers are exposed to grave risks in their jobs and have little or no ability to safely defend themselves from abuse. Often flower workers live in rural areas where few income-earning jobs are available. They may be landless peasants or unemployed city dwellers who migrate from other parts of the country to work in the flower industry. While they may have farming experience, flower workers generally lack the training and education that could qualify them for other types of work. For many, floriculture is their first entry into the paid labour force.

Producing flowers in the massive quantities needed for commercial export can have negative effects for workers and the environment.

In the countries of the South, between 65 and 70 percent of flower workers are women. Although they are the majority, they are often subject to discrimination. In some African countries, women can be seen working without any protective clothing next to male workers who wear full protective uniforms. One of the biggest problems women flower workers face is sexual harassment.

They must often submit to a pregnancy test before being hired. Women who are heads of families must work in situations where it is impossible for them to take time off to attend to their children's school or medical needs.

Flower workers are employed on short, fixed-term contracts, or as "temporary" or "casual" workers in most parts of Latin America and Africa, even though they may work at the same farm for years. They have no job security and often do not receive any benefits.They are usually paid only minimum wage—which in many countries is extremely low—and are expected to work overtime when required.

They sometimes work excessively long shifts—fifty or sixty hours per week at peak seasons. Flower workers have little choice but to accept these conditions if they want to keep their jobs. They are under strict surveillance in the workplace, and those who cannot keep up with the pace of the work, or who complain about conditions, are fired.

Few flower workers belong to unions and in most countries the industry does everything it can to prevent workers from organizing. In some countries, the owners themselves have set up "company" unions, which exist on paper but do little to help workers.

Flowers are a Health Hazard

The greatest danger flower workers face is exposure to harmful pesticides and other chemicals in the industry. The countries that import flowers insist these flowers be completely free of insects and disease. One hint of either can cause an entire shipment to be sent back. To meet these standards (which are called phytosanitary regulations) growers rely on heavy use of toxic pesticides, fungicides and other chemicals.

Each crop of flowers begins with workers using steam and chemical fumigation to disinfect the soil in which the seedlings will be planted. This process kills off all of the bacteria in the soil. Later, artificially produced fertilizers are pumped into the soil in order to provide nutrients for the growing plants.

During the planting process, workers kneel on the damp ground of raised beds, quickly setting out seedlings imported from the Netherlands, Germany or the United States. The planting is computer-monitored and calculated right down to the ideal date for supplying specific market destinations. Once the flowers begin to grow, workers string netting to support the plants, pinch back lateral buds," comb" the plants to ensure uniform growth, and cover them with plastic to prevent early blooming. After the flowers are harvested, selection and classification for shipping is done in refrigerated packing rooms. Many flower workers experience physical and emotional problems related to their jobs.

They are subject to drastic temperature changes, a polluted environment, a heavy workload and psychological stress. The most common health complaints are respiratory illness, skin diseases, varicose veins and ulcers. Since floriculture requires people to spend long periods of time working in awkward and uncomfortable positions, many suffer chronic back and muscle pain. Kneeling for hours on damp ground causes a variety of leg and knee problems.

Most flower workers are not given protective clothing or masks, and companies provide little or no training on how to decrease risk when handling and applying pesticides. One-fifth of the chemicals used in flower production in the countries of the South have been proven to produce cancer and have been restricted or banned in Europe and North America. Yet, in some cases, workers do not even have a place to wash their hands before eating lunch.

Dust from fumigation causes dizziness, headaches, cramps and other long-term symptoms. Workers are also vulnerable to severe respiratory infections due to exposure to temperature changes and high humidity levels in the greenhouses.

Women are even more at risk than men. One study in Tanzania found that pregnant and nursing mothers were exposed to the same chemical risks as other workers. The same study found that most workers did not know what chemicals they were handling, or the dangers associated with them.

Even in the Netherlands, where there is control over pesticide use, it was found that 43 percent of sprayers do not use safety protection when they work, and that 23 percent eat, drink or smoke while preparing toxic mixtures.

Despite an expanding flower industry, Bolivia has no regulations on pesticide use. In Tipicaya, the centre of the Bolivian flower industry, 3.79 percent of babies born in 2000 had some form of birth defect. That same year, over 8 percent of the patients at the local hospital were women with miscarriages.

Researchers and activists at the local university, Nuestra Señora de Paz, are demanding strict regulations on the most dangerous pesticides. However, without support from lawmakers and the public, this will be hard to achieve.

In Colombia, the flower industry uses 200 kilos of pesticide per hectare of land, which is double the amount used in the Netherlands. A study of over 8000 workers on flower plantations near Bogota found that the workers were exposed to 127 different pesticides, three of which the World Health Organization has considered to be extremely toxic. About 20 percent of the pesticides are banned or not registered for use in the United States or Canada because they are carcinogenic and extremely toxic.

According to one report, nearly two-thirds of Colombian flower workers suffer from one or more floriculture-related health problems, including headaches, nausea, impaired vision, conjunctivitis, rashes, asthma, stillbirths, miscarriages, congenital malformations and respiratory and neurological problems.

Interviews with fumigation workers in Colombia revealed that although they were using masks, they had not been informed that the mask filters need to be changed frequently. Without this step, the equipment provides only minimal protection.

Eyewitnesses have testified that greenhouses are not always completely cleared of people while fumigation is going on, and that workers sometimes return to greenhouses too early after fumigation has taken place. Workers do not receive enough instruction on how to handle pesticides properly; nor are they always informed about the types of pesticides being used. Not all farms keep toxicological records of stored pesticides, and the quantities used are not always registered. Secure showers are not available at the workplace, so workers cannot wash off pesticide residues before going home. Many companies do not have a doctor for emergencies or the capacity to respond to accidents at the worksite.

Doctors in flower-producing areas of Colombia report up to five cases of acute pesticide poisoning per day.

Flowers and the Environment

This heavy use of pesticides also creates a danger for the communities where the plantations are located. Although some municipalities have rulings prohibiting greenhouses from being placed close to neighbourhoods where people live, this rule is often violated. In most flower producing countries in Africa, workers live on the plantations, where they are exposed to even greater risk.

The high levels of toxic chemicals used in the flower industry contaminate soil and water in flower-growing regions. When not properly treated, the run-off water from the irrigation process is laced with pesticides and as it soaks into the ground it creates a zone of toxic soil. In Colombia, pesticides have contaminated the water table and the subsoil in the flower producing zone. Nearby rivers are also seriously affected. In Ecuador, pollution is causing the extinction of some species of flora and fauna, destabilizing the ecological balance and posing a risk for the food farmers remaining in the region.

Disposal of solid waste generated by the industry is also a problem. Plant stocks, rejected flowers and pesticide containers are all contaminated and need to be safely disposed.

Until recently, farmers in Colombia were feeding their cattle rejected carnation stalks that were sold or given away by the plantations. The contaminated milk from these cows was sold in the surrounding towns and in the capital city. Although the practice is now illegal, it is still going on in some places.

Retailers' desire to market new and special varieties of flowers has led some growers to experiment with genetic alterations. Flower genes are being manipulated to make the plants resistant to insects and disease, to increase the output of blooms and to make flowers last longer after they are cut. Producing flowers in colours not found in nature is also a goal for vendors who want to have something different to sell.

There is growing concern everywhere about the environmental and health effects of manipulating life forms in this way. In 2000, at an international meeting in Colombia, over forty workers, human rights and environmental organizations signed a statement protesting the introduction of genetic alterations in the flower and food industries.

In May 2000, the first genetically modified flower was approved for the Colombian flower industry.

The Blue Carnation contains genes derived from petunias, which provide its new artificial colour.

Genetically altered plants are good business for the companies that develop them because they receive royalties each time the plants are used. However, many people around the world are concerned about the short-term and long-term effects of transgenetic agriculture, and flower workers have joined in this movement.

Cultural Values Threatened

Ecuador began to produce cut flowers in the early 1980s. By 1998, flowers comprised 3.5 percent of the country's total exports. Ecuador is becoming known for its excellent roses, which fetch the best market prices. Many of the flower companies in Ecuador belong to Colombians who have brought their expertise, along with their labour practices, to this neighbouring country.

In the regions where flowers are grown, floriculture has replaced farming for food crops. The flower growing zone of Colombia was once prime farmland. Today food must be imported. The same is true for Ecuador.

Bibliography

Alex Laurie and Victor H. Ries : *Floriculture : Fundamentals and Practices*, Agrobios, Delhi, 2008.

Arundhati P. : *Annapurna : A Bunch of Flower of Indian Culture*, Concept, Delhi, 2001.

Baker J.G. : *Flora of Mauritius and The Seychelles : A Description of the Flowering Plants and Ferns of Those Islands*, AES, Delhi, 1999.

Bhaskar P. : *Floriculture and Plant Biotechnology*, Manglam Pub, Delhi, 2011.

Bhattacharjee S.K. and Chandran Lakshman De : *Advanced Commercial Floriculture*, Aavishkar Pub, Delhi, 2010.

Bose T.K. : *Floriculture and Landscaping*, Naya Prokash, Delhi, 1999.

Chattopadhyay S.K. : *Commercial Floriculture*, Gene Tech Books, Delhi, 2007.

Choudhary Dewasish : *Flower Crops : Cultivation and Management*, Oxford Book Company, Delhi, 2010.

Crozier, William, and Peter Henderson : *How the Farm Pays: The Experiences of Forty Years of Successful Farming and Gardening*, New York: P. Henderson &, 1884.

Datta S.K. and Chakrabarty D. : *Floriculture : Role of Tissue Culture and Molecular Techniques*, Pointer, Delhi, 2010.

Grimshaw, John, and Bobby J. Ward : *The Gardener's Atlas: the Origins, Discovery, and Cultivation of the World's Most Popular Garden Plants*, London: Little, Brown, 2001.

Guha Sudeep : *Organic Methods for Ornamental Gardening and Floriculture*, Dominant, Delhi, 2011.

Gupta Ramesh : *Baba Amte's: Flames and Flowers (Collected and revised version)*, Indian Pub, Delhi, 2001.

Henderson, Peter : *Gardening for Profit*, New York: Orange Judd, 1867.

Henderson, Peter : *Practical Floriculture*, Orange Judd, 1874.

Kala C.P., Rawat G.S. : *Ecology and Conservation of the Valley of Flowers National Park, Garhwal Himalaya*, Wildlife Institute of India, 1998.

Kumar Virendra & Subramaniam B. : *Chromosome Atlas of Flowering Plants of the Indian Subcontinent*, Botanical Survey of India, 1988-89.

Leapman, Michael : *The Ingenious Mr. Fairchild: the Forgotten Father of the Flower Garden,* New York: St. Martin's, 2001.

Mondal S. : *Technology Gap in Floriculture,* , Agrotech, Delhi, 2009.

Nowroz N.S. Jahan : *A Field Guide to Bangladeshi Flowers,* Asiatic Society of Bangladesh, 2005.

Pavord, Anna : *The Tulip,* New York: Bloomsbury, 1999.

Polunin Oleg and Stainton Adam : *Concise Flowers of the Himalaya,* Oxford University Press, Delhi, 2010.

Prasad S. and Kumar U. : *A Handbook of Floriculture,* Agrobios, Delhi, 2010.

Raj Desh : *Floriculture in Hills,* Agrotech Publishing Academy, Delhi, 2005.

Reddy K. Venkata : *Flowering of Indian Drama : Growth and Development,* Prestige Books, Delhi, 2004.

Richardson Boedler Cornelia : *Applying Bach Flower Therapy To The Healing Profession Of Homoeopathy,* B. Jain Pub., Delhi, 2003.

Ronald J. Biondo and Dianne A. Noland : *Floriculture : From Greenhouse Production to Floral Design,* International Book, Delhi, 2006.

Savur Manorama : *And the Bamboo Flowers in the Indian Forests : What Did the Pulp and Paper Industry Do?,* Manohar, Delhi, 2003.

Singh Neeraj Pratap : *Landscaping, Irrigation and Floriculture Terminology,* IBDC Publishers, Delhi, 2010.

Sivadasan M and Mathew Philip : *Biodiversity, Taxonomy and Conservation of Flowering Plants,* Mentor Books, Delhi, 1999.

Smith, Jane S. : *The Garden of Invention: Luther Burbank and the Business of Breeding Plants,* New York: Penguin, 2009.

Somani L.L. : *Floriculture and Landscaping : At A Glance,* Agrotech Pub, Delhi, 2010.

Sridhara Shakunthala and Rajendran T.P. : *Bamboo Flowering and Rodent Outbreaks,* Scientific Pub, Delhi, 2009.

Stewart, Amy : *Flower Confidential,* Chapel Hill: Algonquin, 2007.

Sud Romesh Kumar and Kumar Sudhir : *Flowers and Vegetables of India,* Scientific, Delhi, 2004.

Ward, Bobby J. : *The Plant Hunter's Garden: the New Explorers and Their Discoveries,* Portland: Timber, 2004.

Wells, Diana : *100 Flowers and How They Got Their Names,* Chapel Hill, NC: Algonquin of Chapel Hill, 1997.

Willis J C : *A Manual and Dictionary of Flowering Plants and Ferns,* , Cosmo, 2006.

WrightR. Temple : *Flowers and Gardens : A Manual for Beginners,* Biotech Books, Delhi, 2004.

Wulf, Andrea : *The Brother Gardeners: Botany, Empire, and the Birth of an Obsession,* New York: Alfred A. Knopf, 2009.

Index